TREES AND HILLS:
Methodology for Maximizing Functions
of Systems of Linear Relations

annals of discrete mathematics

NORTH-HOLLAND – AMSTERDAM • NEW YORK • OXFORD

NORTH-HOLLAND
MATHEMATICS STUDIES **96**

Annals of Discrete Mathematics (22)

General Editor: Peter L. Hammer

Rutgers University, New Brunswick, NJ, U.S.A.

TREES AND HILLS:
Methodology for Maximizing Functions of Systems of Linear Relations

Rick GREER
AT & T Bell Laboratories

1984

NORTH-HOLLAND – AMSTERDAM • NEW YORK • OXFORD

ISBN: 0 444 87578 6

Publishers:
ELSEVIER SCIENCE PUBLISHERS B.V.
P. O. BOX 1991
1000 BZ AMSTERDAM
THE NETHERLANDS

Sole distributors for the U.S.A. and Canada:
ELSEVIER SCIENCE PUBLISHING COMPANY, INC.
52 VANDERBILT AVENUE
NEW YORK, N.Y. 10017
U.S.A.

Library of Congress Cataloging in Publication Data

Greer, Rick, 1950-
Trees and hills.

(Annals of discrete mathematics ; 22) (North-Holland mathematics studies ; 96)
Bibliography: p.
Includes index.
1. Maxima and minima--Data processing. 2. Functions --Data processing. 3. Trees (Graph theory)--Data processing. I. Title. II. Series. III. Series: North-Holland mathematics studies ; 96.
QA315.G74 1984 511'.66 84-13557
ISBN 0-444-87578-6

PRINTED IN THE NETHERLANDS

to my parents,
John and Margaret Greer

Preface

The tree algorithm described in this monograph is an algorithm which maximizes functions of systems of linear relations subject to constraints. Typical problems in this class are concerned with identifying all of those vectors which satisfy or don't satisfy given linear equalities or inequalities in such patterns as will maximize certain functions of interest. For example, consider the problem of identifying all of those vectors which satisfy as many of an *inconsistent* system of linear inequalities as possible. For another example, consider two overlapping multidimensional clouds of O's and ×'s; in this setting, the problem is to determine all quadratic hypersurfaces which best separate the clouds in the sense of having the fewest number of O's on the × side of the surface and vice-versa. Also, as very special cases, this class includes the problems of solving linear programs and systems of linear equations.

The tree algorithm will solve many problems in this class, including all of the ones mentioned above. It is also able to solve problems of this type when the solution vectors are constrained to lie in designated linear manifolds or polyhedral sets or are required to solve other problems of this type.

These problems are typically NP-complete. Existing algorithms for solving problems from this class are essentially complete enumeration algorithms since the order of their time complexity is essentially that associated with enumerating the values of the criterion function on all equivalence classes of vectors. On the other hand, as compared to complete enumeration algorithms, the order of the tree algorithm's time complexity is geometrically better as the number of variables increases and polynomially better as the number of linear relations increases. Furthermore, as with the complete enumeration algorithms, the tree algorithm will identify *all* solution equivalence classes. Four examples given in this monograph show the tree algorithm to be from 50 to 30,000 times faster than complete enumeration. A fast approximate version of the tree algorithm is seen to be from 6,000 to 55,000 times faster in these examples.

-- *acknowledgements* --

This monograph extends part of my Ph.D. dissertation at Stanford University. I wish to thank my adviser, Persi Diaconis, for his constant enthusiasm and encouragement which meant a great deal to me.

I would also like to thank Jerry Friedman for many helpful discussions concerning the classification problem and for making it possible for me to use the computation facilities at the Stanford Linear Accelerator Center.

Thanks also go to Bill Brown for providing the biostatistics data used in Chapter 9 and to Eric Grosse for introducing me to Householder transformations and thereby to the world of stable numerical methods. In addition, it is a pleasure to acknowledge several helpful and stimulating conversations with Scott Olmsted, Friedrich Pukelsheim, and Mike Steele.

I am also grateful to AT&T Bell Laboratories for its rewarding and stimulating research environment.

This monograph was phototypeset at AT&T Bell Laboratories. I greatly appreciate both the help of Patrick Imbimbo and Carmela Patuto who did most of the typing and the help of Jim Blinn who explained to me many of the intricacies of that mixed blessing, the TROFF phototypesetting language.

Rick Greer

TREES AND HILLS

Table Of Contents

Notational Conventions

A convention widely used in this monograph is that scalars are denoted by lower-case Greek letters such as α, vectors are denoted by lower case English letters such as x and the coefficients of a vector's representation with respect to some fixed basis $\{b_1, \ldots, b_d\}$ are denoted by using the corresponding Greek letter as, for example, $x = \sum_{1}^{d} \xi_i b_i$. The vector of coefficients in R^d is denoted by the appropriate English letter underlined, as with $\underline{x} = (\xi_1, \ldots, \xi_d) \in \mathsf{R}^d$. This convention necessitates the following forced correspondence between the English and Greek alphabets.

a	α	p	π
b	β	q	χ
c	γ	r	ρ
d	δ	s	σ
e	ϵ	t	τ
f	ϕ	u	Υ
h	θ	v	ψ
i	ι	w	ω
k	κ	x	ξ
l	λ	y	η
m	μ	z	ζ
o	o		

The following notational examples illustrate certain notational conventions that are used subsequently.

$A := B$ — A is defined to be B. The symbol nearest the colon is the one which is being defined.

LHS, *RHS*	Symbols which refer to the left-hand side or the right-hand side of an equation, equivalence, inequality, etc.
$\square$	Symbol indicating the end of a proof
$\Rightarrow$, $\Leftarrow$, $\Leftrightarrow$	Implication arrows
A^c	The complement of the set A
A w.o. B	$A \cap B^c$ (read "A without B")
$A - B$	$\{a-b: a \in A, b \in B\}$
$A \parallel B$	$A \cap B = \varnothing$ (read "A is disjoint from B")
$\#A$	The cardinality of the set A
int A	The interior of the set A
rel int A	The relative interior of the set A
$\overline{A}$	The closure of A
∂A	The boundary of A
$\times_1^n B_i$	The Cartesian product of the sets B_i
R	The real numbers
R^d	The usual vector space over R consisting of vectors of the form $(\alpha_1, \ldots, \alpha_d)$ for $\alpha_i \in \mathrm{R}$
$\mathrm{sgn}(\alpha)$	The sign of $\alpha \in \mathrm{R}$ which is equal to -1, 0, 1 depending on whether α is $<$, $=$, or $>$ 0, respectively
$1\{x=y\}$	The indicator function which is 1 if $x = y$ and 0 if not
δ_{ij}	The Kronecker δ which is equal to $1\{i=j\}$

$\langle a \rangle$	The open ray $\{\alpha a : \alpha > 0\}$
I_k	$\{i \in I : \langle a_i \rangle = \langle a_k \rangle\}$ relative to some set of points $\{a_i : i \in I\}$
$I_j(\pi)$	$\{i \in I : \langle \pi_i y_i \rangle = \langle \pi_j y_j \rangle\}$ relative to some set of points $\{y_i : i \in I\}$ and some $\pi_i \in \{-1, 1\}$ for $i \in I$
$(a : b)$	The open line segment between a and $b = \{\lambda a + (1 - \lambda) b : \lambda \in (0, 1)\}$
(a, b)	For vectors $a, b \in \mathsf{R}^d$, this is the usual Euclidean inner product $a^T b$
$\Vert a \Vert$	$\sqrt{(a, a)}$
$\tilde{v}$	A linear functional in the dual space of the vector space under consideration
$[x, \tilde{v}]$	The value of the linear functional $\tilde{v}$ at the point x, i.e., $\tilde{v}(x)$
$\underline{x}$	The vector of coefficients yielding the representation of the vector x according to some fixed basis
$\underset{\sim}{A}$	The matrix representing a linear transformation A
x^T	The transpose of the column vector $x \in \mathsf{R}^d$
$R \oplus S$	The direct sum of the subspaces R and S
$P[\cdot \mid R, S]$	The projection operator onto R along S
$P[\cdot \mid R]$	The orthogonal projection operator onto R
$S^{\perp}$	Depending on the context, the annihilator of the set S or the subspace orthogonal to the set S

$\tilde{v}\big\vert_R$	The restriction of the linear functional $\tilde{v}$ to the subspace R
ψ	The vector space isomorphism that maps $\tilde{u} \in S^{\perp}$ to $\tilde{u}\big\vert_R \in \tilde{R}$ for specified subspaces R and S such that $R \oplus S = X$
$f \circ g$	The function f composed with the function g
$\theta(nd)$	An otherwise unspecified function which is bounded from below by $\delta_1 nd$ and from above by $\delta_2 nd$ for some δ_1, $\delta_2 > 0$

Chapter 1:

Introduction and Synopsis

A problem of continuing interest in mathematical programming is that of solving the system of linear inequalities $\{a_i^T x \geqslant \mu_i\}_1^m$ for given $\mu_i \in \mathsf{R}$ and $a_i \in \mathsf{R}^d$. Probably the most well-known and efficient method for solving such a system of linear inequalities when a solution exists is that provided by the Phase I method of linear programming. And, in fact, the duality theory of linear programming can be used to show the converse, namely, that any procedure for solving systems of linear inequalities of the form $\{a_i^T x \geqslant \mu_i\}_1^m$ will be able to solve linear programs of the form: maximize $c^T x$ subject to $\underset{\sim}{B} x \geqslant e$ where $c \in \mathsf{R}^d$, $\underset{\sim}{B}$ is an $m \times d$ matrix, and $e \in \mathsf{R}^m$. Other methods for solving $\{a_i^T x \geqslant \mu_i\}_1^m$ do exist; the more well-known ones include Fourier elimination and Motzkin-Schoenberg relaxation.

The tree algorithm described in this monograph is also a procedure which solves $\{a_i^T x \geqslant \mu_i\}_1^m$ when a solution exists. But it does this almost incidentally. More generally, consider the set of linear relations $\{a_i^T x \; R_i \; \mu_i\}_1^m$ where $R_i \in \{<, \leqslant, =, \neq, \geqslant, >\}$. The tree algorithm is the only known non-enumerative algorithm for determining all of those vectors $x \in \mathsf{R}^d$ which satisfy or don't satisfy elements of this set of linear relations in such patterns as will extremize certain functions of interest.

For example, in order to find vectors x which solve $\{a_i^T x \geqslant \mu_i\}_1^m$, one can begin by associating an indicator function of the form $1\{a_i^T x \geqslant \mu_i\}$ with each linear inequality in the system. It then suffices to use the tree algorithm to identify all of those vectors $x \in \mathsf{R}^d$ which maximize $f(x) = \sum_1^m 1\{a_i^T x \geqslant \mu_i\}$. By maximizing f, the tree algorithm will identify all $x \in \mathsf{R}^d$ which satisfy as many of the linear inequalities as possible. If the system is consistent, then the tree algorithm will produce a representative x_0 from the relative interior of the single equivalence class of vectors satisfying all of the linear inequalities; furthermore, it will announce the consistency of the system by asserting that $f(x_0) = m$. If the system is inconsistent, then the tree algorithm will assert this by producing representative vectors with f values $< m$ from the relative interiors of all those equivalence classes whose members satisfy as many of the linear inequalities as possible.

-- *historical context* --

In fact, it would appear that all previous work in this area of maximizing functions of systems of linear inequalities can be characterized as work which sought solutions to special cases of the problem of maximizing $\sum_J \sigma_i \, 1\{a_i^T x > \mu_i\} + \sum_K \sigma_i \, 1\{a_i^T x \geqslant \mu_i\}$ over $x \in \mathsf{R}^d$. Here $\sigma_i, \mu_i \in \mathsf{R}$, J and K are index sets such that $J \cup K \neq \emptyset$ and without loss of generality, all a_i are assumed non-zero. To the author's knowledge, this previous work falls into two categories. The first, which was essentially just discussed, occurs when all $\sigma_i > 0$ and the underlying system is consistent, i.e., when there exists some x_0 which satisfies all of the linear inequalities.

The second category is concerned with maximizing this function when the underlying system is homogeneous (i.e., all $\mu_i = 0$) and inconsistent. Warmack and Gonzalez (1973) present an algorithm for maximizing $\sum_1^m 1\{a_i^T x > 0\}$ when $\{a_i\}_1^m$ is in general position (i.e., for all $J \subset \{1, \ldots, m\}$ such that $\#J$ or the cardinality of J is d, $\{a_i, i \in J\}$ is linearly independent). This monograph was inspired by the Warmack and Gonzalez paper. It greatly extends their basic ideas to the development of the tree algorithm which solves a much larger class of problems than that of maximizing $\sum_1^m 1\{a_i^T x > 0\}$. It also offers rigorous proofs of the validity of the tree algorithm whereas the main algorithm proofs in Warmack and Gonzalez (1973) are incomplete and incorrect as will be seen in section 3.3.

Johnson and Preparata (1978) show that the problem of maximizing $\sum_J \sigma_i \, 1\{a_i^T x > 0\} + \sum_K \sigma_i \, 1\{a_i^T x \geqslant 0\}$ is NP-complete when the system of all of the linear inequalities is inconsistent. They refer to this problem as the Weighted Closed, Open, or Mixed Hemisphere problem depending upon whether $J = \varnothing$, $K = \varnothing$, or $J \neq \varnothing$ and $K \neq \varnothing$, respectively. The rationale behind these mnemonically attractive names is the following: If a norm is introduced on R^d and all a_i are required to be of norm 1, then when $J = \varnothing$ (or $K = \varnothing$), the problem becomes one of identifying all of those closed (or open) hemispheres of the unit sphere which collect the greatest sum total reward for the points they contain.

The algorithms Johnson and Preparata offer for the solution of these problems are complete enumeration algorithms. To see how this is the case, observe that the set of hyperspaces $\{a_i^\perp : i \in J \cup K\}$ where $a_i^\perp := \{x \in \mathsf{R}^d : a_i^T x = 0\}$ divides up the solution space into a union of polyhedral convex cones: Each vector $y \in \mathsf{R}^d$ is in a set of the form $\{x \in \mathsf{R}^d : a_i^T x > 0 \text{ for } i \in L_1,\ a_i^T x < 0 \text{ for } i \in L_2,\ a_i^T x = 0 \text{ for } i \in L_3\}$. Such a set is the relative interior of a polyhedral convex cone. Intuitively speaking, the edges of these cones are the one-dimensional rays which make up their "ribs" or frame. The Johnson-Preparata Weighted Closed Hemisphere (WCH) algorithm enumerates the values of the criterion function on all of the edges of which there are on the order of n^{d-1} where $n = \#(J \cup K)$.

The Johnson-Preparata WOH and WMH algorithms enumerate the values of the criterion function on all of the edges as well as on the order of at most 2^{d-2} more rays. In the case of the WOH problem where the set of all solution vectors is the union of a finite number of interiors of fully-dimensional polyhedral cones, the Johnson-Preparata WOH algorithm enumerates the values of the criterion function on at least all of the edges and all of the interiors of fully-dimensional polyhedral cones in the solution space. When $\{a_i\}_1^m$ is in general position, there are more of these cones than there are edges as will be seen in Chapter 7.

The tree algorithm avoids complete enumeration on this scale by relying upon an observation that all solution vectors to the Weighted Hemisphere problems (as well as many other problems) are in the relative interiors or other faces of certain special polyhedral cones called hills. These hills, which may or may not be fully-dimensional, play the roles of relative maxima in these problems. What the tree algorithm does is to enumerate the hills by constructing a tree of vectors with the property that when the vectors in this tree are perturbed slightly in a prescribed manner, the resulting set of vectors contains at least one representative from the relative interior of every hill. Fortunately, there are typically far fewer hills than there are polyhedral cones in the solution space. In fact, when the system of linear inequalities in a Weighted Hemisphere problem is consistent and in pointed position (cf., (2.3.34)), then the problem defines precisely one hill.

-- the class of problems that the tree algorithm solves --

More formally now, the tree algorithm solves many problems in a large class of problems which are introduced here as problems of extremizing functions of systems of linear relations subject to constraints. This class of problems provides a unifying framework for the research that has been done on finding procedures to produce vectors which satisfy systems of linear inequalities in certain desired patterns. To be more specific, H is said to be a function of the system of linear relations $\{a_i^T x \; R_i \; \mu_i\}_1^m$, where $R_i \in \{<, \leqslant, =, \neq, \geqslant, >\}$ and where $x \in \mathsf{R}^d$ if and only if there is a function $g : \mathop{\times}\limits_1^m \{0, 1\} \to \mathsf{R}$ such that for all $x \in \mathsf{R}^d$,

$$H(x) = g(1\{a_1^T x \; R_1 \; \mu_1\}, \ldots, 1\{a_m^T x \; R_m \; \mu_m\}).$$

The problem is to maximize (or minimize) H over $x \in \mathsf{R}^d$

(i) subject to requiring the maximizing vectors to lie in some designated linear manifold or polyhedral set

(ii) or subject to maximizing another function H_2 of a system of linear relations

(iii) or subject to maintaining the value of yet another function H_3 of a system of linear relations greater than some preset constant

(iv) or any or none of the above constraints.

From the previous discussion, it is easy to see that linear programming and the Weighted Hemisphere problems fall into this category of problems of extremizing functions of systems of linear relations. For that matter, so also do problems of solving systems of linear equations like $\underset{\sim}{A}x = b$.

(Whether or not the tree algorithm is particularly efficient in solving such special purpose problems as solving linear programs and systems of linear equations remains to be seen. In fact, it seems likely that there are many linear programs which could be solved faster with existing linear programming methodology than by the tree algorithm.)

In spite of the apparent complexity of the general case, all problems of extremizing functions of linear relations with or without constraints are equivalent to certain other unconstrained problems in a simple homogeneous canonical form. To define this, the concepts of nondecreasing and nonincreasing variables are needed. The jth variable of $g : \bigtimes_1^m \{0, 1\} \rightarrow \mathsf{R}$ is nondecreasing if and only if for all choices $\xi_1, \ldots, \xi_{j-1}, \xi_{j+1}, \ldots, \xi_m \in \{0, 1\}$,

$$g(\xi_1, \ldots, \xi_{j-1}, 0, \xi_{j+1}, \ldots, \xi_m) \leqslant g(\xi_1, \ldots, \xi_{j-1}, 1, \xi_{j+1}, \ldots, \xi_m).$$

The jth variable of g is nonincreasing if and only if the jth variable $-g$ is nondecreasing. The jth variable of g is constant if and only if it is nondecreasing and nonincreasing. g is a nondecreasing function if and only if all of its variables are nondecreasing.

It will be shown that for every problem of extremizing a function H of a system of linear relations subject to constraints, there is a homogeneous system of linear inequalities $\{b_i^T x \; R_i \; 0\}_1^n$ where $R_i \in \{ >, \geqslant \}$ and a positive function g_2 with no nonincreasing variables such that any vector y which solves the original problem can be obtained from some vector x which maximizes $g_2(1\{b_1^T x \; R_1 \; 0\}, \ldots, 1\{b_n^T x \; R_n \; 0\})$ and vice versa.

Once a problem has been reduced to homogeneous canonical form, then the tree algorithm can solve it if the appropriate g_2 function is nondecreasing. In all practical situations the author has seen to date, the g_2 functions of problems reduced to homogeneous canonical form have all been nondecreasing; consequently, the nondecreasing g_2 function requirement does not seem to affect the utility of the tree algorithm in practice. This section continues with a discussion of a number of specific problems that the tree algorithm solves.

-- *applications in operations research* --

Problems of maximizing functions of systems of linear relations arise in the fields of economics and operations research when there is a need to determine those vectors $x \in \mathrm{R}^d$ which satisfy as many of a system of linear inequalities as possible. It may even be desired to attach more weight to the solution of some inequalities than to others. The associated criterion function is

$$H(x) = \sum_J \sigma_i \; 1\{a_i^T x > \mu_i\} + \sum_K \sigma_i \; 1\{a_i^T x \geqslant \mu_i\}$$

where the $\sigma_i \in \mathrm{R}$ are the weights and J, K are finite index sets with $J \cup K \neq \emptyset$. (Note that this is not expressed as a Weighted Hemisphere problem since the μ_i are not necessarily 0). It is easy to see that this problem is no less general than the one obtained by letting the relations ">" and "$\geqslant$" in H above be any relations in $\{ <, \leqslant, =, \neq, \geqslant, > \}$. The tree algorithm solves these problems.

-- *statistical classification and the tree algorithm* --

Also, in terms of applications, the tree algorithm enables one to solve a longstanding problem in the field of statistical classification. In 1954, Stoller published a complete enumeration algorithm for solving a version of the one-dimensional two-class Bayes loss classification problem. Under certain restrictions, Stoller's algorithm produces consistent estimates of best half-line classification rules. The tree algorithm is the first non-enumerative algorithm for solving not only the multidimensional version of Stoller's problem, but also any of a much larger class of statistical classification problems as well. This class is concerned with estimating linear classification rules that are best according to any of a wide variety of criteria.

In brief, the goal of these problems is to produce good rules for estimating which one of two arbitrary unknown distributions F_1 and F_2 on R^p is responsible for producing the observation vector $x \in \mathsf{R}^p$.

For each subset A of R^p, define a rule d_A which classifies x as class 2 if and only if $x \in A$. Consider only sets of decision regions A of the form $\{x \in \mathsf{R}^p : g(x) > 0\}$ where g is an element of a fixed known vector space of real-valued functions on R^p which includes the identity function. Such regions are known as linear decision regions. The set of all polynomials in the coordinates of x of degree $\leqslant k$ for some $k \geqslant 1$ is such a vector space of functions. The set of all open halfspaces would then be an allowable class of linear decision regions.

In order to define the worth of rules d_A, let the loss function L assign to each rule d_A a real number $L(F_1(A), F_2(A^c), \tau_1)$ where $F_1(A)$, $F_2(A^c)$ are d_A's two conditional misclassification probabilities and τ_1, which may or may not be present, is the probability that the next individual sampled will be from class 1. $L(\cdot, \cdot, \cdot)$ is assumed to be a continuous function that is a nondecreasing function of its first two arguments.

In order to gain information about F_1 and F_2, samples are taken from each and the usual empirical measures $\hat{F}_1$ and $\hat{F}_2$ are formed. If τ_1 is used in the loss function, then an estimate $\hat{\tau}_1$ of τ_1 is also obtained. It can then be shown that those $A = \{x \in \mathsf{R}^p : g(x) > 0\}$ from a given class of linear decision regions which minimize $L(\hat{F}_1(A), \hat{F}_2(A^c), \hat{\tau}_1)$ are in fact consistent estimates of best linear decision regions for the class of regions under consideration. This means that the true losses associated with the empirically best d_A converge almost surely to the best loss function values possible for the class of rules under consideration as the sample sizes become infinite.

In order to identify the empirically best d_A, it can be shown that it is sufficient to find all $a \in \mathsf{R}^d$ which minimize

$$L\left(\sum_{i=1}^{n_1} \sigma_{1i}\ 1\{a^T y_{1i} \leqslant 0\},\ \sum_{j=1}^{n_2} \sigma_{2j}\ 1\{a^T y_{2j} \leqslant 0\},\ \hat{\tau}_1\right)$$

where the σ_{1i}, σ_{2j} are weights and the y_{1i}, $y_{2j} \in \mathsf{R}^d$ are transformed data points. The tree algorithm will perform this minimization.

-- *classification using the Bayes and Kullback loss functions* --

As a specific example, for the Bayes loss function $\lambda_1 \tau_1 F_1(A) + \lambda_2(1 - \tau_1)F_2(A^c)$ with $\lambda_k > 0$, the above empirical objective function is a positive multiple of

$$\sum_1^{n_1} \lambda_1\ 1\{a^T y_{1i} \leqslant 0\} + \sum_1^{n_2} \lambda_2\ 1\{a^T y_{2j} \leqslant 0\}$$

Minimizing this function is equivalent to maximizing the WOH criterion function

$$\sum_1^{n_1} \lambda_1\ 1\{a^T y_{1i} > 0\} + \sum_1^{n_2} \lambda_2\ 1\{a^T y_{2j} > 0\}$$

When $\lambda_1 = \lambda_2$, it can be seen that the Bayes empirical minimization problem is that of finding all allowable classification rules which make the fewest number of errors on the data.

As another example of a specific loss function, consider the empirical minimization problem for Kullback's I(1:2) loss function. Here the task is to find all vectors a which minimize

$$\left((1-\hat{\psi}_1) \log \frac{\hat{\psi}_2}{1-\hat{\psi}_1} + \hat{\psi}_1 \log \frac{1-\hat{\psi}_2}{\hat{\psi}_1} \right) \cdot 1\{\hat{\psi}_1 + \hat{\psi}_2 \leqslant 1\}$$

where for $k = 1, 2$

$$\hat{\psi}_k = \frac{1}{n_k} \sum_{1}^{n_k} 1\{a^T y_{ki} \leqslant 0\}$$

The tree algorithm will solve this problem as well.

(For more detail on these statistical applications, see Chapter 8, and for much more detail, see Greer (1979).)

-- *a pictorial classification example* --

In terms of a pictorial example of what the tree algorithm can do in this statistical classification setting, Figure (1.1.1) shows a cloud of 30 ×'s and 30 O's in the plane which is dichotomized by an ellipsoidal classification rule into a class × region and a class O region. Note that this rule makes a total of 3 errors, where an error is said to occur when there is a × in the O region or vice-versa. Of all of the ways of dichotomizing these 60 points using quadratic curves, the tree algorithm identified the pictured ellipsoidally induced dichotomy as one of the two minimum-error dichotomies existing for this data set. Consequently, the ellipsoid rule shown in Figure (1.1.1) is a consistent estimate of a best Bayes quadratic rule when $\lambda_1 = \lambda_2$ and the usual estimate $\hat{\tau}_1 = \dfrac{n_1}{n_1 + n_2}$ is used.

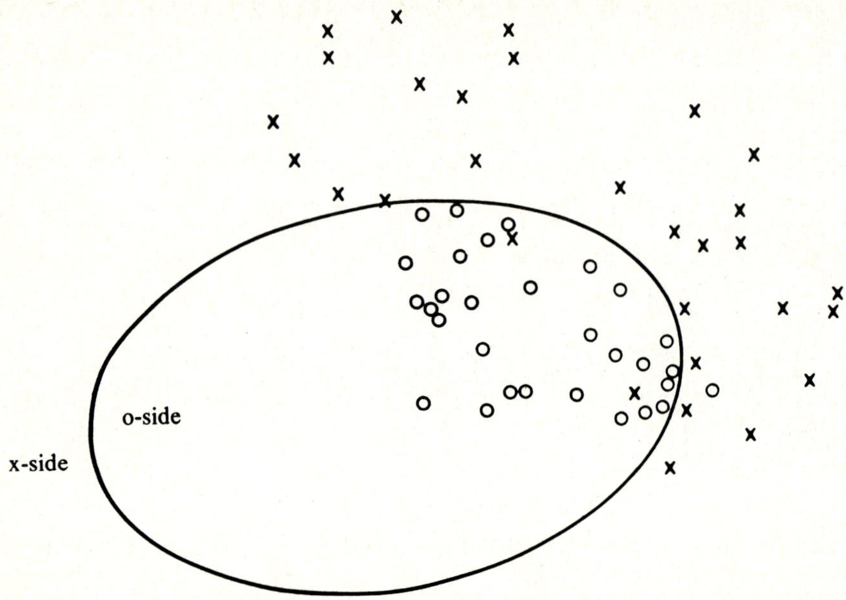

(1.1.1) Figure: One of two minimum-error quadratic curve dichotomies for a set of 30 ×'s and 30 O's in the plane.

-- imputation and the tree algorithm --

As an example of a problem of extremizing a function of a system of linear relations subject to a constraint, consider the following problem from the field of linear numeric editing and imputation. Suppose there is a database consisting of vectors in R^d each of which is known to be incorrect if it fails the consistency test of being in some prespecified polytope $\{x \in \mathsf{R}^d : Ax \leqslant b\}$. Given a vector y which has failed this set of linear edits by not being in the polytope, it is of interest to find the smallest number of components of y which could be change in order to place the modified vector in the polytope. If z is defined by $z := (\zeta_1, \ldots, \zeta_d)$, then the associated mathematical programming problem is to minimize $\sum_1^d 1\{\zeta_i \neq 0\}$ such that $A(y + z) \leqslant b$. The tree algorithm will do this.

-- equal hemispheric partitions of points on a sphere --

As another example of a constrained problem of this kind, consider an open problem posed in Johnson and Preparata (1978), namely, determine a procedure for finding a hemisphere of the unit sphere in R^d which most equally partitions the set $\{a_i\}_1^m$ on the surface of the sphere. This can be expressed in symbols by asking which x minimize $|\sum_1^n 1\{a_i^T x > 0\} - \sum_1^n 1\{a_i^T x < 0\}|$. Note that since the value of this criterion function at x is the same as it is at $-x$, attention may be restricted to those x such that $\sum_1^n 1\{a_i^T x > 0\} \geqslant \sum_1^n 1\{a_i^T x < 0\}$. Consequently, this problem can be solved by using the tree algorithm to minimize $\sum_1^n 1\{a_i^T x > 0\} + \sum_1^n 1\{a_i^T x \geqslant 0\}$ subject to keeping $\sum_1^n 1\{a_i^T x > 0\} + \sum_1^n 1\{a_i^T x \geqslant 0\} \geqslant n$.

-- the time complexity of the tree algorithm --

Chapter 6 discusses the computational complexity of the tree algorithm for maximizing a function $H = g \circ f$ of a system of linear relations when H is in homogeneous canonical form with a nondecreasing g_2 function. In this case, for $x \in \mathsf{R}^d$,

$$H(x) = g_2(1\{b_1^T x \; R_1 \; 0\}, \ldots, 1\{b_n^T x \; R_n \; 0\})$$

where $R_i \in \{ >, \geqslant \}$. Let $\alpha = \inf \{\#\{i : b_i^T x < 0\} : x \neq 0\}$ and suppose $g_2 : \times_1^m \{0, 1\} \rightarrow \mathsf{R}$ can be computed in time of order n. Then, if $\alpha \geqslant 2$, a version of the tree algorithm is shown to have time complexity of order greater than $dn\frac{\alpha^d - 1}{\alpha - 1}$ and less than $dn^d \,/\, 2^{d-1}$. In practice, the lower bound is much more indicative of the tree algorithm's time complexity than the upper bound is. The exponential character of the lower bound comes as no surprise considering the NP-complete nature of the problem.

By way of contrast, the complete enumeration procedure of Johnson and Preparata for solving the WMH problem has time complexity of order between $dn^{d-1}\log n$ and $d2^{d-2}n^{d-1}\log n$. A complete enumeration algorithm extended from one suggested in conversation by Mike Steele is generally faster for solving the WOH problem than the Johnson-Preparata algorithm and has time complexity of order $nd\binom{n}{d-1}$.

A fast approximate tree algorithm was developed which greedily explores subsets of a sequence of trees with the objective of quickly finding vectors with large criterion function values. This algorithm cannot be guaranteed to produce optimal vectors but it has been found to be very successful in practice in producing good if not optimal vectors very quickly.

-- *computer trials* --

As regards the behavior of the tree algorithm in practice, the examples of Chapter 9 describe the results of using a sophisticated WOH tree algorithm to estimate best linear classification rules for four data sets. In these examples the WOH tree algorithm examined only a small fraction ranging from .000034 to .02 of the number of vectors that would have been examined by the modified Steele edge enumeration procedure. In particular, for the Fisher iris data where $\alpha = 1$, $d = 5$, and $n = 100$, the WOH tree algorithm's computer program examined only 128 candidate solution vectors before stopping with the two best solution equivalence classes whereas the complete enumeration procedures would have had to examine at least 3,764,376 candidate vectors.

The fast approximate WOH tree algorithm also did very well in these examples. The version of the fast approximate algorithm that was used here produced vectors that were optimal in 3 out of the 4 examples and only 1 error away from being optimal in the fourth. It accomplished this by examining at most 403 candidate solution vectors in these problems where the complete enumeration procedures would have had to examine millions of vectors. In summary, the fast approximate WOH tree algorithm used in these examples was between 6,000 and 55,000 times faster than the modified Steele edge enumeration procedure.

-- *solving consistent systems of linear equations* --

Even though the tree algorithm's time complexity is, in general, exponential in d, the tree algorithm actually provides a polynomial time method for solving the consistent linear system $\underset{\sim}{A}x = b$. As the discussion in Chapter 8 will indicate, by using prior knowledge that the tree algorithm does not have in general (namely that $\underset{\sim}{A}x = b$ is assumed to be consistent), the tree algorithm can be slightly modified so as to obtain an apparently new way to solve $\underset{\sim}{A}x = b$ which has a time complexity of the same order as Gaussian elimination. This new algorithm will produce as a particular solution the minimum norm solution for any given inner-product norm and, if asked, will go on to identify the entire linear manifold of solutions.

-- *what is to come* --

As a brief synopsis of what is to come, the next chapter will introduce the reader to that subset of the theory of polyhedral convex cones which is needed to understand the nature of the tree algorithm. The tree algorithm is developed in two stages. First, in Chapter 3, a tree algorithm for solving the WOH problem is presented. Then after discussing in Chapter 4 how problems of extremizing functions of systems of linear relations subject to constraints may be reduced to a homogeneous canonical form, the general tree algorithm is presented in Chapter 5. The WOH problem is done first because of the great benefit this provides in understanding the considerably more complicated general situation.

The computational complexity of the tree algorithm is discussed in Chapter 6. Other methodology for extremizing functions of systems of linear relations is compared and contrasted with the tree algorithm in Chapter 7. Various applications of the tree algorithm are discussed in Chapter 8. The tree algorithm's behavior in estimating best linear classification rules for four data sets is presented and analyzed in Chapter 9.

The last chapter, Chapter 10, complements Chapter 1 in summarizing this monograph; in particular, it contains a detailed geometrically oriented summary description of how and why the tree algorithm works. The reader may wish to browse through Chapter 10 from time to time since it contains in one place all of the simple ideas underlying all of the details in this monograph. In short, Chapter 10 provides a good way to see the forest without thinking about the trees.

For the reader's convenience, a list of notational conventions is provided after the Table of Contents. Also, summaries of the more involved sections and chapters are given at the end of each for the reader who wishes to browse.

Chapter 2:

A Tutorial On Polyhedral Convex Cones

In order to understand the proofs validating tree algorithms for maximizing functions of systems of linear relations, it is necessary to know quite a bit about the theory of polyhedral convex cones. Inasmuch as the literature on this subject is somewhat scattered, this chapter was written to develop the necessary theory in an essentially self-contained way.

A substantial portion of the following is based on Gerstenhaber (1951), Goldman and Tucker (1956), and Stoer and Witzgall (1970). Much of the material in this chapter has not appeared in print before.

Those who have some familiarity with polyhedral cones will probably wish to just browse through this chapter on their way to Chapter 3 and beyond. This browsing may be facilitated by the summaries that follow each section in this chapter. Then, when reading subsequent chapters, these readers may wish to make use of this chapter, the notational convention list, and the index to resolve any particular questions that may arise.

It should be noted, however, that this treatment of polyhedral cones does differ in several fundamental ways from preceding treatments. Subsequent tree algorithm proofs depend greatly on these differences. Here is a list of some of them:

(1) All of the polyhedral cone theory is done in a coordinate-free fashion for an arbitrary finite-dimensional vector space over the reals. Strong use is made of the distinction between vectors and their representations according to some fixed basis.

(2) In keeping with (1), the dual space of linear functionals is used extensively instead of the usual transposed vectors from R^d.

(3) All of this theory is developed using purely vector space notions without imposing any norms or metrics on the space as previous authors have almost uniformly done. One noticeable consequence of this is that projectors which project one subspace along another complementary subspace are used instead of the more common inner product based orthogonal projectors which project a subspace along its orthogonal complement.

(4) Polyhedral cones are thought of as being the convex hulls of open rays just as polyhedrons are the convex hulls of points. Consequently, frames of polyhedral cones necessarily consist of open rays and not points.

(5) The concept of (convexly) isolated subsets is introduced. Isolated open rays are found to work quite nicely and naturally with the definition of a frame of a polyhedral cone.

(6) Special indexing notation, I_k and later, $I_k(\pi)$, is introduced for indexing the generators of a polyhedral cone. This notation greatly facilitates subsequent tree algorithm proofs.

(7) A nonstandard definition of face is needed and used.

The first section of this chapter develops and reviews the particular form of basic vector space geometry which will be needed subsequently. The second section introduces some helpful topological considerations to this basic vector space theory. The third section introduces polyhedral convex cones while the fourth section discusses the relationships between these cones and their duals. Since some of the theorems in this chapter are used as lemmas in subsequent tree algorithm proofs, they may seem to be somewhat unmotivated and out of place here. They are placed in this chapter however because they would break up the flow of ideas if placed elsewhere.

Section 2.1: Vector Space Preliminaries

Most problems of maximizing functions of systems of linear relations which are encountered in practice are expressed using vectors in R^d. There is, however, a certain technical reason for couching all of the following theory in the context of an arbitrary abstract d-dimensional vector space X over R. The proof that the tree algorithm works is based on an induction on the dimensionality of the problem, i.e., the d-dimensional version of the problem can be solved for $d > 2$ if certain $d - 1$ dimensional versions can be solved. The reason why X is preferred to R^d is because a subspace of X is a vector space whereas a proper subspace of R^d is not R^p for $p < d$. This should become clearer later. It is of course safe to visualize X as being R^d since all d-dimensional vector spaces over R are isomorphic to R^d.

As a final comment, the computer programs which implement the various algorithms to be discussed are totally insensitive to what X is since they work with the representations of vectors according to some pre-set basis instead of the vectors themselves.

Much of the following presumes a solid understanding of basic vector space theory which may be obtained, if need be, from Halmos (1974) and Nering (1963). The material in this section establishes notation, lists standard definitions, and presents several special interest theorems.

With regard to notation, Greek letters $\alpha, \beta, \gamma, \ldots$ are used to represent elements of R, the underlying field. For the most part, the only exceptions to this rule are the letters d,i,j,k,ℓ,m,n,p,q which are used to represent the positive integers used for indices.

All vectors are denoted by small English letters. The $d \times 1$ matrix which represents the vector x with respect to a basis $B = \{b_1, \ldots, b_d\}$ for X is written as $\underline{x} = (\xi_1, \ldots, \xi_d)$ where $x = \sum_1^d \xi_i b_i$. Matrices which are not column or row vectors are denoted by capital English letters with tildes underneath, as with $\underset{\sim}{A} = [\alpha_{ij}]$. The transpose of $\underline{x}$ or $\underset{\sim}{A}$ is written $\underline{x}^T$ or $\underset{\sim}{A}^T$.

X is not considered to be an inner product space. In fact, no metric or norm is assumed to be associated with X. Extensive use is made however of $\tilde{X}$, the dual space of X (i.e., the space of all linear functionals on X). Elements of the dual space are denoted by small English letters with tildes on top, e.g., $\tilde{v}$. Following Halmos, $[x, \tilde{v}]$ is defined to be $\tilde{v}(x)$ which, of course, is equal to $\underline{\tilde{v}}^T \underline{x}$ where the representation of $\tilde{v}$ is with respect to the dual basis. As will become increasingly evident, explicit use of the dual space is most helpful in keeping straight which vectors are associated with data points and which vectors are associated with hyperspaces.

For $A, B \subset X$, $A + B$ is $\{a+b : a \in A, b \in B\}$. $\{a_0\} + B$ is also denoted by "$a_0 + B$". Similarly, $A - B$ is $A + (-B)$ where $-B = \{-b : b \in B\}$. Note that $A - B$ is distinct from $A \cap B^c$ where B^c is the complement of the set B. $A \cap B^c$ will usually be denoted by "A w.o. B" (read "A without B").

$A \parallel B$ indicates that set A is disjoint from set B, i.e., $A \cap B = \varnothing$.

$\#A$ denotes the cardinality of a set A.

A list of notational conventions follows the table of contents.

In what follows, proofs of standard, tangential, or easy results may be omitted.

-- *segments, rays, convex sets, cones, subspaces, and manifolds* --

(2.1.1) Definitions: Take $x, y \in X$. $(x:y)$ is *the open line segment between* x *and* y i.e., $\{(1-\alpha)x + \alpha y : \alpha \in (0, 1)\}$. *The closed line segment between* x *and* y is $[x:y] := \{(1-\alpha)x + \alpha y : \alpha \in [0, 1]\}$. $(x:y]$ and $[x:y)$ are defined similarly.

Notice that $(x:y) = [x:y]$ w.o. $\{x, y\}$ if and only if $x \neq y$. If $x = y$, then $(x:y) = \{x\} \neq \varnothing$.

(2.1.2) Definition: *The open half-line* or *ray* originating at 0 and passing through $x \in X$ is $\langle x \rangle := \{\alpha x : \alpha > 0\}$.

(2.1.3) Definitions: Let $\emptyset \neq A \subset X$. A is a *convex set* if and only if for all $x, y \in A$ such that $x \neq y$, $(x:y) \subset A$. A is a *cone* if and only if for all $x \in A$, $\{\alpha x : \alpha \geqslant 0\} \subset A$. A is a *convex cone* if and only if A is convex and a cone.

(2.1.4) Theorem: Let $\emptyset \neq A \subset X$. A is a convex cone if and only if for all $x, y \in A$ and all $\alpha, \beta \geqslant 0$, $\alpha x + \beta y \in A$.

(2.1.5) Definition: Let $\emptyset \neq A \subset X$. A is a *subspace* if and only if for all $x, y \in A$ and all $\alpha, \beta \in \mathrm{R}$, $\alpha x + \beta y \in A$.

(2.1.6) Definition: $T \subset X$ is a *linear manifold* if and only if $T = x_0 + S$ for some $x_0 \in X$ and subspace $S \subset X$.

The next theorem shows that the subspace associated with a linear manifold is unique.

(2.1.7) Theorem: Let T be a linear manifold and suppose that $T = t_1 + S_1 = t_2 + S_2$ where $t_1, t_2 \in T$ and S_1, S_2 are subspaces. Then $S_1 = S_2$. Note t_1 need not equal t_2.

Proof: First note that $t_2 - t_1 \in S_1 \cap S_2$. Now take $s_1 \in S_1$. Then $t_1 + s_1 \in t_2 + S_2$ and so $s_1 \in (t_2 - t_1) + S_2 \subset S_2$. Similarly, $S_2 \subset S_1$. □

-- *hulls* --

The four types of subsets of X of the greatest interest here are convex sets, convex cones, subspaces, and linear manifolds. For an arbitrary nonempty subset A of X, it will prove useful to have a notion for the smallest set of each of the above types which contains A. Here a smallest set with a property P is defined to be a set R_0 with property P such that for all R with property P, $R_0 \subset R$.

(2.1.8) Definitions: Let $\emptyset \neq A \subset X$.

(a) The *convex hull* of A, denoted by "$H(A)$" and also called the *convex span* of A, is the smallest convex set containing A.

(b) The *convex conical hull* of A, denoted by "$C(A)$" and also called the *positive span of* A, is the smallest convex cone containing A.

(c) The *linear hull* of A, denoted by "$L(A)$" and also called the *linear span of* A, is the smallest subspace containing A. $L(\emptyset) := \{0\}$.

(d) The *linear manifold hull* of A, denoted by "$M(A)$" and also called the *dimensionality space of* A, is the smallest linear manifold containing A.

(2.1.9) Theorem: Let $\emptyset \neq A \subset X$. Then each of the four hulls defined in (2.1.8) exists. In fact,

(a) $H(A) = \cap \{K\colon K$ is convex and $K \supset A\}$

(b) $C(A) = \cap \{C\colon C$ is a convex cone and $C \supset A\}$

(c) $L(A) = \cap \{S\colon S$ is a subspace and $S \supset A\}$

(d) $M(A) = \cap \{T\colon T$ is a linear manifold and $T \supset A\}$

Proof: All intersections above are well-defined since X is itself a convex cone and a subspace containing A. Since A is contained in all of the intersections, none are empty.

Clearly if each intersection above has the desired property, then it is the smallest such set with that property. The fact that arbitrary intersections of convex sets, convex cones, and subspaces retain their respective properties is immediate. The analogous result for linear manifolds follows directly from the following lemma. □

Lemma: Let $\{x_i + S_i : i \in I\}$ be an arbitrary set of linear manifolds $\subset X$. Suppose there exists $z_0 \in \bigcap_I x_i + S_i$. Then $\bigcap_I x_i + S_i = z_0 + \bigcap_I S_i$.

Proof of Lemma: First of all, note that for each i, $z_0 = x_i + r_i$ for some $r_i \in S_i$. Now, for each i, take $z = x_i + s_i$ for some $s_i \in S_i$ and observe that $z - z_0 \in S_i$ for all i. For the other inclusion, take $s \in \bigcap_I S_i$ and observe that $z_0 + s = x_i + (r_i + s)$ for all i. □

See Figure (2.1.10) for examples of these hulls.

Next is a characterization of these four different kinds of hulls.

(2.1.11) Definitions: Let $a_1, \ldots, a_n \in X$ and $\gamma_1, \ldots, \gamma_n \in \mathsf{R}$. $\sum_1^n \gamma_i a_i$ is a *linear combination.* A linear combination is called:

(a) a *positive combination* if and only if $\gamma_i \geqslant 0$ for all i.

(b) an *affine combination* if and only if $\sum_1^n \gamma_i = 1$.

(c) a *convex combination* if and only if $\gamma_i \geqslant 0$ for all i and $\sum_1^n \gamma_i = 1$.

The combination is *strictly convex* when $\gamma_i > 0$ for all i.

(2.1.12) Theorem: Let $\varnothing \neq A \subset X$. Then:

(a)

$$\begin{aligned} \mathsf{H}(A) &= \{\text{convex combinations of elements of } A\} \\ &= \{\sum_1^n \gamma_i a_i : n \geqslant 1,\, a_i \in A,\, \gamma_i \geqslant 0,\, \sum_1^n \gamma_i = 1\} \end{aligned}$$

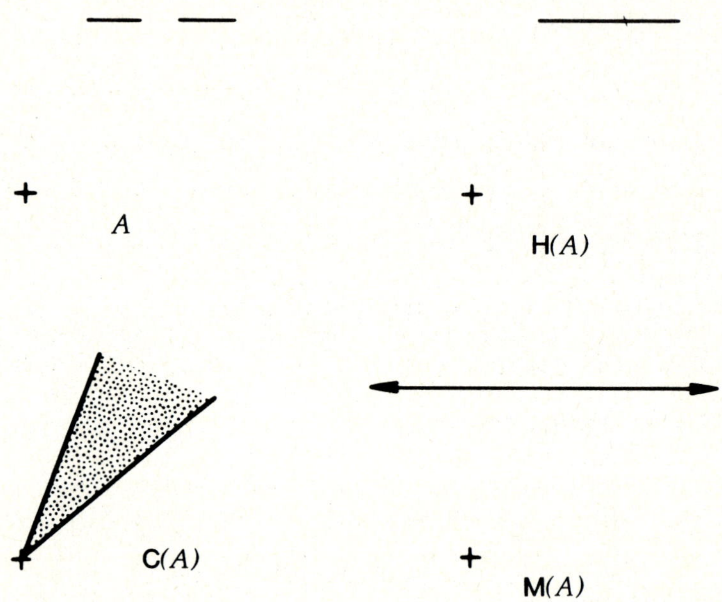

(2.1.10) Figure: $A \subset \mathsf{R}^2$ and three of its associated hulls. $\mathsf{L}(A)$ is the plane itself. The origin is denoted by +.

(b)

$$\mathsf{C}(A) = \{\text{ positive combinations of elements of } A\}$$
$$= \{\sum_1^n \gamma_i a_i : n \geqslant 1,\, a_i \in A,\, \gamma_i \geqslant 0\}$$

(c)

$$\mathsf{L}(A) = \{\text{ linear combinations of elements of } A\}$$
$$= \{\sum_1^n \gamma_i a_i : n \geqslant 1,\, a_i \in A,\, \gamma_i \in \mathsf{R}\}$$

(d)

$$\begin{aligned} \mathsf{M}(A) &= \{\text{affine combinations of elements of } A\} \\ &= \{\sum_1^n \gamma_i a_i : n \geq 1, a_i \in A, \gamma_i \in \mathsf{R}, \sum_1^n \gamma_i = 1\} \end{aligned}$$

Proof: (a) and (c) are shown in many standard texts such as Nering [33].

If (2.1.4) is used, then the *RHS* of (b) is easily seen to be a convex cone which contains A and so $\mathsf{C}(A) \subset$ *RHS* of (b). On the other hand, if C is a convex cone containing A then by (2.1.4), $\sum_1^n \gamma_i a_i$ must be in C for any $n \geq 1$, $\gamma_i \geq 0$, $a_i \in A$.

To show (d), first set T equal to the *RHS* of (d). Now, since clearly $A \subset T$, to show $\mathsf{M}(A) \subset T$, it will suffice to show that T is a linear manifold. Take $t_0 = \sum_1^m \beta_j a_j' \in T$ where $a_j' \in A$ and $\sum_1^m \beta_j = 1$. It remains to show that $T - t_0$ is a subspace. With regard to closure under scalar multiplication, take $\delta \in \mathsf{R}$ and note that

$$\delta\left(\sum_1^n \gamma_i a_i - \sum_1^m \beta_j a_j'\right) = \left(\sum_1^n \delta\gamma_i a_i + \sum_1^m (1-\delta)\beta_j a_j'\right) - \sum_1^m \beta_j a_j' .$$

Closure under addition follows easily now that closure under scalar multiplication has been established.

To show $\mathsf{M}(A) \supset T$, take a linear manifold $z + S$ containing A and take $\sum_1^n \gamma_i a_i \in T$. Since $a_i \in z + S$ for each i, $a_i = z + s_i$ for some s_i. Observe that $\sum \gamma_i a_i = z + \sum \gamma_i s_i \in z + S$. □

Here are a few corollaries:

(2.1.13) Theorem: Let $\emptyset \neq A \subset X$. Then:

(a) $\mathsf{C}(A) = \mathsf{C}(\mathsf{H}(A)) = \mathsf{H}(\mathsf{C}(A))$

(b) $\mathsf{L}(A) = \mathsf{L}(\mathsf{H}(A)) = \mathsf{H}(\mathsf{L}(A))$

(c) $\mathsf{M}(A) = \mathsf{M}(\mathsf{H}(A)) = \mathsf{H}(\mathsf{M}(A))$

(d) $\mathsf{L}(A) = \mathsf{L}(\mathsf{C}(A)) = \mathsf{C}(\mathsf{L}(A))$

(e) Let A' be the set A modified by multiplying arbitrarily selected elements by -1. Then $\mathsf{L}(A) = \mathsf{L}(A')$.

-- *the dimension of a set* --

A non-standard definition of linear independence will lead naturally into a definition of the dimension of a set $A \subset X$.

(2.1.14) Definition: Let I be a nonempty index set and $W = \{x_i : i \in I\} \subset X$. W is *linearly independent* if and only if for all $i \in I$, $x_i \notin \mathsf{L}\{x_j : j \in I, j \neq i\}$.

The following theorem shows how to construct linearly independent sets and will be used as a lemma shortly.

(2.1.15) Theorem: Let $W = \{x_i : i \in I\}$ be linearly independent and let $x_k \in X$ be such that $k \notin I$. If $x_k \notin \mathsf{L}(W)$, then $\{x_i : i \in I \cup \{k\}\}$ is linearly independent.

Proof: It is necessary to show for each $i \in I$, $x_i \notin \mathsf{L}\{x_j, j \in I \cup \{k\} \text{ w.o. } i\}$. Suppose, to the contrary, that $x_i = \sum_{j \in I \text{ w.o. } i} \alpha_j x_j + \alpha_k x_k$. Now $x_i \neq 0$ for each $i \in I$ since W is linearly independent. Consequently not all α_j, $j \in I \cup \{k\}$ w.o. i, are 0. Now if $\alpha_k = 0$, then this contradicts the assumed linear independence of W whereas if $\alpha_k \neq 0$, then that contradicts $x_k \notin \mathsf{L}(W)$. □

The dimension of a set can now be defined. Remembering that a basis for a finite dimensional vector space is any linearly independent set which linearly spans the space and that all bases have the same cardinality, consider:

(2.1.16) Definitions: The *dimension of a subspace* S is the cardinality of one of its bases. The *dimension of a linear manifold* is the dimension of its unique associated subspace. For $\emptyset \neq A \subset X$, the *dimension of* A, denoted by "$\mathrm{dim}\ A$", is $\mathrm{dim}\ \mathrm{M}(A)$. A *hyperspace* is a subspace of dimension $d - 1$. A *hyperplane* is a linear manifold of dimension $d - 1$.

It will be convenient to know that a basis for $\mathrm{L}(A)$ can always be chosen from A itself.

(2.1.17) Theorem: Let $\emptyset \neq A \subset X$. Suppose $A \neq \{0\}$. Then there exists a basis B for $\mathrm{L}(A)$ such that $B \subset A$. In fact, B can be taken to be any linearly independent subset of A of the largest possible size.

Proof: Since X is finite dimensional and $A \neq \{0\}$, there is a finite integer $k \geqslant 1$ such that no indexed subset of A containing more than k indices is linearly independent and there is an indexed set $B \subset A$ with k elements such that B is linearly independent. Now $B \subset A$, so $\mathrm{L}(B) \subset \mathrm{L}(A)$.

$\mathrm{L}(A) \subset \mathrm{L}(B)$ if $A \subset \mathrm{L}(B)$. To see the latter, take $a_0 \in A$. If $a_0 \notin \mathrm{L}(B)$, then $B \cup \{a_0\}$ (suitably indexed) is linearly independent by (2.1.15). This contradicts the choice of B. □

This has the following corollary:

(2.1.18) Theorem: Let $\emptyset \neq A \subset X$ where $A \neq \{0\}$. Then $\mathrm{dim}\ \mathrm{L}(A) \geqslant p$ if and only if there is a linearly independent set $\{a_i\}_1^p \subset A$.

-- *general position* --

An assumption frequently made about points derived in some specified fashion from a system of linear inequalities is that the set of points be in general position. The general position assumption requires that a set of points have as few linear dependencies as possible. In other words,

(2.1.19) Definition: Let I be an arbitrary index set. The set of vectors $W := \{x_i, i \in I\} \subset X$ is in *general position* in the d-dimensional vector space X if and only if for all $J \subset I$ of cardinality d, $\{x_i, i \in J\}$ is linearly independent.

-- linear manifolds --

Next, linear manifolds are discussed in more detail.

(2.1.20) Theorem: Let $a_0 \in A \subset X$. Then $\mathsf{M}(A) = a_0 + \mathsf{L}(A - a_0)$. If $0 \in A$, then $\mathsf{M}(A) = \mathsf{L}(A)$.

Proof: Observe

$$a_0 + \sum_1^m \gamma_i (a_i - a_0) = (1 - \sum_1^m \gamma_i) a_0 + \sum_1^m \gamma_i a_i \in \mathsf{M}(A)$$

by (2.1.12). So $\mathsf{M}(A) \supset a_0 + \mathsf{L}(A - a_0)$. The other inclusion follows from the fact that $a_0 + \mathsf{L}(A - \{a_0\})$ is a linear manifold containing A. □

There is an interesting the relationship between linear manifolds and elements of the dual space.

(2.1.21) Definition: Let $\varnothing \neq A \subset X$. The *annihilator of A* is $A^\perp := \{\tilde{x} \in \tilde{X} : [a, \tilde{x}] = 0 \text{ for all } a \in A\}$.

(2.1.22) Theorem: Suppose T is a linear manifold of dimension k. Then $T = t_0 + S$ for some $t_0 \in T$ and subspace S of dimension k. Let $\{\tilde{s}_1, \ldots, \tilde{s}_{d-k}\}$ be a basis for $S^\perp$. Let $\sigma_i = [t_0, \tilde{s}_i]$ for all i. Then $T = \{x : [x, \tilde{s}_i] = \sigma_i \text{ for } i = 1, \ldots, d-k\}$.

Proof: Clearly $LHS \subset RHS$. Now take x such that $[x, \tilde{s}_i] = \sigma_i$ for all i. Then $x - t_0 \in (S^\perp)^\perp = S$. □

There is a converse to (2.1.22), namely:

(2.1.23) Theorem: Let $\tilde{V} = \{\tilde{v}_1, \ldots, \tilde{v}_m\}$ be a nonempty subset of $\tilde{X}$. Let $\nu_i \in \mathsf{R}$ for $i = 1, \ldots, m$ and set $A := \{a \in X : [a, \tilde{v}_i] = \nu_i \text{ for } i = 1, \ldots, m\}$. Suppose $A \neq \varnothing$ and take $a_0 \in A$. Then $A = a_0 + (\tilde{V})^\perp$ and is consequently a linear manifold.

Proof: Clearly $LHS \supset RHS$. Now take $a \in A$ and note that $a - a_0 \in (\tilde{V})^\perp$. □

Linear manifolds of dimension $d - 1$ in X, i.e., hyperplanes, provide convenient ways to divide X into two pieces.

(2.1.24) Definition: Let $0 \neq \tilde{v}_0 \in \tilde{X}$ and take $\nu \in R$. $\{x : [x, \tilde{v}_0] = \nu\}$ is a hyperplane which determines four *halfspaces* : $\{x : [x, \tilde{v}_0] < \nu\}$, $\{x : [x, \tilde{v}_0] \leqslant \nu\}$, $\{x : [x, \tilde{v}_0] \geqslant \nu\}$, and $\{x : [x, \tilde{v}_0] > \nu\}$. The first two are called *negative halfspaces* while the last two are called *positive halfspaces*. The first and fourth are called *open halfspaces* while the second and third are called *closed halfspaces*.

-- *direct sum projection* --

The concept of projection used here is the basic vector space one.

(2.1.25) Definitions: Let R and S be subspaces of X such that $R + S = X$ and $R \cap S = 0$. This situation is denoted by writing $R \oplus S = X$ and saying that X is the *direct sum* of subspaces R and S. When $X = R \oplus S$, for each vector x, there is unique $r \in R$ and $s \in S$ such that $x = r + s$.

The *projector on R along S*, denoted by $P[\cdot \mid R, S]$, is a function which maps the point $x = r + s$, $r \in R$ and $s \in S$, onto r. $P[x \mid R, S]$ is said to be the *projection of x on R along S*.

Figure (2.1.26) shows that this concept of projection is *not* identical with the usual Euclidean projection operation.

-- *the dual spaces of subspaces* --

One of the central proof techniques used in the next chapter is to recurse on the dimensionality of the problem. In order to do this in a rigorous manner, it is necessary to establish a connection between $\tilde{R}$, the dual space of a subspace $R \subset X$, and $\tilde{X}$, the dual space of X. This is done via the following

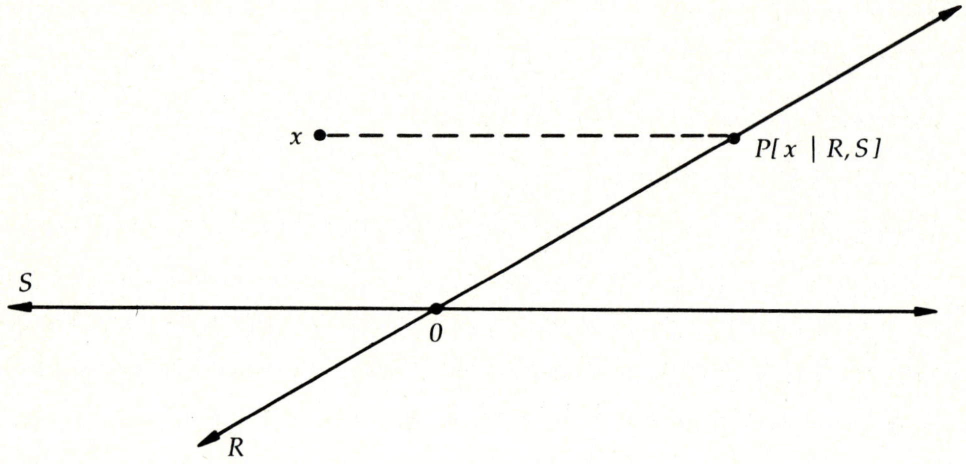

(2.1.26) Figure: Geometrical construction of the projection of the point $x \in \mathbb{R}^2$ on the subspace R along the subspace S.

technical lemma:

(2.1.27) Theorem: Let R be a subspace of X of dimension $k \geqslant 1$. Let S be any subspace such that $R \oplus S = X$. For any $\tilde{u} \in \tilde{X}$, $\tilde{u}|_R$ denotes the restriction of the function $\tilde{u}$ to R.

(a) $S^\perp|_R := \{\tilde{u}|_R : \tilde{u} \in S^\perp\}$ is a vector space with addition defined via $\tilde{u}|_R + \tilde{v}|_R := (\tilde{u}+\tilde{v})|_R$ and scalar multiplication defined via $\alpha \cdot \tilde{u}|_R := (\alpha\tilde{u})|_R$.

(b) Let $\{\tilde{u}_i\}_1^k$ be a basis for $S^\perp$. Then $\{\tilde{u}_i|_R\}_1^k$ is a basis for $S^\perp|_R$.

(c) $S^\perp$ is in one-to-one correspondence with $S^\perp|_R$ via the vector space isomorphism ψ which maps $\tilde{u}$ onto $\tilde{u}|_R$.

(d) $\tilde{R} = S^\perp|_R = \tilde{X}|_R$

(e) For $\tilde{r} \in \tilde{R}$, $\psi^{-1}(\tilde{r})$ is defined via $\psi^{-1}(\tilde{r})(r+s) = \tilde{r}(r)$ where $r \in R$ and $s \in S$.

In other words, the set of linear functionals on R may be obtained by taking one of a certain class of subspaces of $\tilde{X}$ and restricting the domain of each linear functional in that subspace to be R. A useful one-to-one correspondence then exists between $\tilde{R}$ and a subspace of $\tilde{X}$.

Proof: (b): To show $\{\tilde{u}_i|_R\}_1^k$ is linearly independent, suppose $\sum_1^k \alpha_i \tilde{u}_i|_R = 0$. Then for all $r \in R$, $s \in S$,

$$0 = \sum_1^k \alpha_i \tilde{u}_i|_R\,(r) = \sum_1^k \alpha_i \tilde{u}_i(r) = \sum_1^k \alpha_i \tilde{u}_i(r+s).$$

$\{\tilde{u}_i|_R\}_1^k$ is clearly a linear spanning set for $S^\perp|_R$.

(c): ψ is easily seen to preserve vector space operations. ψ is onto by virtue of the fact that it maps a basis of $S^\perp$ onto a basis of $S^\perp|_R$ and is easily seen to be $1:1$.

(d): Since $S^\perp|_R$ is a set of linear functionals mapping R into X, $S^\perp|_R \subset \tilde{R}$. Since $S^\perp|_R$ is a vector space of dimension k, equality holds.

Clearly $S^\perp|_R \subset \tilde{X}|_R$. To show the other inclusion, begin by showing $R^\perp \oplus S^\perp = \tilde{X}$. Note that $R^\perp \cap S^\perp = \{0\}$ by virtue of $R + S = X$. Since $\dim R^\perp + \dim S^\perp = d$, $R^\perp \oplus S^\perp = \tilde{X}$ follows. Now take $\tilde{x} = \tilde{t} + \tilde{u}$ for $\tilde{t} \in R^\perp$ and $\tilde{u} \in S^\perp$. Note $\tilde{x}|_R = \tilde{u}|_R$.

(e): For fixed $\tilde{r} \in \tilde{R}$, define $\tau(\tilde{r}) \in \tilde{X}$ via $\tau(\tilde{r})(r+s) = \tilde{r}(r)$ for $r \in R$, $s \in S$. Note that $\psi(\tau(\tilde{r})) = \tilde{r}$. Hence $\tau(\tilde{r}) = \psi^{-1}(\tilde{r})$. □

-- *lineality spaces* --

The next concept is one which is used a great deal in the study of convex cones.

(2.1.28) Definition: Let $0 \in A \subset X$. The *lineality space of A* is $\text{Lin}\, A := \text{H}(\cup\{S : S \subset A \text{ and } S \text{ is a subspace }\})$.

(2.1.29) Theorem: $\text{Lin}\, A \supset \{0\}$ and is a subspace. If $0 \in A$, then A is a subspace if and only if $A = \text{Lin}\, A$.

Proof: Take two convex combinations $\sum_1^m \alpha_i x_i$, $\sum_1^n \beta_i y_i \in \text{Lin}\, A$ where the x_i and y_i are elements of subspaces contained in A. Observe for all real δ, $\delta \sum_1^m \alpha_i x_i = \sum_1^m \alpha_i (\delta x_i) \in \text{Lin}\, A$. Also, note that

$$2\left(\sum_1^m \frac{\alpha_i}{2} x_i + \sum_1^n \frac{\beta_i}{2} y_i\right) \in \text{Lin}\, A. \ \square$$

(2.1.30) Theorem: If K is convex and $0 \in K$, then $\text{Lin}\, K \subset K$ and consequently $\text{Lin}\, K$ is the largest subspace contained in K.

See Figure (2.1.31) for examples.

Proof: Since $\cup\{S : S \subset K \text{ and } S \text{ is a subspace}\} \subset K$, $\text{Lin}\, K \subset \text{H}(K) = K$. $\square$

(2.1.32) Theorem: If C is a convex cone, then $\text{Lin}\, C = C \cap (-C)$.

See Figure (2.1.31).

-- *extreme and isolated subsets* --

The last topics for this section are the related ideas of extreme and isolated subsets.

(2.1.33) Definition: Let $\emptyset \neq W \subset A \subset X$. W is an *extreme subset of A* if and only if for all $a_1, a_2 \in A$, if $W \cap (a_1 : a_2) \neq \emptyset$ then $a_1, a_2 \in W$.

(2.1.34) Definition: Let $\emptyset \neq W \subset A \subset X$. W is an *isolated subset of A* if and only if it is *not* the case that there exist $a_1, a_2 \in A$ w.o. W such that $W \cap (a_1 : a_2) \neq \emptyset$. This is also equivalent to the statement that for all

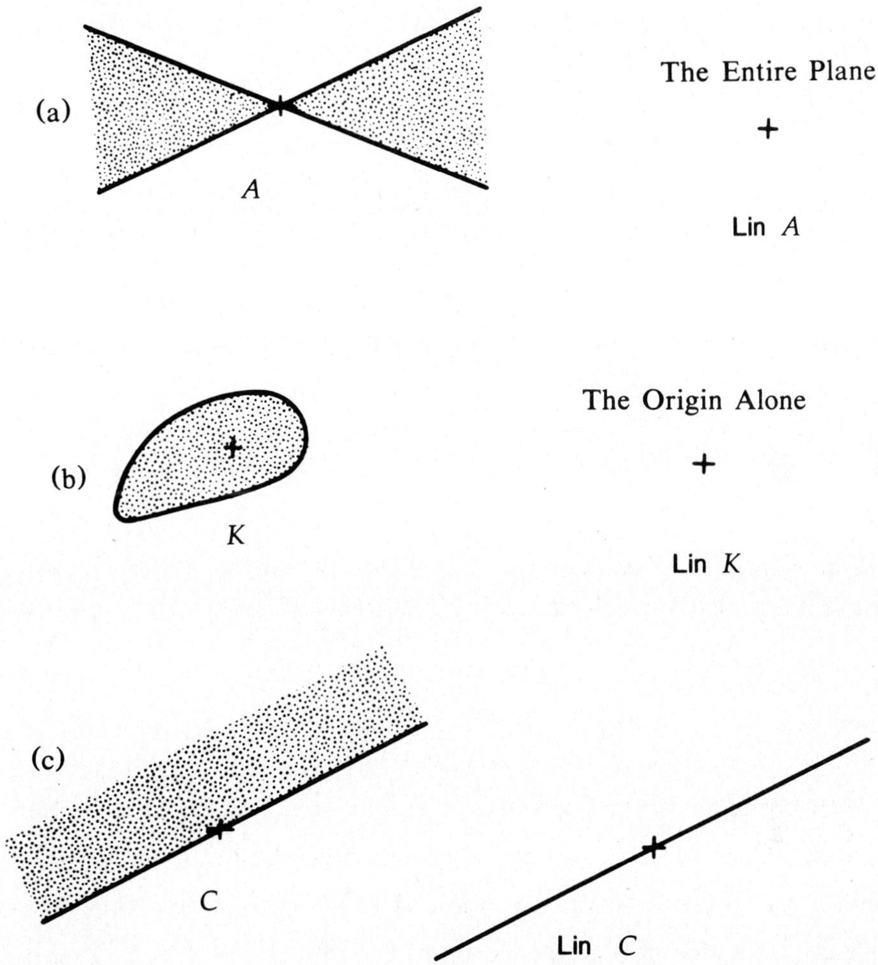

(2.1.31) Figure: Examples of lineality spaces in R^2.

$a_1, a_2 \in A$, if $W \cap (a_1: a_2) \neq \emptyset$ then either $a_1 \in W$ or $a_2 \in W$.

Note that every extreme subset is isolated but, as Figure (2.1.35) shows, not every isolated subset is extreme.

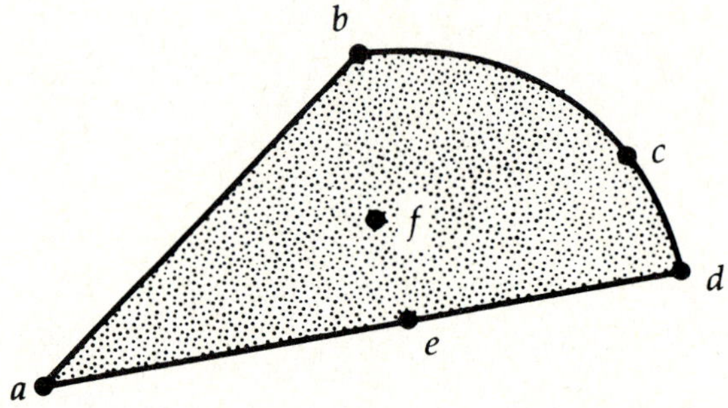

(2.1.35) Figure: Examples of extreme and isolated subsets of a convex set in R^2. $\{a\}$, $\{b\}$, $\{c\}$, $[a:b]$, and the closed and open arcs from b to c are extreme subsets of the figure. $(a:b]$ and $[a:b]$ are isolated but not extreme. The sets $\{e\}$, $\{f\}$, $(a:b)$, and $(a:e]$ are not isolated.

The definition of extreme subset is in wide use. The basic idea behind it, as the next theorem shows, is that W is an extreme subset of A if and only if whenever any point of W can be expressed as a strictly convex combination of points in A, then all of those points must be in W.

(2.1.36) Theorem: Let $\varnothing \neq W \subset A \subset X$. W is an extreme subset of A if and only if for all $\{a_i\}_1^n \subset A$, if there exists $\{\lambda_i\}_1^n$, $\lambda_i > 0$, and $\sum_1^n \lambda_i = 1$ such that $\sum_1^n \lambda_i a_i \in W$, then $\{a_i\}_1^n \subset W$.

Proof: The "if" direction follows from the definition. Suppose that W is extreme. Observe that

$$\sum_1^n \lambda_i a_i = \beta \left[\sum_1^{n-1} \frac{\lambda_i}{\beta} a_i \right] + \lambda_n a_n \text{ where } \beta = \sum_1^{n-1} \lambda_i . \quad \square$$

The definition of isolated subset generalizes Goldman and Tucker's (1956) definition of extreme face and its use is apparently confined to this monograph at this time. A few comments on the nature of isolated subsets might be helpful.

One can think of an isolated subset W of A as one whose members can never be reached by walking along the line segment connecting two points in A but not in W. In fact, the next theorem shows that a subset of a convex set is isolated if and only if it is disjoint from the convex hull of the points remaining after its removal from the convex set. This is reminiscent of the topological notion of isolated where W is a topologically isolated subset of A if and only if $\overline{A \text{ w.o. } W} \parallel W$ where $\bar{S}$ is the closure of S.

The idea is that if one is seeking to find a subset of a convex set whose convex hull is that convex set, then the isolated subsets which are not in turn generated by smaller isolated subsets will have to be included in this subset because there is no way to generate them from the other points.

(2.1.37) Theorem: Let $K \subset X$ be convex. If W is an isolated subset of K, then K w.o. W is convex and so $W \parallel \mathrm{H}(K \text{ w.o. } W)$. Conversely, if $\emptyset \neq W \subset K$ is such that $W \parallel \mathrm{H}(K \text{ w.o. } W)$, then W is an isolated subset of K.

Proof: ($\Rightarrow$): Take $k_1, k_2 \in K$ w.o. W. To show $(k_1 : k_2) \subset K$ w.o. W, first observe that $(k_1 : k_2) \subset K$ since K is convex. If $W \cap (k_1 : k_2) \neq \emptyset$ then either $k_1 \in W$ *or* $k_2 \in W$, which contradicts the choice of k_1 and k_2. $\square$

The usual definition of an extreme point of a convex set follows from the definition of an isolated singleton. The term extreme point (instead of isolated point) is used here in deference to common usage.

(2.1.38) Definition: Let $K \subset X$ be convex. k_0 is an *extreme point of K* if and only if $\{k_0\}$ is an isolated subset of K.

(2.1.39) Theorem: Let $K \subset X$ be convex. The following are equivalent:

(a) k_0 is an extreme point of K

(b) $\{k_0\}$ is an extreme subset of K

(c) it is not the case that there exist $k_1, k_2 \in K$ such that $k_1 \neq k_0$, $k_2 \neq k_0$, and $k_0 \in (k_1 : k_2)$.

Note that neither isolated nor extreme subsets are necessarily composed of extreme points (cf., Figure (2.1.35)).

Summary For Section 2.1

This section contained a potpourri of necessary background vector space information. It started with a discussion of the basic geometrical objects needed by this monograph, namely, line segments, rays, convex sets, convex cones, subspaces, and linear manifolds.

Four different types of smallest sets containing a given set were described. The convex hull, the convex conical hull, the linear hull, and the linear manifold hull will permeate the rest of this chapter and the next three.

The dimension of a set in X is the dimension of the unique subspace associated with the smallest linear manifold containing the set. This concept will be of value in visualizing subsequent results.

The dual space of X makes its introduction in providing an alternate representation of linear manifolds as the intersection of a finite number of level sets of linear functionals. The later sections of this chapter will involve quite a bit of hopping back and forth between the original space and the dual space. A useful correspondence was established between the dual space of a subspace of X and certain subspaces of $\tilde{X}$.

The lineality space of a convex set $K \subset X$ is the largest subspace contained in K. The lineality space concept is essential for an understanding of polyhedral convex cones. In fact, the lineality space of a polyhedral cone in X is closely connected with the dimensionality space of another cone in $\tilde{X}$ as will be seen in section 2.4.

An isolated subset of a convex set in X is one which can in no way be generated in a convex fashion by the other points of the set. An extreme subset of a convex set is one whose points can be generated in a strictly convex fashion from other points of the set only if all of those other points are in the extreme subset.

Section 2.2:

Topological Considerations

All of the essential theorems leading up to and justifying the algorithms of the next chapter are purely algebraic in character. However, one's intuition as to what should be true in a d-dimensional vector space X is greatly enhanced by attempting to see the geometry of R^d in suitably constructed two and three dimensional pictures. Everyone has a natural feeling for the concepts of boundary, interior, relative interior, and dimension. It would be a false economy not to provide the mathematical structure (i.e., the topological considerations) which makes these notions rigorous.

This section shows how to generate in a natural, constructive, and purely vector space fashion a topology for any subset of a vector space over R which coincides with the topology induced on the set by the usual topology on R^d when the vector space is R^d. This is aesthetically pleasing because no inner product, norm, metric, or any other structure is needed to generate this topology. It also provides characterizations of open sets and relative interiors which are very convenient for use with polyhedral and other convex sets.

-- *the* Γ_W *topology* --

Using only vector space concepts, the next definition defines what will later prove to be the natural topology for a set W in the vector space X. The basic idea here is that a set G is open relative to the Γ_W topology if and only if for every point $g \in G$ there is a polyhedron of the same dimensionality as W which when intersected with W both contains g in its "middle" and is itself contained in G.

(2.2.1) Definition: Let $W \subset X$. Suppose W consists of at least two distinct points, one of which is w_0. Let $B = \{b_i\}_1^p$ be a basis for $\mathsf{L}\{W-w_0\}$

for some $1 \leqslant p \leqslant d = \dim X$. Let $\Gamma_W := \{G \subset W\colon$ for all $g \in G$ there exists $\alpha > 0$ such that $\mathrm{H}\{g \pm \alpha b_i\}_1^p \cap W \subset G\}$. Let $\Gamma := \Gamma_X$.

(2.2.2) Comments: $\{g \pm \alpha b_i\}_1^p$ is $\{g - \alpha b_i, g + \alpha b_i\}_1^p$. At first glance, it may seem that Γ_W is dependent on the choice of w_0. To see why it is not, take $w_1 \in W$ such that $w_1 \neq w_0$. Then $\mathrm{M}(W) = w_0 + \mathrm{L}\{W - w_0\} = w_1 + \mathrm{L}\{W - w_1\}$ and so by (2.1.7), $\mathrm{L}\{W - w_0\} = \mathrm{L}\{W - w_1\}$.

Also at this point, it may seem that Γ_W is dependent on the choice of basis B. That this is not the case will be seen shortly when, for any B, Γ_W is shown to be precisely the same as the topology generated by any norm on W. For an example of $G \in \Gamma$ in R^2, see Figure (2.2.3).

Using Kelley (1955) as a reference if need be, the reader will find the proof of the next theorem straightforward.

(2.2.4) Theorem: Γ_W as in (2.2.1) is a topology for W.

(2.2.5) Example: Definition (2.2.1) will be used to show that for $\tilde{v} \neq 0$, $\{x \in X\colon [x, \tilde{v}] > 0\} \in \Gamma$, i.e. is open in X. Since this set is easily shown to be convex, it is only necessary to show that for all $x_0 \in X$ with $[x_0, \tilde{v}] > 0$, there exists $\alpha_0 > 0$ such that for $i = 1, \ldots, d$, $[x_0 \pm \alpha_0 b_i, \tilde{v}] > 0$, i.e., $[x_0, \tilde{v}] > \pm\, \alpha_0 [b_i, \tilde{v}]$. Since there is no constraint on α_0 if $[b_i, \tilde{v}] = 0$, it suffices to select α_0 such that

$$0 < \alpha_0 < \min\{[x_0, \tilde{v}] \,/\, |[b_i, \tilde{v}]|\colon [b_i, \tilde{v}] \neq 0,\ i = 1, \ldots, d\}.$$

The next theorem is used to establish the equivalence of Γ_W to any norm-induced topology on W.

(2.2.6) Theorem: Let $W \subset X$. Suppose W consists of at least two distinct points, one of which is w_0. Let $B = \{b_i\}_1^p$ be a basis for $\mathrm{L}\{W - w_0\}$ for some $1 \leqslant p \leqslant d$. Define a norm $\|\cdot\|$ on $\mathrm{L}\{W - w_0\}$ via $\|y\| := \sum_1^p |\eta_i|$ for $y = \sum_1^p \eta_i b_i \in \mathrm{L}\{W - w_0\}$. Fix $a_0 \in A \subset W$. Then the following two statements are equivalent:

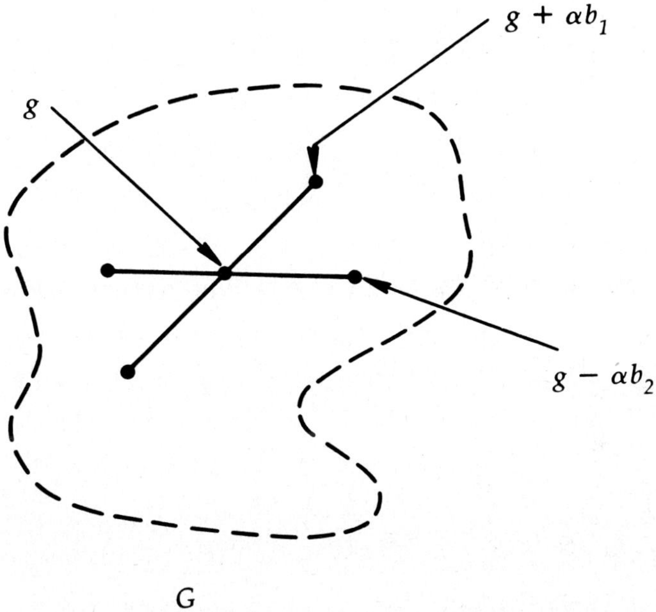

(2.2.3) Figure: Example of $G \in \Gamma$ in R^2. The dashed line indicates that $\partial G \cap G = \emptyset$.

(a) There exists $\epsilon > 0$ such that $\{w \in W: \|a_0 - w\| < \epsilon\} \subset A$

(b) There exists $\alpha > 0$ such that $W \cap \mathsf{H}\{a_0 \pm \alpha b_i\}_1^p \subset A$.

Proof: $\|\cdot\|$ is easily seen to be a norm. Also, note that $\|a_0 - w\| = \|a_0 - w_0 - (w - w_0)\|$ is well defined.

$((a) \Rightarrow (b))$: Let $\alpha = \epsilon/2$. By (2.1.12), any element of $\mathsf{H}\{a_0 \pm \alpha b_i\}_1^p$ has the form $\sum_1^p \lambda_i \ (a_0 + \beta_i b_i)$ where $\lambda_i \geqslant 0$, $\sum \lambda_i = 1$, and $|\beta_i| \leqslant \alpha$ for all i.

Observe that

$$\|a_0 - \sum_1^p \lambda_i (a_0 + \beta_i b_i)\| = \|\sum_1^p \lambda_i \beta_i b_i\| = \sum_1^p \lambda_i |\beta_i| \leqslant \alpha < \epsilon .$$

$((b) \Rightarrow (a))$: Suppose $\alpha > 0$ is such that $W \cap \mathsf{H}\{a_0 \pm \alpha b_i\}_1^p \subset A$. Let $\epsilon = \alpha$. Take $w \in W$ such that $\|w - a_0\| < \alpha$. Now, $w - a_0 = \sum_1^p \eta_i b_i$ for some η_i and so $\sum_1^p |\eta_i| < \alpha$. Observe

$$\begin{aligned} w &= a_0 + \sum_1^p \eta_i b_i \\ &= \sum_1^p \frac{|\eta_i|}{\alpha} (a_0 + (\mathrm{sgn}\ \eta_i)\alpha b_i) + \frac{\beta}{2}(a_0 - \alpha b_1) + \frac{\beta}{2}(a_0 + \alpha b_1) \end{aligned}$$

where $\beta = 1 - \sum_1^p \frac{|\eta_i|}{\alpha}$. So, $w \in W \cap \mathsf{H}\{a_0 \pm \alpha b_i\}_1^p \subset A$. □

Recall the definition of relative topology.

(2.2.7) Definition: Let (S, T) be a topological space. Let $R \subset S$. Then (R, U) is a topological space where $\mathsf{U} = \{T \cap R : T \in \mathsf{T}\}$ is the *relative topology* or the *relativization of* T *to* R.

Any norm which makes X into a normed linear space makes X into a topological vector space (see Rudin (1973) for definitions). It turns out that all topologies which make X into a topological vector space are the same (see p. 15-16, Rudin (1973)). This fact coupled with the preceding theorem make the next theorem easy.

(2.2.8) Theorem: For any choice of basis B for X, Γ is the same as the only vector topology X can have and consequently is the same as any norm-induced topology for X.

For any $W \subset X$ with at least two distinct points and any choice of basis B, Γ_W is the same as the relativization of Γ to W, the relativization of the unique vector topology on X to W, and the topology for W generated by the set of W-open balls $\{\{w \in W: \|w - a\| < \epsilon\}: a \in W, \epsilon > 0\}$ for any norm on W.

Proof: Using the information from Rudin and the fact that the metric-induced topology for a subset of a metric space is the same as the relativization of the parent metric topology to that subset, it will suffice to show for any nonempty, non-singleton $W \subset X$ that Γ_W is the same as the topology induced on W by some convenient norm. Using the norm in (2.2.6), observe $\Gamma_W = \{G \subset W\colon$ for all $g \in G$ there exists $\epsilon > 0$ such that $\{w \in W\colon \|g-w\| < \epsilon\} \subset G\}$. By extending $\{b_i\}_1^p$ to a basis $\{b_i\}_1^d$ for X if necessary, the norm defined for $\mathsf{L}\{W-w_0\}$ may be extended to a norm defined for all of X via $\|x\| := \sum_1^d |\xi_i|$. Then $\Gamma_W = \{G \subset W\colon$ for all $g \in G$ there exists $\epsilon > 0$ such that $W \cap \{x \in X\colon \|g-x\| < \epsilon\} \subset G\}$. □

-- *relative closures, interiors, and boundaries* --

Closed set, interior, closure, and boundary in the standard way as in Kelley [25].

When one tries to visualize a lower dimensional set $A \subset \mathsf{R}^d$ with respect to analyzing its internal geometry, it is natural to picture this set A relative to its dimensionality space, $\mathsf{M}(A)$. For example, in R^3 where the point x has coordinates (ξ_1, ξ_2, ξ_3), picture a closed circular disk lying in the $\xi_3 = 1$ hyperplane. Clearly this disk has no interior with respect to the usual topology on R^3. However it has a rather natural looking interior consisting of the open disk when considered with respect to the usual R^3 topology relativized to the $\xi_3 = 1$ hyperplane.

(2.2.9) Definition: Let $\emptyset \neq A \subset \mathsf{R}^d$. rel int A, i.e., the *relative interior* of A, is the interior of A relative to $\Gamma_{\mathsf{M}(A)}$. The *relative closure* of A is the closure of A relative to $\Gamma_{\mathsf{M}(A)}$. rel ∂A, i.e., *relative boundary* of A, is the boundary of A relative to $\Gamma_{\mathsf{M}(A)}$. In short, when the word relative is used in a topological context without specifying the background space W, then W is understood to be the dimensionality space of whatever set is being discussed. Note rel int A = int A if $\mathsf{M}(A) = \mathsf{R}^d$.

The advantages of working with linear manifolds W will become apparent in the next few theorems.

(2.2.10) Theorem: Suppose W is a non-singleton linear manifold. Let $w_0 \in W$ and $B = \{b_i\}_1^p$ be a basis for $\mathsf{L}(W - w_0)$. Then $\Gamma_W = \{G \subset W:$ for all $g \in G$ there exists $\alpha > 0$ such that $\mathsf{H}\{g \pm \alpha b_i\}_1^p \subset G\}$.

Proof: It suffices to show that for all $w_1 \in W$ and all $\alpha > 0$, $\mathsf{H}\{w_1 \pm \alpha b_i\}_1^p \cap W = \mathsf{H}\{w_1 \pm \alpha b_i\}_1^p$. Note that since $\pm \alpha b_i \in \mathsf{L}(W - w_0) = \mathsf{L}(W - w_1)$,

$$\{w_1 \pm \alpha b_i\}_1^p \subset w_1 + \mathsf{L}(W - w_1) = \mathsf{M}(W) = W.$$

Hence, $\mathsf{H}\{w_1 \pm \alpha b_i\}_1^p \subset \mathsf{H}(W) = W$. □

The next theorem characterizes the relative interior of a set.

(2.2.11) Theorem: Let $a_0 \in A \subset X$ with A non-singleton. Let $B = \{b_i\}_1^p$ be a basis for $\mathsf{L}(A - a_0)$. Then the following statements are equivalent for each $a_1 \in A$:

(a) $a_1 \in \text{rel int } A$

(b) there exists $\alpha > 0$ such that $\mathsf{H}\{a_1 \pm \alpha b_i\}_1^p \subset A$.

Note that if A is convex, then (b) may be replaced by (b'):

(b') there exists $\alpha > 0$ such that $\{a_1 \pm \alpha b_i\}_1^p \subset A$.

Proof: ($\Rightarrow$) Since rel int A is nonempty and open in $\mathsf{M}(A)$, (2.2.10) can be used to obtain $\alpha > 0$ such that $\mathsf{H}\{a_1 \pm \alpha b_i\}_1^p \subset \text{rel int } A \subset A$.

($\Leftarrow$) By using the norm defined in (2.2.6) suitably extended to X as in the proof of (2.2.8), one obtains $\epsilon > 0$ such that

$$\{x \in X: \|a_1 - x\| < \epsilon\} \cap \mathsf{M}(A) \subset A.$$

Since a_1 is an element of an $\mathsf{M}(A)$ -open ball contained in A, $a_1 \in \text{rel int } A$. □

(2.2.12) Example: Consider showing that, for $\tilde{v} \neq 0$, int $\{x \in X: [x, \tilde{v}] \geqslant 0\} = \{x: [x, \tilde{v}] > 0\}$. The RHS was shown to be open in (2.2.5). Any larger open set contained in the closed halfspace would have to contain x_0 such that $[x_0, \tilde{v}] = 0$. Let $\{b_i\}_1^d$ be a basis for X. There must be b_j such that $[b_j, \tilde{v}] \neq 0$. Note that for all $\alpha > 0$, $[x_0 \pm \alpha b_j, \tilde{v}] = \pm \alpha [b_j, \tilde{v}]$.

As further examples of the utility of Theorem (2.2.11), see (2.3.37) and (2.4.9).

The next theorem says that a convex set has an interior relative to a linear manifold W which contains it if and only if they are of the same dimension. It also proves that the relative interior of nonempty convex sets is always nonempty.

(2.2.13) Theorem: Let $\varnothing \neq K \subset \mathsf{M}(K) \subset W \subset X$ where K is convex and W is a non-singleton linear manifold. Then $\mathsf{int}\, K \neq \varnothing$ (relative to Γ_W) if and only if $W = \mathsf{M}(K)$.

This theorem has two special cases, one where $W = X$ which speaks for itself and the other where $W = \mathsf{M}(K)$ which leads to the conclusion that $\mathsf{rel\ int}\, K \neq \varnothing$ for nonempty convex K, the case for singleton K being trivial.

Proof: ($\Rightarrow$): Let $k_0 \in \mathsf{int}\, K$ relative to Γ_W. Since $\varnothing \neq \mathsf{int}\, K \in \Gamma_W$, choose a basis $\{b_i\}_1^p$ for $\mathsf{L}(W - k_0)$ and obtain via (2.2.10) an $\alpha > 0$ such that $\mathsf{H}\{k_0 \pm \alpha b_i\}_1^p \subset \mathsf{int}\, K \subset K$. Now $\mathsf{L}(K - k_0) \subset \mathsf{L}(W - k_0)$, so, in order to show $\mathsf{M}(K) = W$, it suffices to show $\dim \mathsf{L}(K - k_0) \geqslant p$. This follows from (2.1.18) since $\{\alpha b_i\}_1^p \subset K - k_0$.

($\Leftarrow$): Since $\mathsf{M}(K) = W$, $\mathsf{L}(K - k_0) = \mathsf{L}(W - k_0)$. By (2.1.17), there exists a basis $\{k_i - k_0\}_1^p \subset K - k_0$ for $\mathsf{L}(K - k_0)$ and hence for $\mathsf{L}(W - k_0)$. Note that $\mathsf{H}\{k_i\}_0^p \subset K$ and is a simplex. Let $\bar{k} = \sum_0^p \frac{1}{p+1} k_i$ be the centroid of this simplex.

To show $\bar{k} \in \mathsf{int}\, K$, begin by showing that there is an $\alpha > 0$ such that $\{\bar{k} \pm \alpha(k_i - k_0)\}_1^p \subset K$. Taking $0 < \alpha < \frac{1}{p+1}$, observe that, for $i = 1, \ldots, p$,

$$\bar{k} \pm \alpha(k_i - k_0) = \sum_{j \neq 0,i} \frac{1}{p+1} k_j + (\frac{1}{p+1} \pm \alpha) k_i + (\frac{1}{p+1} \mp \alpha) k_0 \in K$$

Now use (2.2.11). □

(2.2.14) Comment: Note that although int $K \neq \emptyset$ implies $\mathsf{M}(K) = W$ even when K is not convex, the converse is not true. To see this, let $W = \mathsf{R}^2$ for $d = 2$ and consider three points not all on a line. This three point set has no interior relative to R^2 yet its dimensionality space is R^2.

Summary for Section 2.2

The usual topology on R^d and more generally, the unique vector topology for any finite-dimensional vector space can be obtained without using a metric, norm, or inner product. This can be accomplished by defining a set $G \subset W \subset X$ to be open relative to W if and only if for every point in G, there is a polyhedron of the same dimensionality as W which when intersected with W both contains that point in its "middle" and is itself contained in G.

This discussion provided the tools for introducing the relative topology for a set A in R^d, namely the above topology relative to $\mathsf{M}(A)$. This led to defining the concepts of relative interior and relative boundary.

Relative interior points were characterized. Lastly, it was shown that a convex set has an interior relative to a containing linear manifold W if and only if they are of the same dimension.

Section 2.3:

Polyhedral Convex Cones

This is the section which introduces and develops the basic characteristics of polyhedral convex cones. The first topic, however, is indexing.

-- indexing --

Let $I := \{0,1,2, \ldots, n\}$ for some n. Consider the set $A = \{a_i, i \in I\}$. By convention, $a_0 := 0$. For each $j \in I$, $I_j := \{i \in I : \langle a_i \rangle = \langle a_j \rangle\}$ where, recall, $\langle x \rangle := \{\alpha x : \alpha > 0\}$. Consequently, for fixed $j \in I$, $\{a_i, i \in I_j\}$ is the set of vectors in A which generate the same open ray as a_j. The care taken in this chapter to force 0 into A and to keep track of vectors in A which generate the same open ray, in fact, to survive all of this bookkeeping, will greatly simplify matters in the next chapter.

-- polyhedral and finite cones --

The next two definitions define two types of cones which, in fact, turn out to be identical.

(2.3.1) Definition: Let $\emptyset \neq C \subset X$. C is a *polyhedral cone* if and only if $C = \{x : [x, \tilde{b}_j] \geqslant 0, j = 1, \ldots, n\}$ for some $\{\tilde{b}_j\}_1^n \subset \tilde{X}$. Polyhedral cones are also called polyhedral convex cones.

A polyhedral cone is the intersection of a finite number of closed halfspaces whose bounding hyperspaces pass through the origin. Such an object is easily seen to be a convex cone.

(2.3.2) Definition: Let $\emptyset \neq C \subset X$. C is a *finite* (or finitely generated) *cone* if and only if there is a finite set $\{a_i\}_1^m$ such that $C = \mathrm{C}\{a_i\}_1^m$.

An easy consequence of these two definitions is:

(2.3.3) Theorem: A finite sum (using vector addition) of finite cones is a finite cone. A finite intersection of polyhedral cones is a polyhedral cone.

(2.3.4) Theorem: (Minkowski-Weyl): Every finite cone is a polyhedral cone and vice-versa. More precisely, for each $A = \{a_i\}_1^m \subset X$, there exists $\tilde{B} = \{\tilde{b}_j\}_1^n \subset \tilde{X}$ and for each such $\tilde{B}$, there exists A such that $\mathsf{C}(A) = \{x \in X: [x, \tilde{b}_j] \geqslant 0 \text{ for } j = 1, \ldots, n\}$.

Proof: See Nering (1963), Goldman and Tucker (1956), or Stoer and Witzgall (1970). □

Even though these two types of cones are equivalent, the appropriate name is useful when emphasizing how certain cones are generated.

-- *examples of polyhedral cones* --

The following examples of finite/polyhedral convex cones serve to illustrate this theorem as well as other concepts later in this chapter and the next three.

(2.3.5) Example: Let $a_1 \in \mathsf{R}^d$. Then $\mathsf{C}\{a_1\} = \{\alpha a_1: \alpha \geqslant 0\}$ is a finite cone. Note that $\mathsf{C}\{a_1\}$ is the closed half-line or ray originating at 0 and passing through a_1 and as such is equal to $\langle 0 \rangle \cup \langle a_1 \rangle$.

To see that $\mathsf{C}\{a_1\}$, $a_1 \neq 0$, is also polyhedral, let $\{\tilde{v}_i\}_1^{d-1}$ be a basis for $a_1^\perp$ and let $\tilde{y}$ be such that $[a_1, \tilde{y}] > 0$. Then $\mathsf{C}\{a_1\} =$

$$\{x \in \mathsf{R}^d: [x, \tilde{y}] \geqslant 0, [x, \tilde{v}_i] \geqslant 0, [x, -\tilde{v}_i] \geqslant 0, i = 1, \ldots, d-1\}.$$

(2.3.6) Example: Any subspace S in R^d is a polyhedral convex cone since for proper $S \subset \mathsf{R}^d$,

$$S = \{x \in \mathsf{R}^d: [x, \tilde{v}_i] \geqslant 0 \text{ and } [x, -\tilde{v}_i] \geqslant 0 \text{ for } i = 1, \ldots, k\}$$

where $\{\tilde{v}_i\}_1^k$ is a basis for $S^\perp$.

To see that it is also a finite cone, let $\{b_1, \ldots, b_q\}$ be a basis for $S \neq \langle 0 \rangle$. Let $b_{q+1} = -\sum_1^q b_i$. The claim is that $S = \mathsf{C}\{b_i\}_1^{q+1}$. Note that

$$\mathsf{C}\{b_i\}_1^{q+1} \subset \mathsf{L}\{b_i\}_1^{q+1} = S.$$

So, take $s = \sum_1^q \sigma_i b_i$. Let $\gamma = \sup\{|\sigma_i| : \sigma_i < 0\}$. Observe

$$s = \sum_1^q (\sigma_i + \gamma) b_i + \gamma b_{q+1} \in \mathsf{C}\{b_i\}_1^{q+1}.$$

(2.3.7) Example: Every closed halfspace through the origin $\{x \in \mathsf{R}^d : [x, \tilde{a}] \geqslant 0\}$ with $\tilde{a} \neq 0$ is a finite cone. The argument here begins by denoting the $d-1$ dimensional subspace $\{x : [x, \tilde{a}] = 0\}$ by $S = \mathsf{C}\{y_i\}_1^d$ for suitable y_i. Now take any $x_0 \notin S$ and set y_{d+1} equal to either $-x_0$ or x_0 so that $[y_{d+1}, \tilde{a}] > 0$.

The assertion is that $\mathsf{C}\{y_i\}_1^{d+1} = \{x : [x, \tilde{a}] \geqslant 0\}$. Clearly $LHS \subset RHS$. For the other direction, take x such that $[x, \tilde{a}] \geqslant 0$. Let $\lambda = [x, \tilde{a}] / [y_{d+1}, \tilde{a}] \geqslant 0$. Then $x - \lambda y_{d+1} \in S$.

(2.3.8) Example: Consider next the polyhedral cone $C = \{x \in \mathsf{R}^d : [x, \tilde{a}_1] \geqslant 0 \text{ and } [x, \tilde{a}_2] \geqslant 0\}$ where $\{\tilde{a}_1, \tilde{a}_2\}$ is linearly independent. The stated conditions on the $\tilde{a}_i$ imply that there exist y_1, y_2 such that $[y_1, \tilde{a}_1] = 0$, $[y_1, \tilde{a}_2] > 0$, $[y_2, \tilde{a}_1] > 0$, and $[y_2, \tilde{a}_2] = 0$. C is a finite cone since for any $x \in C$, if one sets $\lambda_1 = [x, \tilde{a}_2] / [y_1, \tilde{a}_2] \geqslant 0$ and $\lambda_2 = [x, \tilde{a}_1] / [y_2, \tilde{a}_1] \geqslant 0$, then

$$x - \lambda_1 y_1 - \lambda_2 y_2 \in \{x : [x, \tilde{a}_1] = 0, [x, \tilde{a}_2] = 0\}.$$

Figure (2.3.9) shows C when $X = \mathsf{R}^2$. For $i = 1, 2$, the + signs indicate which of the two halves of R^2 defined by $\tilde{a}_i^{\perp}$ should be considered the positive halfspace $\{x \in \mathsf{R}^2 : [x, \tilde{a}_i] > 0\}$.

In R^3, $C = \{x \in \mathsf{R}^3 : [x, \tilde{a}_1] \geqslant 0 \text{ and } [x, \tilde{a}_2] \geqslant 0\}$ looks like a wedge. This wedge is a very useful example of a polyhedral cone which is not a

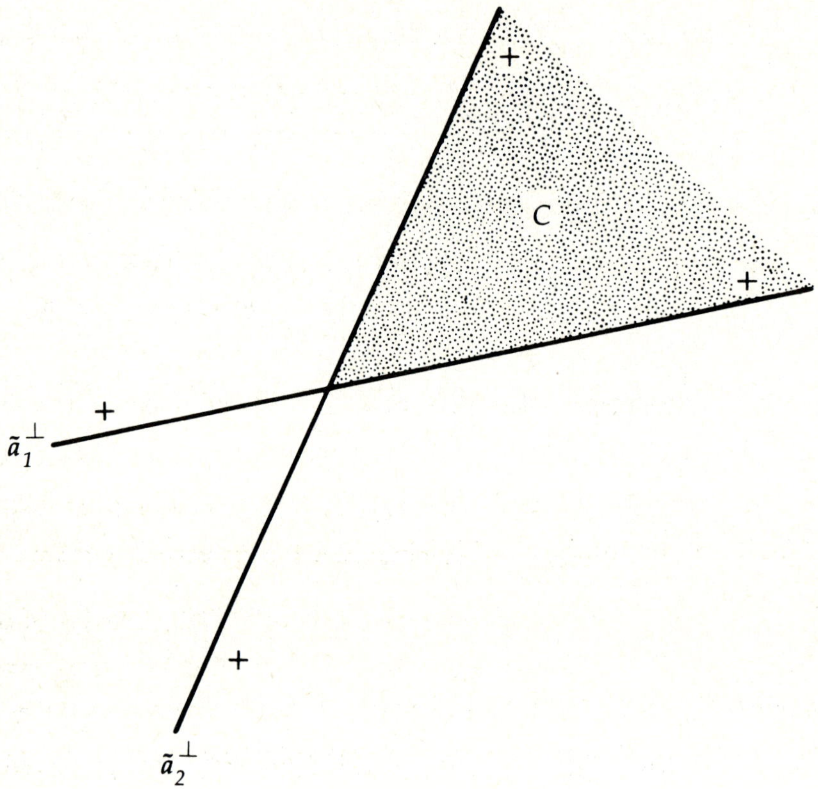

(2.3.9) Figure: A polyhedral cone C in R^2.

subspace but yet has a non-trivial lineality space (which in this case is $\tilde{a}_1^\perp \cap \tilde{a}_2^\perp$).

(2.3.10) Example: In the context of R^3, let $a_1 = (1, 0, 1)$, $a_2 = (0, 1, 1)$, $a_3 = (-1, 0, 1)$, and $a_4 = (0, -1, 1)$. Consider the finite cone $\mathsf{C}\{a_i\}_1^4$. After visualizing this cone in R^3, one can see that it is the intersection of the appropriate halfspaces associated with the planes generated by each pair of adjacent a_i.

-- rays as points --

Finite cones have a number of interesting properties. The first one is that they may be viewed as being the convex hulls of rays in the same way as bounded polyhedra are considered to be convex hulls of points. In short, a useful way of viewing $\mathrm{C}\{x_i\}_1^n$ is as $\mathrm{H}(\langle 0\rangle \cup \bigcup_1^n \langle x_i\rangle)$. Since, by the conventions made at the start of this section, $a_0 = 0$, it is possible to write more compactly for $\{a_i, i \in I\}$, $\mathrm{C}\{a_i, i \in I\} = \mathrm{H}(\bigcup_I \langle a_i\rangle)$. The notation will be slightly abused subsequently when the last expression is written as $\mathrm{H}\{\langle a_i\rangle, i \in I\}$. This is done to emphasize the idea that the open rays $\langle a_i\rangle$ for $i \in I$ may be thought of as "points" for which

$$\mathrm{C}\{a_i, i \in I\} = \bigcup\{\sum_I \lambda_i \langle a_i\rangle : \lambda_i \geqslant 0, \sum_I \lambda_i = 1\}$$

is just the convex hull of these "points". In short, the reader will discover as he reads further, particularly if he tries to do it the other way, that the basic objects constructing $\mathrm{C}\{a_i, i \in I\}$ are not the a_i but rather the $\langle a_i\rangle$. However, even though this is the case, it will at times be notationally convenient to work with the a_i instead of the $\langle a_i\rangle$.

-- isolated rays of finite cones --

The next theorem follows easily from the definition of isolated subset and is needed in order to define the frame of a finite cone.

(2.3.11) Theorem: Let the finite cone $B = \mathrm{H}\{\langle a_i\rangle, i \in I\} \subset X$ and suppose $\langle z\rangle \subset B$. Then the following are equivalent:

(a) $\langle z\rangle$ is an isolated subset of B.

(b) it is not the case that there exists $y_1, y_2 \in B$ such that $z = y_1 + y_2$ and for all $\beta > 0$, $y_1 \neq \beta z$ and $y_2 \neq \beta z$.

(c) it is not the case that there exist $\langle y_1\rangle, \langle y_2\rangle \subset B$ such that $\langle z\rangle \subset \langle y_1\rangle + \langle y_2\rangle$ and $\langle y_1\rangle \neq \langle z\rangle$ and $\langle y_2\rangle \neq \langle z\rangle$.

The reader may want to algebraically verify the following visually obvious examples:

(2.3.12) Example: For $\mathsf{C}\{a_1\}$, $a_1 \neq 0$, both $\langle a_1\rangle$ and $\langle 0\rangle$ are isolated subsets of $\mathsf{C}\{a_1\}$.

(2.3.13) Example: For every $\langle 0\rangle \neq \langle x\rangle \subset S$, where S is a subspace of X, $\langle x\rangle$ is an isolated subset of S if and only if $\dim S = 1$.

(2.3.14) Example: Consider a wedge in R^3 (cf., Example (2.3.8)). If $\mathsf{L}\{x_0\} = \tilde{a}_1^{\perp} \cap \tilde{a}_2^{\perp}$, then $\langle x_0\rangle$ and $\langle -x_0\rangle$ are isolated rays of the wedge whereas no other open ray in the wedge is isolated.

(2.3.15) Example: Consider the cone of Example (2.3.10). $\langle 0\rangle$ and $\langle a_i\rangle$, $i = 1, \ldots, 4$, are the only isolated open rays of this cone.

Part (c) of Theorem (2.3.11) is of interest because it is the result of the formal substitution of open rays for points in the definition of extreme point (2.1.39). This is further evidence that the open rays of a cone should be thought of as "points".

The treatment of polyhedral cones in this chapter differs from that of Gerstenhaber in two ways. The first is that Gerstenhaber works with closed rays $\{\alpha x : \alpha \geqslant 0\}$ instead of open rays $\langle x\rangle$. Open rays are used in this presentation because any point of the ray can be used to generate the ray whereas 0 cannot be used to generate $\{\alpha x : \alpha \geqslant 0\}$ for $x \neq 0$. Thus, in some sense, the open ray is a more homogeneous set of points than the closed ray. The open ray is also compatible with Goldman and Tucker's faces (to be discussed later) which are here preferred over Gerstenhaber's facets, again for reasons of homogeneity.

Second, the Gerstenhaber definition of an extreme closed ray does not agree with the Theorem-Definition (2.3.11) of isolated open ray. For a counter-example, note that the rays $\langle x_0 \rangle$, $\langle -x_0 \rangle \neq 0$ contained in the line contained in the 3-dimensional wedge of (2.3.8) are isolated whereas, for those familiar with Gerstenhaber's paper, neither $\{\alpha x_0 \colon \alpha \geqslant 0\}$ nor $\{-\alpha x_0 \colon \alpha \geqslant 0\}$ are extreme closed rays in this wedge by Gerstenhaber's definition.

The next theorem gives a necessary and sufficient condition for $\langle 0 \rangle$ to be an isolated ray of a polyhedral cone.

(2.3.16) Theorem: Let $C \subset X$ be a convex cone. Then $\langle 0 \rangle$ is isolated if and only if $\text{Lin}\ C = \{0\}$.

Proof: ($\Rightarrow$) Suppose $\text{Lin}\ C \neq \{0\}$. Then there exists $x_0 \in \text{Lin}\ C$, $x_0 \neq 0$, such that $0 = x_0 + (-x_0)$ and so $\langle 0 \rangle$ is not isolated.

($\Leftarrow$) Suppose $\langle 0 \rangle$ is not isolated. Then there exist $y_1, y_2 \in C$, $y_1, y_2 \neq 0$, such that $0 = y_1 + y_2$. Hence $y_2, -y_2 \in C$ and $\dim \text{Lin}\ C \geqslant 1$. □

(2.3.17) Theorem: Let $A = \{\langle a_i \rangle, i \in I\} \subset X$ where A has at least two distinct rays. If $\langle a_j \rangle$ is an isolated ray of $\text{H}(A)$, then $\langle a_j \rangle \parallel \text{H}\{\langle a_i \rangle, i \in I \text{ w.o. } I_j\}$. ("$\parallel$" is read "is disjoint from ".)

Proof: $\{\langle a_i \rangle, i \in I \text{ w.o. } I_j\} \subset \text{H}(A) \text{ w.o. } \langle a_j \rangle$ which is convex by (2.1.37) and so $\text{H}\{\langle a_i \rangle, i \in I \text{ w.o. } I_j\} \subset \text{H}(A) \text{ w.o. } \langle a_j \rangle \parallel \langle a_j \rangle$. □

To see that the converse does not hold, let $a_1 = (1, 0, 0)$, $a_2 = (0, 1, 0)$, $a_3 = (-1, -1, 0)$, and $a_4 = (0, 0, 1)$ and consider the halfspace $\text{H}\{\langle a_i \rangle\}_1^4 \subset \text{R}^3$. Note that $\langle a_1 \rangle$ is not isolated yet $\langle a_1 \rangle \parallel \text{H}\{\langle a_2 \rangle, \langle a_3 \rangle, \langle a_4 \rangle\}$. A partial converse exists. See (2.3.32).

-- *frames of finite cones* --

Now for the definition of frame.

(2.3.18) Definitions: Let $A = \{\langle a_i \rangle, i \in I\} \neq \varnothing$ since $0 \in I$ and let C be a convex cone in X.

(a) A is *conically independent* if and only if for all $j \in I$, $\langle a_j \rangle \parallel \mathsf{H}\{\langle a_i \rangle, i \in I \text{ w.o. } j\}$.

(b) A is a *conical spanning set* for C if and only if $\mathsf{H}(A) = C$.

(c) A is a *frame* (or conical basis) for C if and only if A is a conically independent conical spanning set for C.

(2.3.19) Examples: All of the conical spanning sets given in Examples (2.3.5) to (2.3.10) are frames.

(2.3.20) Remarks: Note that for a conically independent set, different indices correspond to different rays. Note also the similarity to linear independence as defined in (2.1.14).

The theory for bases of vector spaces is only incompletely paralleled here for frames of finite cones. For example, while it is easy to see that every conical spanning set of smallest cardinality is conically independent, it turns out that there are frames which are not minimal in size. (See Example (2.3.23)). For more details, see Davis (1954). Also, as another difference, certain rays must be in any frame:

(2.3.21) Theorem: Let $C \subset X$ be a convex cone. Let $A = \{\langle a_i \rangle, i \in I\}$ be a conical spanning set for C and let $\langle y \rangle$ be an isolated ray of C. Then $\langle y \rangle \in A$.

Proof: For some index $k \notin I$, set $\langle a_k \rangle = \langle y \rangle$. Suppose $\langle a_k \rangle \notin A$. Since $\{\langle a_i \rangle, i \in I \cup \{k\}\} \subset C$, $\mathsf{H}\{\langle a_i \rangle, i \in I \cup \{k\}\} \subset C$. On the other hand, $C = \mathsf{H}(A) \subset \mathsf{H}\{\langle a_i \rangle, i \in I \cup \{k\}\}$. Consequently, $C = \mathsf{H}\{\langle a_i \rangle, i \in I \cup \{k\}\}$ and by (2.3.17), $\langle a_k \rangle \parallel \mathsf{H}(A) = C$, which is a contradiction. □

The next theorem shows that every finite cone has a frame and shows how one may be obtained. Appropriate analogues of the procedure given here will produce a basis for $\mathsf{L}\{a_i\}_1^n$ and the set of extreme points for $\mathsf{H}\{a_i\}_1^n$.

(2.3.22) Theorem: Let $C = \mathsf{H}\{\langle a_i \rangle,\ i \in I\}$ be a finite cone. Then a frame for C exists and may be obtained through the following procedure:

(This procedure is written as an algorithm in a hopefully self-explanatory hybrid of Fortran, BASIC, and English which will also be used in subsequent theorems.)

Set $K_{-1} = I = \{0, \ldots, n\}$.

For $j = 0, \ldots, n$ do:

If $\langle a_j \rangle \subset \mathsf{H}\{\langle a_k \rangle,\ k \in K_{j-1}$ w.o. $j\}$ then set $K_j = K_{j-1}$ w.o. j.

Else set $K_j = K_{j-1}$.

next j;

The set $\{\langle a_i \rangle,\ i \in K_n\}$ is a frame.

Proof: Claim: For $j = 0, \ldots, n$, $\mathsf{H}\{\langle a_k \rangle,\ k \in K_{j-1}\} = \mathsf{H}\{\langle a_k \rangle,\ k \in K_j\}$.

Fix j. If $K_j = K_{j-1}$, then the claim follows. Suppose then that $\langle a_j \rangle \subset \mathsf{H}\{\langle a_k \rangle,\ k \in K_{j-1}$ w.o. $j\}$. To show $\mathsf{H}\{\langle a_k \rangle,\ k \in K_{j-1}$ w.o. $j\} = \mathsf{H}\{\langle a_k \rangle,\ k \in K_{j-1}\}$, first note that the *LHS* $\subset$ *RHS* inclusion is trivial. For the other inclusion, since $a_j = \sum_{K_{j-1} \text{ w.o. } j} \lambda_k a_k$ for $\lambda_k \geqslant 0$ but not all 0, it is easy to see that any positive combination of a_k, $k \in K_{j-1}$, is a positive combination of a_k, $k \in K_{j-1}$ w.o. j. So, $\{\langle a_i \rangle,\ i \in K_n\}$ is a conical spanning set.

To show that it is conically independent, suppose for $j \in K_n$, $\langle a_j \rangle \subset \mathsf{H}\{\langle a_k \rangle,\ k \in K_n$ w.o. $j\}$. Then since $K_j \subset K_{j-1}$ for all j, $\langle a_j \rangle \subset \mathsf{H}\{\langle a_k \rangle,\ k \in K_{j-1}$ w.o. $j\}$, which implies $j \notin K_n$, a contradiction. □

(2.3.23) Example: See Figure (2.3.24).

(2.3.25) Remark: The problem of determining whether or not $\langle b \rangle \subset \mathsf{H}\{\langle a_i \rangle,\ i=1, \ldots, p\}$ can be solved using linear programming. Note that $\langle b \rangle \subset \mathsf{H}\{\langle a_i \rangle,\ i=1, \ldots, p\}$ if and only if there exist $\xi_i \geqslant 0$, not all 0, such that $b = \sum_1^p \xi_i a_i$. The case where $b = 0$ is deferred to a more

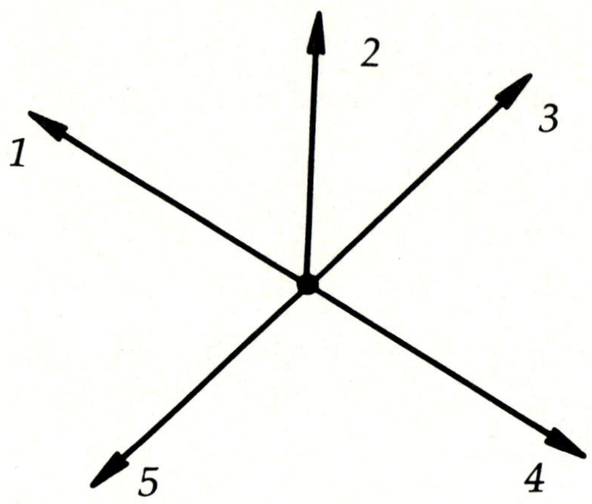

(2.3.24) Figure: Five rays in R^2. Note that {2, 4, 5} and {1, 3, 4, 5} are both frames for the subspace (namely R^2) conically spanned by the five rays.

appropriate time, namely, (2.3.33). When $b \neq 0$, then the condition that not all ξ_i equal 0 can be dropped. Writing $\underset{\sim}{A} = [\underline{a}_1 \cdots \underline{a}_p]$ and $\underline{x} = (\xi_1, \ldots, \xi_p)$, the problem reduces to that of finding whether or not the standard linear programming problem, maximize $\underline{c}^T \underline{x}$ subject to $\underset{\sim}{A}\underline{x} = \underline{b}$ for $\underline{x} \geqslant 0$, is feasible.

For a more efficient way of finding the frame, see Wets and Witzgall (1967).

-- *pointed cones* --

It is important to know when a cone looks like the common conception of a cone.

(2.3.26) Definition: A convex cone C is *pointed* if and only if $\mathsf{Lin}\ C = \{0\}$.

(2.3.27) Examples: The cones in Examples (2.3.5), (2.3.6) when $\dim S = 0$, (2.3.7) when $d = 1$, (2.3.8) when $d = 2$, and (2.3.10) are pointed. See Figure (2.3.9) for a picture of a pointed cone and Figure (2.1.31) (c) for a picture of a cone which is not pointed.

Gordan's Theorem is useful for determining whether or not polyhedral cones are pointed.

(2.3.28) Theorem: (Gordan): Let $\{b_j\}_1^m \subset X$. The following statements are equivalent:

(a) There is $\tilde{x} \in \tilde{X}$ such that $[b_j, \tilde{x}] > 0$ for $j = 1, \ldots, m$.

(b) There does not exist $\{\lambda_j\}_1^m$ with $\lambda_j \geqslant 0$, not all 0, such that $0 = \sum_1^m \lambda_j b_j$.

Proof: See Gale (1960) or Stoer and Witzgall (1970). Note that (a) easily implies (b). □

Here is the connection between Gordan's Theorem and pointed cones:

(2.3.29) Theorem: Let $C = \mathsf{H}\{\langle a_i \rangle, i \in I\} \subset X$. (Remember $I = \{0, 1, 2, \ldots, n\}$ for some n and $a_0 = 0$). Suppose $C \neq \langle 0 \rangle$. The following statements are equivalent:

(a) C is pointed.

(b) There is no $x \neq 0$ in C such that $-x \in C$.

(c) $\langle 0 \rangle$ is isolated.

(d) $\langle 0 \rangle \parallel \mathsf{H}\{\langle a_i \rangle, i \in I \text{ w.o. } I_0\}$

(e) There does not exist $\lambda_i \geqslant 0$, not all 0, for $i \in I$ w.o. I_0 such that $0 = \sum_{I \text{ w.o. } I_0} \lambda_i a_i$.

(f) There is $\tilde{x} \in \tilde{X}$ such that $[a_i, \tilde{x}] > 0$ for $i \in I$ w.o. I_0.

(g) $\mathsf{H}\{\langle a_i \rangle, i \in I \text{ w.o. } I_0\}$ lies in the interior of some closed halfspace whose boundary passes through the origin.

Proof: (b) easily implies (a). (a) and (c) are equivalent by (2.3.16). (c) implies (d) by (2.3.17). (d) and (e) rephrase each other. (e) and (f) are equivalent by Gordan's Theorem.

To show that (e) implies (b), suppose there exists $x = \sum_{I \text{ w.o. } I_0} \alpha_i a_i = -\sum_{I \text{ w.o. } I_0} \beta_i a_i \neq 0$ where $\alpha_i, \beta_i \geqslant 0$, not all $\alpha_i = 0$, and not all $\beta_i = 0$. Then $\sum_{I \text{ w.o. } I_0} (\alpha_i + \beta_i) a_i = 0$.

Clearly (g) implies (f). Assume (f) holds and consider $\sum_{I \text{ w.o. } I_0} \lambda_i a_i$ where $\lambda_i \geqslant 0$, not all 0. Observe $[\sum_{I \text{ w.o. } I_0} \lambda_i a_i, \tilde{x}] > 0$ and use (2.2.10). □

(2.3.30) Example: In Counter-example (3.5.6) of the next chapter, it will be necessary to show a certain cone in R^3 is pointed. It may be instructive to do that here. Let $\underset{\sim}{A} = [\underline{a}_1\ \underline{a}_2\ \underline{a}_3\ \underline{a}_4\ \underline{a}_5] = \begin{bmatrix} -1 & 0 & 1 & 0 & 0 \\ 1 & 1 & 1 & 1 & 0 \\ 0 & 0 & 0 & 1 & 1 \end{bmatrix}$ and suppose there exists $\underline{e} = (\lambda_1, \lambda_2, \lambda_3, \lambda_4, \lambda_5)$ with all $\lambda_i \geqslant 0$ and such that $\underline{0} = \underset{\sim}{A}\underline{e}$. Then $\lambda_3 = \lambda_1$, $\lambda_4 + \lambda_5 = 0$, and $\lambda_1 + \lambda_2 + \lambda_3 + \lambda_4 = 0$. Hence all $\lambda_i = 0$ and $\mathsf{C}\{a_i\}_1^5$ is pointed.

-- *frames of pointed cones* --

Pointed cones have frames that are unique upto the indexing of their elements.

(2.3.31) Theorem: Let $C \neq \langle 0 \rangle$ be a pointed finite cone in X. Then there is an essentially unique frame for C in the sense that if $\{\langle b_i \rangle, i \in J\}$ is a frame for C, then every $\langle b_i \rangle$ is an isolated ray of C and if $\langle c \rangle$ is an isolated ray of C, then there exists $i \in J$ such that $\langle c \rangle = \langle b_i \rangle$.

Proof: Let $B = \{\langle b_i \rangle, i = 0, \ldots, p\}$ be a frame for C. By (2.3.21), all isolated open rays of C are in B. Since C is pointed, $\langle 0 \rangle$ is isolated. Without loss of generality, let $\langle b_0 \rangle = \langle 0 \rangle$. The next step is to show that for $k = 1, \ldots, p$, $\langle b_k \rangle \neq \langle 0 \rangle$ is isolated.

Suppose for some $k = 1, \ldots, p$, $\langle b_k \rangle$ is not isolated. Then there exist $\langle y_1 \rangle, \langle y_2 \rangle \subset C$ such that $b_k = y_1 + y_2$ and $\langle b_k \rangle \neq \langle y_1 \rangle$ and $\langle b_k \rangle \neq \langle y_2 \rangle$. Observe that for $i = 1, 2$, there exists $\lambda_{ij} \geqslant 0$ such that $y_i = \sum_{j=1}^{p} \lambda_{ij} b_j$. Now, if $p = 1$, then $b_1 = (\lambda_{11} + \lambda_{21}) b_1$ and since $b_1 \neq 0$, it must be that $\lambda_{11} + \lambda_{21} = 1$. This implies either $\langle b_1 \rangle = \langle y_1 \rangle$ or $\langle b_1 \rangle = \langle y_2 \rangle$, a contradiction. So, suppose $p \geqslant 2$. For each $i = 1, 2$, there is $j \neq k$ such that $\lambda_{ij} > 0$ since $\langle b_k \rangle \neq \langle y_i \rangle$. Observe $b_k = (\lambda_{1k} + \lambda_{2k}) b_k + \sum_{j \neq 0,k} (\lambda_{1j} + \lambda_{2j}) b_j$. If $\lambda_{1k} + \lambda_{2k} < 1$, then $\langle b_k \rangle \subset \mathsf{H}\{\langle b_j \rangle, j \neq 0,k\}$. If $\lambda_{1k} + \lambda_{2k} \geqslant 1$, then C is not pointed. □

One can in fact show that a finite cone C is pointed if and only if it has an essentially unique frame.

The following is a partial converse to (2.3.17).

(2.3.32) Theorem: Let $C = \mathsf{H}\{\langle a_i \rangle, i \in I\} \neq \langle 0 \rangle$. Suppose C is pointed. Fix $j \in I$. The following statements are equivalent:

(a) $\langle a_j \rangle \parallel \mathsf{H}\{\langle a_i \rangle, i \in I \text{ w.o. } I_j\}$

(b) $\langle a_j \rangle$ is an isolated ray of C.

(c) $\langle a_j \rangle$ is in the frame of C.

Proof: (b) and (c) are equivalent by (2.3.31). (2.3.17) shows that (b) implies (a). Suppose (a) holds. Then the frame construction procedure of (2.3.22) will choose the ray $\langle a_j \rangle$ (ignore the index) for the frame after the indices to all duplicate rays have been deleted from the K_i earlier in the procedure. □

The computational problem of determining whether or not $\mathsf{H}\{\langle a_i \rangle, i \in I\}$ is pointed can be solved by using linear programming. The following LP is a useful component of the algorithms to be described in the next chapter.

(2.3.33) Theorem: Let $\{a_i\}_1^p \subset X$, all $a_i \neq 0$. Consider the following LP and its dual:

Primal: maximize γ subject to $\gamma \leqslant 1$ and $\gamma \leqslant \underline{a_i}^T \underline{x}$ for $i = 1, \ldots, p$ and where $\underline{x} \in \mathsf{R}^d$.

Dual: minimize δ subject to $\sum_1^p \zeta_i \underline{a_i} = \underline{0}$, $\sum_1^p \zeta_i + \delta = 1$, $\underline{z} = (\zeta_1, \ldots, \zeta_p) \geqslant 0$, and $\delta \geqslant 0$.

Then the optimal value of the objective function exists and is either 0 or 1. If the value is 0, then $\{\underline{x} : \underline{a_i}^T \underline{x} > 0, \text{ all } i\} = \varnothing$. If the value is 1, then the solution vector x_0 for the primal problem is such that $\underline{a_i}^T \underline{x_0} > 0$ for $i = 1, \ldots, p$.

Proof: Note that since the primal program is feasible with $\underline{x} = \underline{0}$ and $\gamma = 0$ and since the dual is feasible with $\underline{z} = \underline{0}$ and $\delta = 1$, a solution exists. The optimal objective function value must be in $[0, 1]$.

Suppose the optimal value is 0. Then there exist $\zeta_i \geqslant 0$, not all 0, such that $\sum_1^p \zeta_i \underline{a_i} = \underline{0}$. Hence by Gordan's Theorem, $\{\underline{x} : \underline{a_i}^T \underline{x} > 0 \text{ for all } i\} = \varnothing$.

If the optimal value is 1, then x_0 is such that $1 \leqslant \underline{a_i}^T \underline{x_0}$ for all i.

Suppose the optimal value $\gamma^* = \delta^*$ is greater than 0 and less than 1. Then once again, there exist $\zeta_i \geqslant 0$, not all 0, such that $\sum_1^p \zeta_i \underline{a_i} = \underline{0}$. This contradicts the existence of $\underline{x_0}$ such that $0 < \delta^* < \underline{a_i}^T \underline{x_0}$ for all i. □

-- pointed position --

The concept of pointed position will subsequently prove to be a more natural sufficient condition for certain results to hold than the oft-used concept of general position (cf., (2.1.19)).

(2.3.34) Definition: $\{a_i\}_1^p \subset X$ is in *pointed position* if and only if for all nonempty subsets J of $\{1, \ldots, p\}$, if $\{x_i, i \in J\}^\perp \neq \langle 0 \rangle$, then $\mathsf{C}\{x_i, i \in J\}$ is pointed.

Note that every set in general position is in pointed position. Also, should one ever want to prove that a set is in pointed position, the LP of (2.3.33) will do the lion's share of the work.

-- making pointed cones --

The next theorem shows how a collection of nonzero vectors can be made to generate a pointed cone by multiplying certain vectors by -1. This is a crucial lemma for the next chapter.

(2.3.35) Theorem: Let $A = \{a_i\}_1^p$ be a set of nonzero vectors in X. Then there exist $\theta_1, \ldots, \theta_p \in \{-1, 1\}$ and $\tilde{x} \in \tilde{X}$ such that for $i = 1, \ldots, p$, $[\theta_i a_i, \tilde{x}] > 0$.

Proof: The proof follows from induction on $\dim \mathsf{L}(A)$.

Suppose $\dim \mathsf{L}(A) = 1$. Then, $\mathsf{L}\{a_1\} = \mathsf{L}(A)$. Now there exists $\tilde{x}_1 \in \tilde{X}$ such that $[a_1, \tilde{x}_1] \neq 0$, otherwise $a_1 = 0$. For each $i = 1, \ldots, p$, set $\theta_i = 1$ if $[a_i, \tilde{x}_1] > 0$ and $\theta_i = -1$ if $[a_i, \tilde{x}_1] < 0$. Since, for each i, $a_i \in \langle a_1 \rangle$ or $\langle -a_1 \rangle$, $[a_i, \tilde{x}_1] \neq 0$ for all i.

Suppose the theorem holds for all A such that $\dim \mathsf{L}(A) \leqslant k - 1$. Let $\dim \mathsf{L}\{a_i\}_1^p = k$ and suppose $\{a_i\}_1^k$ is a basis for $\mathsf{L}\{a_i\}_1^p$ (cf. (2.1.17)). By (e) of (2.3.29), $\mathsf{C}\{a_i\}_1^k$ is pointed and so there exists $\tilde{x}_1$ such that $[a_i, \tilde{x}_1] > 0$ for $i = 1, \ldots, k$. Set $\theta_i = 1$ for $i = 1, \ldots, k$. Now for $i = k + 1, \ldots, p$, if $[a_i, \tilde{x}_1] > 0$, set $\theta_i = 1$ and if $[a_i, \tilde{x}_1] < 0$, set $\theta_i = -1$. Let

$J = \{i : [a_i, \tilde{x}_1] = 0\}$. If $J = \emptyset$, then the theorem follows. Suppose $J \neq \emptyset$.

Consider $\mathsf{L}\{a_i\}_J \subset \mathsf{L}\{a_i\}_1^p$. Now $\mathsf{L}\{a_i\}_J$ must have dimension $\leqslant k - 1$ because if it had dimension k, then $\mathsf{L}\{a_i\}_J = \mathsf{L}\{a_i\}_1^p$ and $a_1 = \sum_J \alpha_i a_i$ for suitable α_i. This would imply $0 < [a_i, \tilde{x}_1] = \sum_J \alpha_i [a_i, \tilde{x}_1] = 0$.

So by the induction hypothesis, there exists $\tilde{x}_2 \in \tilde{X}$ and θ_j for $j \in J$ such that $[\theta_j a_j, \tilde{x}_2] > 0$ for all $j \in J$. Now set $\tilde{x} = \tilde{x}_1 + \alpha \tilde{x}_2$ where $\alpha > 0$ is to be determined. Clearly for all $j \in J$, $[\theta_j a_j, \tilde{x}] > 0$. Consider for $i \notin J$, $[\theta_i a_i, \tilde{x}_1] + \alpha[\theta_i a_i, \tilde{x}_2]$. If $[\theta_i a_i, \tilde{x}_2] \geqslant 0$, then any $\alpha > 0$ will suffice. Hence take α such that

$$0 < \alpha < \min\left\{ \frac{[\theta_i a_i, \tilde{x}_1]}{-[\theta_i a_i, \tilde{x}_2]} : [\theta_i a_i, \tilde{x}_2] < 0, i \notin J \right\}$$

□

-- characterizing finite cones which are subspaces --

Theorem (2.3.29) presents several conditions characterizing pointed finite cones. Stiemke's theorem can be used to characterize those cones which are subspaces. This knowledge will be used in constructing the most general form of the tree algorithm.

(2.3.36) Theorem: (Stiemke): Let $\{b_j\}_1^m \subset X$. The following statements are equivalent:

(i) There exists $\lambda_j > 0$ such that $0 = \sum_1^m \lambda_j b_j$

(ii) There does not exist $\tilde{x} \in \tilde{X}$ such that $[b_j, \tilde{x}] \geqslant 0$ for all j with at least one strict inequality.

Proof: Note that (i) easily implies (ii). See Stoer and Witzgall (1970) for a proof of the rest. □

To help in understanding Stiemke's theorem, the next theorem shows that condition (i) of Stiemke's theorem is equivalent to saying that $0 \in \text{rel int } C\{b_i\}_1^m$. Note how Theorem (2.2.11) contributes to a simple proof of the following characterization of the relative interior of a finite cone. (The relative interior of a polyhedral cone will be characterized in Theorem (2.4.9).)

(2.3.37) Theorem: Let $\{b_j\}_1^m \subset X$. Then

$$\text{rel int } C\{b_j\}_1^m = \{\sum_1^m \lambda_j b_j : \lambda_j > 0, \text{ for all } j\}.$$

Note how an economy of expression results when $\langle 0 \rangle \cup \{\langle b_j \rangle\}_1^m$ is a conical spanning set of minimum size for $C = C\{b_j\}_1^m$.

Proof: First, if $\langle 0 \rangle = \{b_j\}_1^m$, then since $X \cap \langle 0 \rangle$ is open in the relative topology for $\{b_j\}_1^m$, $\text{rel int } C\{b_j\}_1^m = \langle 0 \rangle$. So suppose $\{b_j\}_1^m \neq \langle 0 \rangle$.

To show $LHS \supset RHS$ (and thus $\text{rel int } C\{b_j\}_1^m \neq \emptyset$ which is known by (2.2.13) anyway), begin by using (2.1.17) to select a basis $\{b(j_k)\}_1^p$ from $\{b_j\}_1^m$ for $L(C\{b_j\}_1^m - 0)$. Next, take $\sum_1^m \lambda_j b_j$ with $\lambda_j > 0$ for all j. Take $0 < \alpha < \min\{|\lambda_j| : j = 1, \ldots, m\}$ and observe that

$$\sum_1^m \lambda_j b_j \pm \alpha b(j_k) \in C\{b_j\}_1^m$$

for $k = 1, \ldots, p$.

To show $LHS \subset RHS$, take $\sum_1^m \beta_j b_j \in \text{rel int } C\{b_j\}_1^m$. Observe that each $b_k \neq 0$ is in some basis for $L(C\{b_j\}_1^m)$ and so therefore there exists $\alpha_k > 0$ and $\gamma_{kj} \geqslant 0$ such that $\sum_{j=1}^m \beta_j b_j = \alpha_k b_k + \sum_{j=1}^m \gamma_{kj} b_j$. Now average both sides over all $b_k \neq 0$ and finish by setting any 0 coefficient for $b_k = 0$ to 1. □

The next theorem, then, characterizes those cones which are subspaces.

(2.3.38) Theorem: Let $C = \mathsf{C}\{a_i\}_1^n \subset X$ be a finite cone. The following statements are equivalent:

(i) C is a subspace

(ii) $C = \text{Lin } C$

(iii) $0 \in \text{rel int } C = \{\sum_1^n \lambda_i a_i : \lambda_i > 0 \text{ for all } i\}$

(iv) There does not exist $\tilde{x} \in \tilde{X}$ such that $[a_i, \tilde{x}] \geqslant 0$ for all i with at least one strict inequality.

Proof: (i) $\Leftrightarrow$ (iii): In the case when $C \neq \langle 0 \rangle$, let $\{x_j, j = 1, \ldots, p\}$ be a basis for $\mathsf{L}(C - 0)$ and use (2.2.11) to establish the first in a sequence of equivalences:

$$0 \in \text{rel int } C \Leftrightarrow \text{there exists } \alpha > 0 \text{ such that } \{\pm \alpha x_j, j = 1, \ldots, p\} \subset C$$

$$\Leftrightarrow \{\langle x_j \rangle, \langle -x_j \rangle, j = 1, \ldots, p\} \subset C$$

$$\Leftrightarrow C = \mathsf{L}(C)$$

To show the " $\Rightarrow$ " direction of the last equivalence, take nonzero $y \in \mathsf{L}(C)$ and observe for some η_j, $y = \sum_1^p |\eta_j| \, (\text{sgn } \eta_j) \, x_j$. □

So, in short, Gordan's theorem helps to characterize those finite cones which are pointed and Stiemke's theorem helps to characterize those finite cones which are subspaces.

The following linear program is used subsequently when an algorithm is needed which will determine whether or not $\mathsf{C}\{a_i\}_1^n$ is a subspace.

(2.3.39) Theorem: Let $\{a_i\}_1^n \subset X$. Consider the following LP and its dual:

Primal: maximize $\left(\sum_1^n \underline{a}_j\right)^T \underline{x}$ subject to $\underline{a}_j^T \underline{x} \geqslant 0$ for $i = 1, \ldots, n$ and $\left(\sum_1^n \underline{a}_j\right)^T \underline{x} \leqslant 1$.

Dual: minimize δ subject to $\sum_1^n (\zeta_i + 1 - \delta)\underline{a}_j = \underline{0}$ and all $\zeta_i, \delta \geqslant 0$.

Then the optimal value of the objective function exists and is either 0 or 1. If the value is 0, then there is no x such that $\underline{a}_j^T \underline{x} \geqslant 0$ for all $i = 1, \ldots, n$ with at least one strict inequality and consequently $\mathrm{C}\{a_i\}_1^n$ is a subspace. If the value is 1, then the solution x_0 is a vector such that $\underline{a}_j^T \underline{x} \geqslant 0$ for all i with at least one strict inequality and so $\mathrm{C}\{a_i\}_1^n$ is not a subspace.

Proof: Note that since the primal program is feasible with $\underline{x} = \underline{0}$ and since the dual is feasible with $\delta = 1$ and all $\zeta_i = 0$, a solution exists. The optimal objective function value must be in $[0, 1]$.

Suppose the optimal value is 0. Then there exist $\lambda_i > 0$ such that $\sum_1^n \lambda_i a_i = 0$. Hence by Stiemke's Theorem, there is no x such that $\underline{a}_j^T \underline{x} \geqslant 0$ for all $i = 1, \ldots, n$ with at least one strict inequality.

Suppose the optimal value $\delta^* = \left(\sum_1^n \underline{a}_j\right)^T \underline{x}^*$ is greater than 0 and less than 1. Then once again, there exist $\lambda_i > 0$ such that $\sum_1^n \lambda_i a_i = 0$. But this contradicts the existence of x^* with $\underline{a}_j^T \underline{x}^* \geqslant 0$ for all i and $\sum_1^n \underline{a}_j^T \underline{x}^* > 0$. □

-- *faces of polyhedral cones* --

The last part of this section looks at finite cones from the viewpoint of polyhedral convex cones. The concept of face is defined and various properties of the lineality space are developed.

(2.3.40) Definition: Let $\tilde{A} = \{\tilde{a}_i , i \in K\} \subset \tilde{X}$ where K is a nonempty finite index set. $\tilde{A}^+ := \{x \in X: [x , \tilde{a}_i] \geqslant 0$ for $i \in K\}$ is a polyhedral convex cone. For each subset J of K, define:

(a) $L_J := \{x \in X: [x , \tilde{a}_i] = 0, i \in J\}$

(b) $O_J := \{x \in X: [x , \tilde{a}_i] > 0, i \in I \text{ w.o. } J\}$

(c) $F_J := O_J \cap L_J$

F_J is called a *face* of $\tilde{A}^+$ and is sometimes written $F_J(\tilde{A}^+)$.

This definition is essentially the same as that of Goldman and Tucker (1956). It is however essentially different from Stoer and Witzgall's (1970) face and Gerstenhaber's (1951) facet in that, for example, a face as defined here is *not* an extreme subset nor is it the intersection of the cone with some supporting hyperplane. A few examples of faces follow the next two theorems.

(2.3.41) Theorem: The set of all faces of $\tilde{A}^+$ partitions $\tilde{A}^+$, i.e., $\tilde{A}^+ = \bigcup_J F_J$ and for $J_1 \neq J_2$, $F_{J_1} \cap F_{J_2} = \emptyset$.

The next theorem is stated without proof because it is not essential to what follows. However it may reinforce the reader's intuition.

(2.3.42) Theorem: Let $F_J \neq \emptyset$ be a face of $\tilde{A}^+$. Then $\mathsf{M}(F_J) = \mathsf{L}(F_J) = L_J$ and so F_J is open in its dimensionality space.

(2.3.43) Examples: Figure (2.3.44) shows the face structure of the cone C in Figure (2.3.9). Observe that $F_{\{1,2\}} = \langle 0 \rangle$, $F_{\{1\}}$ and $F_{\{2\}}$ are the nonzero open rays bounding C, and $F_\emptyset$ is the interior of C.

Now, suppose in Example (2.3.8) that $X = \mathsf{R}^3$ and the wedge $C = \{x \in \mathsf{R}^3: [x , \tilde{a}_1] \geqslant 0, [x , \tilde{a}_2] \geqslant 0\}$ is being considered. C has four faces, namely the line $\tilde{a}_1^\perp \cap \tilde{a}_2^\perp$ which forms the edge of the wedge, each of the two bounding open halfspaces, and the interior of C.

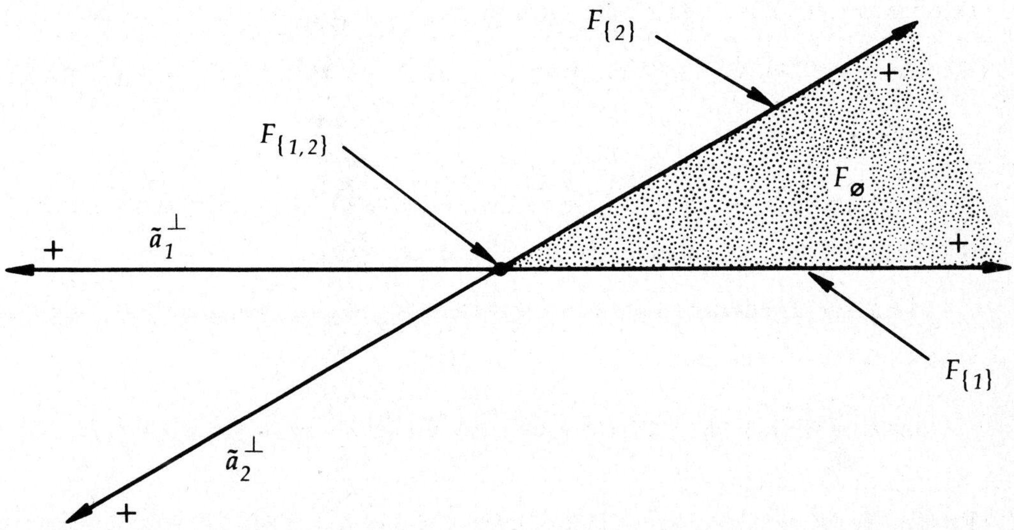

(2.3.44) Figure: The face structure of the cone C depicted in Figure (2.3.9).

As a more abstract example, suppose $\dim \mathsf{L}(\tilde{A}) = \dim X = d \geqslant 2$ and $\tilde{a}_k \neq 0$ for some k. Then $\{x \in X: [x, \tilde{a}_k] > 0, [x, \tilde{a}_j] = 0 \text{ for } j \neq k\}$ is a face and if it is nonempty, an open ray.

-- *the lineality spaces of finite cones* --

The last three theorems of this section concern themselves with the lineality spaces of finite cones.

(2.3.45) Theorem: Let $\tilde{A} = \{\tilde{a}_i, i \in K\} \subset \tilde{X}$ where K is a nonempty finite index set. Then $\operatorname{Lin} \tilde{A}^+ = F_K = \tilde{A}^{\perp}$.

Proof: Take $x \in \operatorname{Lin} \tilde{A}^+ \subset \tilde{A}^+$. Then $-x \in \tilde{A}^+$ and x must satisfy all of the inequalities as equalities. For the other direction, note that since $\tilde{A}^{\perp}$ is a subspace contained in $\tilde{A}^+$, $\tilde{A}^{\perp} \subset \operatorname{Lin} \tilde{A}^+$. □

(2.3.46) Theorem: Let $\tilde{A} = \{\tilde{a}_i, i \in K\} \subset \tilde{X}$. Then $\mathsf{Lin}\ \tilde{A}^+$ is an extreme subset of $\tilde{A}^+$. In other words, if $x_1, x_2 \in \tilde{A}^+$ and $x_1 + x_2 \in \mathsf{Lin}\ \tilde{A}^+$, then $x_1, x_2 \in \mathsf{Lin}\ \tilde{A}^+$. In fact, if $x_1, \ldots, x_k \in \tilde{A}^+$ and $\sum_1^k x_i \in \mathsf{Lin}\ \tilde{A}^+$, then $x_i \in \mathsf{Lin}\ \tilde{A}^+$ for $i = 1, \ldots, k$.

Proof: Suppose $x_1 + x_2 \in \mathsf{Lin}\ \tilde{A}^+$ and $x_1 \notin \mathsf{Lin}\ \tilde{A}^+$. Then there exists j such that $[x_1, \tilde{a}_j] > 0$. Consequently, $x_1 + x_2 \notin \mathsf{Lin}\ \tilde{A}^+$. □

(2.3.47) Theorem: Let $A = \{\langle a_i \rangle, i \in I\}$ be a spanning set for a finite cone $C \subset X$. (Remember $\langle a_0 \rangle = \langle 0 \rangle$). Then

$$\mathsf{Lin}\ C = \mathsf{H}\{\langle a_i \rangle : a_i \in \mathsf{Lin}\ C\} = \mathsf{C}\{a_i : a_i \in \mathsf{Lin}\ C\} = \mathsf{L}\{a_i : a_i \in \mathsf{Lin}\ C\}.$$

Proof: To show the first equality, take $x \in \mathsf{Lin}\ C$. Then $x = \sum_1^n \lambda_i a_i$ for some $\lambda_i \geqslant 0$, not all 0. Since $x \in \mathsf{Lin}\ C$, if $\lambda_i > 0$, then $a_i \in \mathsf{Lin}\ C$. The other direction follows since $\mathsf{Lin}\ C$ is a subspace. □

Summary For Section 2.3

A polyhedral cone is the intersection of a finite number of halfspaces. A finite cone is the convex conical hull of a finite number of vectors. The Minkowski-Weyl theorem states that every polyhedral cone is a finite cone and vice-versa. A number of examples were provided illustrating the diversity of polyhedral cones.

A good way to interpret the convex conical hull of a set of vectors is as the convex hull of an associated set of open rays. Just as polyhedrons (convex hulls of a finite number of vectors) have extreme points, so do polyhedral cones have isolated rays. The analogy fails somewhat, however, in that polyhedral cones are not necessarily the convex hulls of their isolated open rays.

A set of open rays is said to be conically independent if each one is disjoint from the convex hull of the rest. A set of open rays conically spans a convex cone if the convex hull of the set is the convex cone. A frame is a conically independent, conical spanning set for a convex cone and as such is an analog of a basis for a vector space.

A convex cone is pointed if there is some open halfspace whose boundary passes through the origin which contains all nonzero elements of the cone. Pointed finite cones have unique frames consisting of the isolated open rays of the cone and are consequently the convex hulls of their isolated open rays. Linear programming can be used to determine whether or not a given cone is pointed and, if so, to produce an open halfspace through the origin which contains the nonzero elements of the cone.

A fact useful in the next chapter is that any collection of vectors can be made to generate a pointed cone by multiplying certain vectors by -1.

Stiemke's theorem and a characterization of the relative interior of a finite cone help to provide conditions characterizing when a finite cone is a subspace. Linear programming can be used to determine whether or not a given finite cone is a subspace.

The last part of this section looked at finite cones from the alternate viewpoint of being polyhedral cones. Every polyhedral cone can be decomposed into a disjoint collection of sets open in their dimensionality spaces, each of which is called a face of the cone.

The polyhedral cone representation facilitates the proof of several theorems concerning the nature of the lineality space of a finite cone, one of which states that the lineality space of a finite cone is an extreme subset of that cone and another of which says that the lineality space of a finite cone is the conical hull of the original generating points contained in it.

Section 2.4:

Finite Cones And Their Duals

For each finite cone in X, there is an associated polyhedral cone in $\tilde{X}$ and vice-versa. This section examines the inter-relationships of these cones.

(2.4.1) Definitions: Let $\emptyset \neq A \subset X$.

(a) $A^+ := \{\tilde{x} \in \tilde{X} : [a, \tilde{x}] \geqslant 0 \text{ for all } a \in A\}$. A^+ is known as the positive polar of A, the positive conjugate cone of A, or the *dual cone* of A. Note that if A is finite, then A^+ is a polyhedral cone.

(b) $A^- := \{\tilde{x} \in \tilde{X} : [a, \tilde{x}] \leqslant 0 \text{ for all } a \in A\}$. A^- is known as the *negative polar* of A or the negative conjugate cone.

For notational reasons, A^+ will be used almost exclusively in what follows. A^- could have just as easily been used instead of A^+ because, *generally* speaking, since $A^- = (-A)^+$, the same theorems that hold for one hold for the other. The advantage that A^- has over A^+ is that it is easier to draw pictures showing A and A^- than it is to draw pictures showing A and A^+.

(2.4.2) Example: In Figure (2.4.3), a set A, $\mathsf{C}(A)$, and A^- in R^2 are shown. Now even though A and A^- are in different spaces, it is nonetheless convenient to overlay the spaces one on the other in order to picture what is happening. The axes (which are not shown) then simultaneously represent a basis in the original space and its dual basis in the dual space.

-- *standard results about dual cones* --

There are a fair number of straightforward standard results about dual cones which are listed next.

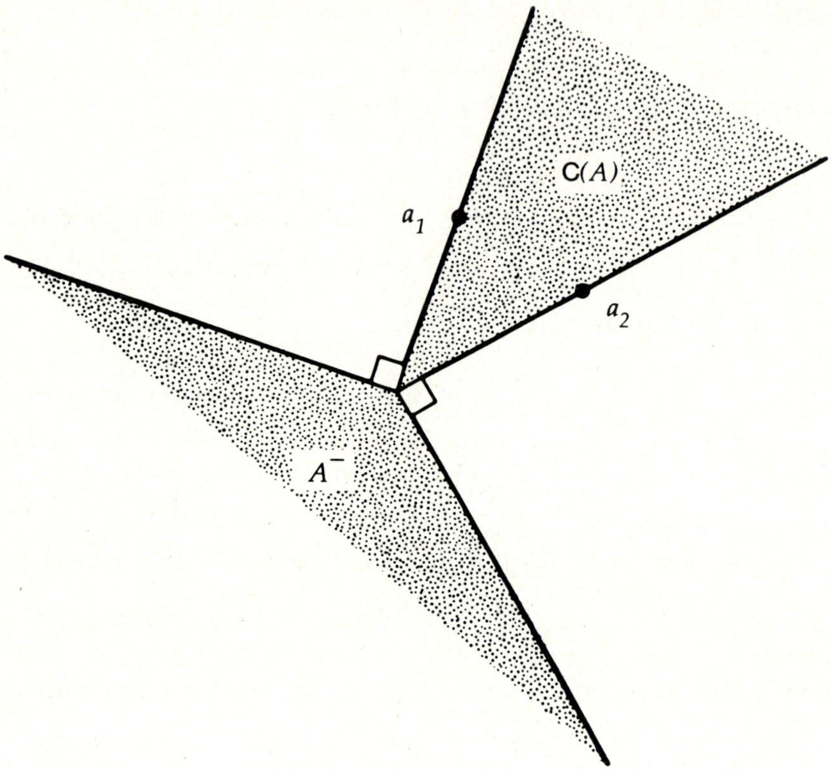

(2.4.3) Figure: A, $\mathsf{C}(A)$, and A^- for A = $\{a_1,a_2\}$ in R^2. The reader may want to visualize $A^+ = -A^-$.

(2.4.4) Theorem: Let $\emptyset \neq A \subset X$. Then:

(a) A^+ is a convex cone.

(b) $\mathsf{C}(A)^+ = A^+$

(c) If A is a subspace, then $A^+ = A^- = A^\perp$.

(d) $A^\perp = A^+ \cap A^-$

(e) If $A \subset B$, then $B^+ \subset A^+$

(f) $A \subset A^{++}$

(g) $A^+ = A^{+++}$

Note that (f) and (g) use the customary identification of X with $\tilde{\tilde{X}}$, the dual of the dual space of X.

As an analog of the fact that $\mathsf{L}(A) = A^{\perp\perp}$, there is:

(2.4.5) Theorem: Let A be a finite subset of X. Then $\mathsf{C}(A) = A^{++}$.

Proof: $\mathsf{C}(A) \subset A^{++}$ since A^{++} is a convex cone containing A. Since $A \subset \mathsf{C}(A)$, $A^{++} \subset \mathsf{C}(A)^{++}$. The proof is complete if it can be shown that if C is a finite cone, then $C = C^{++}$. By the Minkowski-Weyl theorem, $C = B^+$ for some finite B and so $C = B^+ = B^{+++} = C^{++}$. □

(2.4.6) Theorem: Let $\varnothing \neq A_1, A_2 \subset X$. Then:

(a) If $0 \in A_1 \cap A_2$, then $(A_1 + A_2)^+ = A_1^+ \cap A_2^+$.

(b) $A_1^+ + A_2^+ \subset (A_1 \cap A_2)^+$ with equality holding when A_1 and A_2 are finite cones.

Proof: (a) is straight-forward. For (b), $A_1 \supset A_1 \cap A_2$ so that $A_1^+ \subset (A_1 \cap A_2)^+$. Similarly, $A_2^+ \subset (A_1 \cap A_2)^+$. Hence $A_1^+ + A_2^+ \subset (A_1 \cap A_2)^+$. If both A_1 and A_2 are finite cones, then

$$(A_1 \cap A_2)^+ = (A_1^{++} \cap A_2^{++})^+ = (A_1^+ + A_2^+)^{++} = (A_1^+ + A_2^+)$$

using (2.3.3) and (2.4.5) in the last step. □

-- *dimensionality and lineality* --

There is an important connection between the dimensionality space of a finite cone and the lineality space of its dual.

(2.4.7) Theorem: Let $A \neq \emptyset$ be a finite subset of X. Then:

(a) $\mathsf{L}(\mathsf{C}(A)) = (\mathsf{Lin}\ A^+)^\perp$

(b) $\mathsf{L}(A^+) = (\mathsf{Lin}\ \mathsf{C}(A))^\perp$

Proof: To show (a), observe $\mathsf{L}(\mathsf{C}(A))^\perp = \mathsf{C}(A)^\perp = \mathsf{C}(A)^+ \cap \mathsf{C}(A)^- = \mathsf{C}(A)^+ \cap (-\mathsf{C}(A)^+) = \mathsf{Lin}\ \mathsf{C}(A)^+ = \mathsf{Lin}\ A^+$. □

-- *using the lineality space to characterize cones* --

The next theorem shows for a finite cone C how $\mathsf{Lin}\ C$ can be used to provide a useful characterization for C^+.

(2.4.8) Theorem: Let $A = \{a_i,\ i \in I\} \subset X$ where, recall, $I = \{0, 1, \ldots, n\}$ for some n and $a_0 = 0$. Then:

(a) $\mathsf{L}(A^+) = \{\tilde{x} \in \tilde{X}:$

$[a_i, \tilde{x}] = 0$ for all $i \in I$ such that $a_i \in \mathsf{Lin}\ \mathsf{C}(A)\}$

(b) $A^+ = \{\tilde{x} \in \tilde{X}: [a_i, \tilde{x}] \geqslant 0$ for $a_i \notin \mathsf{Lin}\ \mathsf{C}(A)$,

$[a_i, \tilde{x}] = 0$ for $a_i \in \mathsf{Lin}\ \mathsf{C}(A)\}$

Note since $a_0 = 0$, there is always an $a_i \in \mathsf{Lin}\ \mathsf{C}(A)$. If all $a_i \in \mathsf{Lin}\ \mathsf{C}(A)$, then the resulting null condition in (b) is to be dropped.

Proof: With regards to (a), by (2.4.7) and (2.3.47),

$$\mathsf{L}(A^+) = (\mathsf{Lin}\ \mathsf{C}(A))^\perp = (\mathsf{L}\{a_i: a_i \in \mathsf{Lin}\ \mathsf{C}(A)\})^\perp = \{a_i: a_i \in \mathsf{Lin}\ \mathsf{C}(A)\}^\perp.$$

Suppose in (b) there are no $a_i \notin \mathsf{Lin}\ \mathsf{C}(A)$. Then by (2.3.47), $\mathsf{Lin}\ \mathsf{C}(A) = \mathsf{C}(A) = \mathsf{L}(A)$ and so $A^+ = A^\perp$. Suppose then that there are $a_i \notin \mathsf{Lin}\ \mathsf{C}(A)$.

$$\begin{aligned} A^+ &= A^+ \cap \mathsf{L}(A^+) \\ &= \{\tilde{x} \in \tilde{X}: [a_i, \tilde{x}] \geqslant 0 \text{ for } a_i \notin \mathsf{Lin}\ \mathsf{C}(A), [a_i, \tilde{x}] = 0 \text{ for } a_i \in \mathsf{Lin}\ \mathsf{C}(A)\}. \end{aligned}$$

□

The next theorem yields an expression for rel int A^+.

(2.4.9) Theorem: Let $A = \{a_i, i \in I\} \subset X$ where $\mathsf{C}(A) \neq \langle 0 \rangle$. Then rel int $A^+ =$

$$\{\tilde{x} \in \tilde{X}: [a_i, \tilde{x}] > 0 \text{ for } a_i \notin \mathsf{Lin\ C}(A)\ , [a_i, \tilde{x}] = 0 \text{ for } a_i \in \mathsf{Lin\ C}(A)\}$$

where the first condition is omitted if it isn't satisfied.

Note how the number of constraints above can be reduced if $\{\langle a_i \rangle, i \in I\}$ is a minimal spanning set for $\mathsf{C}(A)$. Note also that if $\mathsf{C}(A)$ is pointed, then int $A^+ = \{\tilde{x} \in \tilde{X}: [a_i, \tilde{x}] > 0 \text{ for } a_i \neq 0\} \neq \emptyset$.

Proof: Assume Lin $\mathsf{C}(A) \neq \mathsf{C}(A)$. By (2.4.8),

$$A^+ = \{\tilde{x} \in \mathsf{L}(A^+): [a_i, \tilde{x}] \geqslant 0 \text{ for } a_i \notin \mathsf{Lin\ C}(A)\}.$$

Take $\tilde{x} \in$ rel int $A^+ \neq \emptyset$ by (2.2.13). By (2.2.11), there is $\alpha > 0$ such that $[a_i, \tilde{x}] \geqslant \alpha |[a_i, \tilde{b}_j]|$ for all i such that $a_i \notin$ Lin $\mathsf{C}(A)$ and for all j where $\{\tilde{b}_j\}_1^p$ is a basis for $\mathsf{L}(A^+)$. To conclude then, for each i such that $a_i \notin$ Lin $\mathsf{C}(A)$, there must exist j such that $[a_i, \tilde{b}_j] \neq 0$ or else $a_i \in \{\tilde{b}_j\}_1^{p\perp} =$ Lin $\mathsf{C}(A)$.

Next take $\tilde{x} \in \mathsf{L}(A^+)$ such that $[a_i, \tilde{x}] > 0$ for all i such that $a_i \notin$ Lin $\mathsf{C}(A)$. Fix a basis $\{\tilde{b}_j\}_1^p$ for $\mathsf{L}(A^+)$. It must be shown that there exists $\alpha > 0$ such that for all i such that $a_i \notin$ Lin $\mathsf{C}(A)$, $[a_i, \tilde{x}] \geqslant \alpha |[a_i, \tilde{b}_j]|$ for all $j = 1, \ldots, p$. Take

$$0 < \alpha < \min \left\{ \frac{[a_i, \tilde{x}]}{|[a_i, \tilde{b}_j]|} : [a_i, \tilde{b}_j] \neq 0 \quad , \quad a_i \notin \mathsf{Lin\ C}(A) \right\}$$

□

-- *pointed cones and their duals* --

The next theorem summarizes the relationship between $\mathsf{C}(A)$ and A^+ when $\mathsf{C}(A)$ is pointed.

(2.4.10) Theorem: Let $A = \{a_i, i \in I\} \subset X$ where $\mathsf{C}(A) \neq \langle 0 \rangle$. Then the following are equivalent:

(a) $\dim A^+ = d$

(b) $\mathsf{M}(A^+) = \tilde{X} = \mathsf{L}(A^+)$

(c) $\text{int } A^+ \neq \varnothing$

(d) There exists $\tilde{x}$ such that $[a_i, \tilde{x}] > 0$, $i \in I$ w.o. I_0.

(e) $\mathsf{C}(A)$ is pointed.

If $\text{int } A^+ \neq \varnothing$, then

$$\text{int } A^+ = \{\tilde{x} \in \tilde{X} : [a_i, \tilde{x}] > 0 \text{ for } i \in I \text{ w.o. } I_0\}.$$

The next theorem establishes a needed correspondence between the isolated rays (i.e., one dimensional faces) of a pointed finite cone and the nonempty $d - 1$ dimensional faces of the dual cone. It also establishes the existence of certain cones neighboring the above mentioned dual cone. This theorem is actually an application of Slater's general theorem of the alternative. The tricky part of the proof comes from the papers of Slater (1951) and Gerstenhaber (1951).

(2.4.11) Theorem: Let $A = \{a_i, i \in I\} \subset X$ where $\dim X \geqslant 1$. Suppose $\mathsf{C}(A)$ is pointed. Choose any $\langle a_k \rangle \neq \langle 0 \rangle$ and let

$$F_k := \{\tilde{x} \in \tilde{X} : [a_i, \tilde{x}] > 0 \text{ for } i \in I \text{ w.o. } (I_0 \cup I_k),$$
$$[a_i, \tilde{x}] = 0 \text{ for } i \in I_0 \cup I_k\}$$

Note F_k is a $d - 1$ dimensional face. The following are equivalent:

(a) $F_k \neq \varnothing$

(b) $\langle a_k \rangle \parallel \mathsf{H}\{\langle a_i \rangle, i \in I \text{ w.o. } I_k\}$

(c) $\langle a_k \rangle$ is in the frame of $\mathsf{C}(A)$

(d) $\{\tilde{x} \in \tilde{X} : [a_i, \tilde{x}] > 0$ for $i \in I$ w.o. $(I_0 \cup I_k)$,

$[a_i, \tilde{x}] < 0$ for $i \in I_k$,

$[a_i, \tilde{x}] = 0$ for $i \in I_0\} \neq \varnothing$,

i.e., $\mathsf{H}\{\langle a_i \rangle, i \in I$ w.o. $I_k, \langle -a_k \rangle\}$ is pointed.

Any null conditions above are to be omitted.

(2.4.12) Example: Before the proof of (2.4.11) is presented, an example will serve to illustrate the theorem. Consider the positive orthant in R^3 as spanned by the vectors $a_0 = (0, 0, 0)$, $a_1 = (1, 0, 0)$, $a_2 = (0, 1, 0)$, $a_3 = (0, 0, 1)$, and $a_4 = (1, 1, 0)$. Clearly a frame for $\mathsf{C}\{a_i\}_0^4$ is $\{\langle a_0 \rangle, \langle a_1 \rangle, \langle a_2 \rangle, \langle a_3 \rangle\}$. Visualize now the axes corresponding to the dual basis in the dual space being super-imposed on the coordinate axes in R^3. Observe that $(\mathsf{C}\{a_i\}_0^4)^+$ fills exactly the same space as $\mathsf{C}\{a_i\}_0^4$. Now take a_3 and consider $F_3 \neq \varnothing$. F_3 is that section of the $x - y$ plane which borders the positive orthant in R^3. Note that a certain neighboring cone is nonempty, namely that orthant which is directly below the positive orthant. If $\langle a_4 \rangle$ which is not an isolated ray of $\mathsf{C}\{a_i\}_0^4$ is considered, then it is found that $a_4^\perp$ intersects $(\mathsf{C}\{a_i\}_0^4)^+$ in precisely the upper z-axis which is why $F_4 = \varnothing$.

As a general interpretation of part (d) of the theorem, suppose that a point in the interior of a d-dimensional polyhedral cone is selected and that the $d - 1$ dimensional boundary face $F_{\{k\}}$ is nonempty. Then that point may be moved through the face into a neighboring cone which is different from the original cone only in that it is on the other side of $a_k^\perp$.

Proof: There are two cases depending on whether or not $\{\langle a_i \rangle, i \in I\}$ contains at least two distinct nonzero rays. Suppose it doesn't so that $\mathsf{C}(A) = \langle 0 \rangle \cup \langle a_1 \rangle$ where, without loss of generality, $\langle a_1 \rangle \neq \langle 0 \rangle$. Let $\langle a_k \rangle \neq 0$. Then of course $\langle a_k \rangle = \langle a_1 \rangle$. Note that

$$F_k = \{\tilde{x} \in \tilde{X} : [a_i, \tilde{x}] = 0 \text{ for } i \in I\} \supset \{0\} \neq \varnothing,$$

$\langle a_k \rangle \,\|\, \mathsf{H}\{\langle a_i \rangle, i \in I_0\}$, $\langle a_k \rangle$ is in the frame of $\mathsf{C}(A)$, and $\{\tilde{x} \in \tilde{X} : [a_i, \tilde{x}] < 0 \text{ for } i \in I_k\} \neq \varnothing$. In short, the theorem holds in this special case.

Now suppose that $\{\langle a_i \rangle, i \in I\}$ contains at least two distinct nonzero rays. (b) and (c) are equivalent by (2.3.32). Suppose (a) holds and (b) doesn't. Then there exists $\tilde{x}$ such that $[a_i, \tilde{x}] > 0$ for $i \in I$ w.o. $(I_0 \cup I_k)$ and $[a_k, \tilde{x}] = 0$. There also exists $\lambda_i \geqslant 0$, not all 0, such that $a_k = \sum_{I \text{ w.o. } (I_k \cup I_0)} \lambda_i a_i$. Apply $\tilde{x}$ to both sides of this equation to get a contradiction.

Assume that (b) holds. First note that $\langle -a_k \rangle \,\|\, \mathsf{H}\{\langle a_i \rangle, i \in I \text{ w.o. } I_k\}$ since otherwise $\mathsf{C}(A)$ is not pointed. Consequently,

$$\mathsf{L}\{a_k\} \cap \mathsf{H}\{\langle a_i \rangle, i \in I \text{ w.o. } I_k\} = \{0\}.$$

So, $a_k^\perp + \mathsf{C}\{a_i, i \in I \text{ w.o. } I_k\}^+ = \tilde{X}$. Since a subcone of a pointed cone is pointed, $\mathsf{C}\{a_i, i \in I \text{ w.o. } I_k\}$ is pointed and so $\mathsf{L}(\mathsf{C}\{a_i, i \in I \text{ w.o. } I_k\}^+) = \tilde{X}$. The result of putting these two equations together is

$$\mathsf{L}(\mathsf{C}\{a_i, i \in I \text{ w.o. } I_k\}^+) + a_k^\perp = \mathsf{C}\{a_i, i \in I \text{ w.o. } I_k\}^+ + a_k^\perp.$$

Now select $\tilde{x}_1$ such that $[a_i, \tilde{x}_1] > 0$ for $i \in I$ w.o. $(I_k \cup I_0)$. Taking nonzero $\tilde{x}_2 \in a_k^\perp$, obtain $-\tilde{x}_1 + \tilde{x}_2$, an element of the LHS of the equation immediately above. Then there exists $\tilde{x}_3 \in \mathsf{C}\{a_i, i \in I \text{ w.o. } I_k\}^+$ and $\tilde{x}_4 \in a_k^\perp$ such that $-\tilde{x}_1 + \tilde{x}_2 = \tilde{x}_3 + \tilde{x}_4$. Let $\tilde{x} = \tilde{x}_2 - \tilde{x}_4 = \tilde{x}_1 + \tilde{x}_3$ and observe that $\tilde{x} \in a_k^\perp$ and $[a_i, \tilde{x}] > 0$ for all $i \in I$ w.o. $(I_k \cup I_0)$. Hence $F_k \neq \varnothing$.

Suppose that (d) holds. Then there exists $\tilde{x}$ such that $[a_i, \tilde{x}] > 0$ for $i \in I$ w.o. $(I_0 \cup I_k)$ and $[a_i, \tilde{x}] < 0$ for $i \in I_k$. If (b) does not hold, then there exist $\lambda_i \geqslant 0$, not all 0, such that $a_k = \sum_{I \text{ w.o. } (I_k \cup I_0)} \lambda_i a_i \neq 0$. Now apply $\tilde{x}$ to both sides and get a contradiction.

Suppose (b) holds and (d) doesn't hold, i.e., there exist $\lambda_i \geqslant 0$, not all 0, such that $\lambda_k a_k = \sum_{I \text{ w.o. } (I_0 \cup I_k)} \lambda_i a_i$. If $\lambda_k > 0$ then (b) is contradicted. If $\lambda_k = 0$, then $C(A)$ is not pointed. So (d) must hold. □

Summary of Section 2.4

This section introduces the idea of the dual cone A^+ of a cone $C(A)$ in R^d. The dual cone is the set of all linear functionals which take non-negative values on all of the elements of the original cone. A few basic facts about the operation of forming dual cones are established.

It is shown that the dimensionality space of a finite cone is the annihilator of the lineality space of its dual. This yields a useful representation of A^+ as a function of $\text{Lin } C(A)$. This, in turn, yields an expression for $\text{rel int } A^+$. The nature of the relationship between $C(A)$ and A^+ when $C(A)$ is pointed is also presented.

Slater's theorem relating the solution of homogeneous linear inequalities in one space to the behavior of associated cones in the dual space is implicitly used to develop a correspondence between the isolated rays of a pointed finite cone and the nonempty $d - 1$ dimensional boundary faces of its dual. The value of this is that by knowing the structure of the pointed finite cone, one can determine which $d - 1$ dimensional boundary faces of the dual cone exist and what characterizes the space just beyond each such face.

Chapter 3:

Tree Algorithms For Solving The Weighted Open Hemisphere Problem

This chapter develops and explains tree algorithms for solving the Weighted Open Hemisphere (WOH) problem described in Chapter 1. At the same time, the conceptual groundwork will be laid for the presentation in Chapter 5 of tree algorithms for extremizing functions of systems of linear relations subject to constraints.

The first section of this chapter restates and clarifies by example the WOH problem. The second section details the first steps that can be taken towards understanding and simplifying the problem; it also presents one of the central insights behind why the tree algorithms work. The third section discusses the first or boundary vector collection phase of the generic WOH tree algorithm. The construction of a more sophisticated first phase is discussed in the fourth section and the description of the generic WOH tree algorithm is completed in the fifth section when the second or displacement phase is described.

Section 3.1:

Problem Statement And Preliminaries

Recall from Chapter 1 that the Weighted Open Hemisphere (WOH) problem requests the identification of all vectors $x \in \mathsf{R}^d$ which maximize $\sum_1^n \sigma_i \, 1\{y_i^T x > 0\}$ where the σ_i are positive and the y_i lie on a unit sphere in R^d (i.e., for some norm, $\| y_i \| = 1$ for all i).

In this chapter, an algorithm called the tree algorithm is developed for solving the WOH problem. But first, consider recasting the WOH problem into an equivalent, more natural, and more convenient form.

-- recasting the WOH problem --

For example, observe that solutions to the problem remain unchanged when each y_i is replaced by $\alpha_i \, y_i$ for any desired $\alpha_i > 0$. In other words, the basic objects here are not the y_i but rather the $\langle y_i \rangle$. Hence, it is not necessary to norm the y_i to unit length. In fact, as will be seen, norms and metrics on R^d play only an artificial role in this problem.

Second, by virtue of the isomorphism between sets of vectors and their representations and since $[y_i, \tilde{x}] = \underline{y}_i^T \, \underline{\tilde{x}}$ where $\underline{\tilde{x}}$ is formed with respect to the dual basis, it is clear that for $X = \mathsf{R}^d$, it suffices to find all $\tilde{x} \in \tilde{X}$ which maximize $\sum_1^n \sigma_i \, 1\{[y_i, \tilde{x}] > 0\}$. As commented at the start of section 2.1, the decision to work with an arbitrary finite-dimensional vector space X over R and its dual space $\tilde{X}$ is necessary for subsequent proofs to hold.

In short, modulo the soon-to-be-seen nonrestrictive assumption that $\mathsf{L}\{y_i, i \in I\} = X$, in order to solve the WOH problem, it is sufficient to solve the following Problem (3.1.1):

(3.1.1) Problem: Let X be a d-dimensional vector space over R with $d \geqslant 1$. Let $I = \{0, 1, \ldots, n\}$ for some n. Let $\{y_i, i \in I\} \subset X$ where $y_0 = 0$. Suppose $\mathsf{L}\{y_i, i \in I\} = X$. Let $\sigma_i > 0$ for $i \in I$. Define $h : \tilde{X} \to \mathsf{R}$ via $h(\tilde{x}) := \sum_I \sigma_i \, 1\{[y_i, \tilde{x}] > 0\}$. Find all $\tilde{x}_0 \in \tilde{X}$, if any, such that $h(\tilde{x}_0) = \sup_{\tilde{x} \in \tilde{X}} h(\tilde{x})$.

A good way to visualize this problem is to picture a cloud of points (the y_i) in R^3. Associate with each y_i a reward $\sigma_i > 0$ which is collected if one chooses a hyperplane through the origin, say $\{x : [x, \tilde{x}_0] = 0\}$ for some $\tilde{x}_0$, which has the point y_i on the positive side of this hyperspace, i.e., for which $[y_i, \tilde{x}_0] > 0$. The problem is to find those positive halfspaces resting on the origin which collect the greatest total reward. As mentioned before, this problem was considered by Warmack and Gonzalez (1973) who offer a solution when all $\sigma_i = 1$ and the set $\{y_i\}_1^n$ is in general position, i.e., all size d subsets of $\{y_i\}_1^n$ are linearly independent.

Note that if $d = 1$, then the problem is easy since then there are essentially only two nonzero elements of $\tilde{X}$ to examine.

-- *a non-restrictive assumption* --

The assumption that $\mathsf{L}\{y_i : i \in I\} = X$ is not restrictive in either theory or practice for if $R := \mathsf{L}\{y_i : i \in I\}$ and $\dim R < d$ then in order to find all $\tilde{x} \in \tilde{X}$ which maximize $h_1(\tilde{x}) := \sum_I \sigma_i \, 1\{[y_i, \tilde{x}] > 0\}$, it is sufficient to find all $\tilde{r} \in \tilde{R}$ which maximize $h_2(\tilde{r}) := \sum_I \sigma_i \, 1\{[y_i, \tilde{r}] > 0\}$. This latter problem is a version of Problem (3.1.1). The correspondence between the two problems is such that, using the notation of Theorem (2.1.27),

(i) if $\tilde{x}_0 = \tilde{u}_0 + \tilde{v}_0$ for $\tilde{u}_0 \in R^\perp$ and $\tilde{v}_0 \in S^\perp$ maximizes h_1, then $\psi(\tilde{v}_0) = \tilde{v}_0\big|_R = \tilde{x}_0\big|_R$ maximizes h_2 and

(ii) if $\tilde{r}_0 \in \tilde{R}$ maximizes h_2, then any vector in $\psi^{-1}(\tilde{r}_0) + R^{\perp}$ maximizes h_1.

The proof of these assertions is straightforward and omitted.

-- *an illustrative example* --

(3.1.2) Example: Other aspects of the problem statement are best discussed in the context of an example. Figure (3.1.3) lists sample σ_i and $y_i \in \mathsf{R}^2$, Figure (3.1.4) shows the y_i in R^2, and Figure (3.1.5) shows the situation in the dual space.

i	σ_i	y_i
0	2	(0, 0)
1	1	(1, 1)
2	2	(2, 2)
3	2	(−1, −1)
4	1	(1, 0)
5	2	(0, 1)
6	3	(−1, −2)

(3.1.3) Figure: Data for an example in R^2.

The first thing to note is that $y_0 = 0$ can easily be deleted from the problem without changing it at all since 0 is never in the interior of any halfspace through the origin.

In Figure (3.1.5), the hyperspaces corresponding to the nonzero y_i are pictured as lines through the origin. The + signs indicate which halfspace associated with a given y_i is the positive halfspace. Note that the value of the

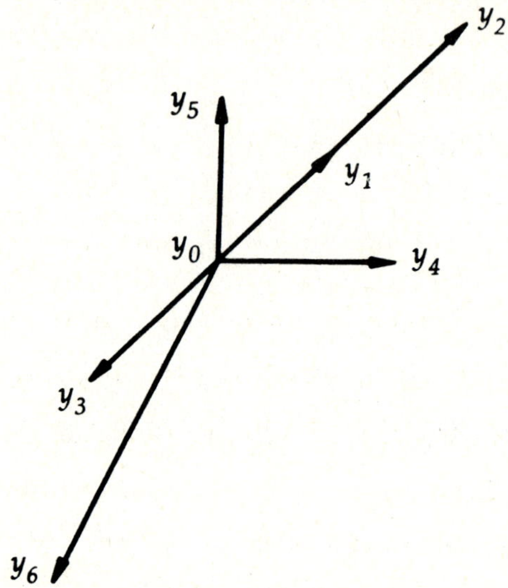

(3.1.4) Figure: The data plotted in R^2.

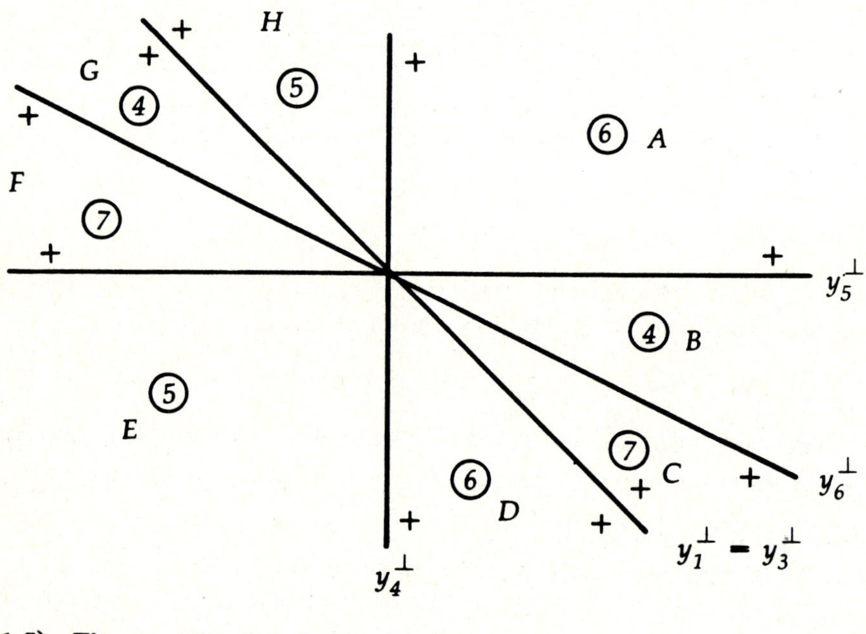

(3.1.5) Figure: The situation in the dual space.

criterion function h is constant on the interiors of the polyhedral cones defined by the lines acting as boundaries. Each interior is labelled by a capital letter and the circled numbers indicate the value assumed by h on the corresponding interior.

Since $\langle y_1 \rangle = \langle y_2 \rangle$, matters could be simplified by deleting y_2 and increasing the reward for y_1 to 3. It is not necessary to consolidate, however, and in fact the result of allowing "conical ties" like this and setting $y_0 = 0$ is a great simplification in the algorithm proofs to come.

One might think that the event $\langle y_3 \rangle = -\langle y_1 \rangle$ is an anomaly which should be ruled out since it yields a hyperspace $y_1^{\perp}$ with two positive sides and two negative sides. This is, however, a situation which naturally occurs in unrestricted problems of maximizing functions of systems of linear relations. Besides, it introduces no real theoretical complications in this chapter.

Looking more closely at the behavior of h in Figure (3.1.5), notice that the value of h is constant on the rays as well. More importantly, observe that for those cones whose interior lies on the positive side of their bounding hyperspaces, the value of h is strictly greater in the interior than it is on the boundary. This is an artifact of working with strict inequalities and positive σ_i. Cones which have interiors which are on the positive sides of their bounding hyperplanes (except for anomalies of the sort $\langle y_1 \rangle = -\langle y_3 \rangle$) will be called *hills* because if a point in the interior of such a cone crosses over a non-anomaly boundary then the criterion function h will decrease.

It is of interest to note that there are two distinct cones (C and F) whose interiors each assume the maximum criterion function value of 7 and both are hills. The presence of multiple maxima like this is one of the more interesting aspects of this kind of problem. The next section will generalize the observations made here to higher dimensional problems.

Section 3.2:

Analyzing The WOH Problem

-- *the role of polyhedral cones* --

The first step in analyzing the problem requires discovering the role that polyhedral cones play. Take any $\tilde{v} \in \tilde{X}$. There exist index sets $J_1, J_2, J_3 \subset I$, one of which must be nonempty, such that

$$[y_i, \tilde{v}] > 0 \text{ for } i \in J_1,\ [y_i, \tilde{v}] < 0 \text{ for } i \in J_2, \text{ and } [y_i, \tilde{v}] = 0 \text{ for } i \in J_3.$$

In other words, there exist $\pi_i \in \{-1, 1\}$ for $i \in I$ such that

$$\tilde{v} \in \{\tilde{x} \in \tilde{X} : [\pi_i y_i, \tilde{x}] > 0 \text{ for } i \in J_1 \cup J_2, [\pi_i y_i, \tilde{x}] = 0 \text{ for } i \in J_3\}.$$

Hence, $\tilde{v} \in \tilde{X}$ is in face F_{J_3} of the polyhedral cone $\{\tilde{x} \in \tilde{X} : [\pi_i y_i, \tilde{x}] \geqslant 0, i \in I\}$ for the π_i just determined. The next definition gives both notation for these cones and a way of identifying those indices i which yield the same open rays $\langle \pi_i y_i \rangle$ for a fixed set $\{\pi_i, i \in I\}$.

(3.2.1) Definition: For $i \in I$ w.o. I_0, let $\pi_i \in \{-1, 1\}$ and for $i \in I_0$, set $\pi_i = 1$. For this set $\{\pi_i, i \in I\}$,

$$C(\pi) := \mathrm{C}\{\pi_i y_i, i \in I\}.$$

For $j \in I$,

$$I_j(\pi) := \{i \in I : \langle \pi_i y_i \rangle = \langle \pi_j y_j \rangle\}.$$

Note for any selection of π_i, $I_0(\pi) = I_0$ and $\dim C(\pi) = d$ by virtue of parts (d) and (e) of (2.1.13) and the assumption that $\mathrm{L}\{y_i, i \in I\} = X$.

The discussion preceding (3.2.1) shows that each $\tilde{v} \in \tilde{X}$ is in a face of $C(\pi)^+$ for suitable $\{\pi_i, i \in I\}$. As examples of $C(\pi)^+$, consider the two optimal cones of Figure (3.1.5). $\mathrm{C}\{y_0, -y_1, -y_2, y_3, -y_4, y_5, y_6\}^+$ is F and

$\mathsf{C}\{y_0, y_1, y_2, -y_3, y_4, -y_5, y_6\}^+$ is C.

-- *h is constant on nonempty faces* --

The next thing to see is that if a face F_J of $C(\pi)^+$ is nonempty for some $\{\pi_i, i \in I\}$, then h assumes the same value for each $\tilde{x}$ in $F_J(C(\pi)^+)$. Now there are only finitely many $C(\pi)^+$ possible, each with only finitely many faces. Since h is constant on nonempty faces of $C(\pi)^+$ for all $\{\pi_i, i \in I\}$ and since each $\tilde{v} \in \tilde{X}$ is in a face belonging to some $C(\pi)^+$, it is easy to see that the problem (3.1.1) of maximizing h is solvable because it is a finite optimization problem where it suffices to enumerate the values of h on all nonempty faces of the $C(\pi)^+$ in order to find all of the solutions.

-- *the solutions occur in the interiors of max-sum cones* --

The next theorem serves as the first step in reducing the number of faces that must be examined.

(3.2.2) Theorem: Let $\tilde{x}_0 \in \tilde{X}$ be such that $h(\tilde{x}_0) \geqslant h(\tilde{x})$ for all $\tilde{x} \in \tilde{X}$. Then there exist $\pi_i \in \{-1, 1\}$ such that $[\pi_i y_i, \tilde{x}_0] > 0$ for $i \in I$ w.o. I_0. Hence, for these π_i, $C(\pi)$ is pointed and $C(\pi)^+$ is a d-dimensional cone with an interior containing $\tilde{x}_0$.

Proof: First, since $\dim X \geqslant 1$ and $\mathsf{L}\{y_i, i \in I\} = X$, there exists some $y_k \neq 0$ and consequently, there exists $\tilde{v}$ such that $[y_k, \tilde{v}] > 0$. Thus, $h(\tilde{x}_0) > \sigma_k$ and $\tilde{x}_0 \neq 0$. There also exists a nonempty set J_1 and corresponding π_i such that $[\pi_i y_i, \tilde{x}_0] > 0$ for $i \in J_1$. Let $J_2 = \{i \in I: [y_i, \tilde{x}_0] = 0\}$. If $J_2 = I_0$, then the conclusion follows. Suppose $J_2 \neq I_0$.

By (2.3.35), there exist $\theta_i \in \{-1,1\}$ for $i \in J_2$ w.o. I_0 and $\tilde{x}_2 \in \tilde{X}$ such that $[\theta_i y_i, \tilde{x}_2] > 0$ for $i \in J_2$ w.o. I_0. It is possible to assume that at least one $\theta_i = 1$ for if not, it suffices to replace $\tilde{x}_2$ with $-\tilde{x}_2$. Observe that $h(\tilde{x}_0 + \alpha\tilde{x}_2) > h(\tilde{x}_0)$ where $\alpha > 0$ is chosen such that

$[\pi_i y_i, \tilde{x}_0] + \alpha[\pi_i y_i, \tilde{x}_2] > 0$ for $i \in J_1$ and $[\theta_i y_i, \tilde{x}_0] + \alpha[\theta_i y_i, \tilde{x}_2] > 0$ for $i \in J_2$ w.o. I_0.

The rest of this theorem follows from (2.4.10). □

This prompts a definition:

(3.2.3) Definition: Consider $C(\pi)$ for some $\{\pi_i, i \in I\}$. $C(\pi)^+$ is a *max-sum cone* if and only if $C(\pi)$ is pointed and int $C(\pi)^+$ contains a vector $\tilde{x}_0$ which maximizes $h(\tilde{x}) = \sum_I \sigma_i \, 1\{[y_i, \tilde{x}] > 0\}$.

Note that max-sum cones exist by virtue of (3.2.2) since a solution vector exists. The optimal cones C and F depicted in Figure (3.1.5) are max-sum cones.

It can be seen now that every solution vector is in the interior of some max-sum cone and any element of the interior of a max-sum cone is a solution. Consequently, in order to identify every solution to the problem, it suffices to identify all max-sum cones.

The following procedure will yield a concise description of the set of solution vectors if such is desired:

(a) By using the tree algorithm, obtain for every max-sum cone, a vector $\tilde{v}$ in its interior.

(b) For each such $\tilde{v}$, compute the max-sum cone $C(\pi)^+$ containing $\tilde{v}$ by computing the sign of $[y_i, \tilde{v}]$ for all $i \in I$.

(c) A concise way of describing each max-sum cone $C(\pi)^+$ is by specifying its $d-1$ dimensional boundary hyperspaces. This can be done by using (2.4.11) to determine the frame of $C(\pi)$.

-- *max-sum cones are hills* --

Returning to the main discussion, at this stage the problem has been reduced to enumerating the values of h on the interiors of fully dimensional $C(\pi)^+$ in the dual space. The next reduction comes about from the fact that every max-sum cone is what will soon be formally defined as a hill. This is based on one of several insights contained in Warmack and Gonzalez (1973).

(3.2.4) Theorem: Let $C(\pi)^+$ be a max-sum cone and suppose that $\{\langle \pi_i y_i \rangle, i \in I^*(\pi)\}$ is the frame for $C(\pi) = \mathrm{C}\{\pi_i y_i, i \in I\}$ where $I^*(\pi) \subset I$. Then for all $k \in I^*(\pi)$, if $\pi_k = -1$ then there exists $j \in I$ such that $\langle -y_k \rangle = \langle y_j \rangle$.

Proof: The basic idea here is simple. Speaking informally, if one has a point in the interior of a d-dimensional cone in $\tilde{X}$ and notices a $d-1$ dimensional boundary face which is generated only by $\pi_k y_k$ where $\pi_k = -1$, then by crossing just over that hyperspace, one collects all of the rewards for those y_k and keeps all of the other rewards he has collected so far. This cannot happen if one starts out in the interior of a max-sum cone.

More formally, suppose there exists $k \in I^*(\pi)$ such that $\pi_k = -1$ and for all $i \in I$, $\langle -y_k \rangle \neq \langle y_i \rangle$. Since $\pi_k = -1$, $\langle -y_k \rangle \neq \langle 0 \rangle$. Consequently, by (2.4.11),

$$\begin{aligned}\{\tilde{x} \in \tilde{X} : [\pi_i y_i, \tilde{x}] > 0 \text{ for } i \in I \text{ w.o. } (I_0 \cup I_k(\pi)),\\ [\pi_i y_i, \tilde{x}] < 0 \text{ for } i \in I_k(\pi),\\ [\pi_i y_i, \tilde{x}] = 0 \text{ for } i \in I_0\} \neq \varnothing.\end{aligned}$$

Now $i \in I_k(\pi)$ if and only if $\langle \pi_i y_i \rangle = \langle -y_k \rangle$ but since $\langle -y_k \rangle \neq \langle y_i \rangle$ for any $i \in I$, $\pi_i = -1$ for $i \in I_k(\pi)$. The second condition above can now be written as $[y_i, \tilde{x}] > 0$ for $i \in I_k(\pi)$. In short, the cone $\mathrm{C}\{\pi_i y_i, i \in I \text{ w.o. } I_k(\pi), y_i, i \in I_k(\pi)\}^+$ which has just been discovered has a nonempty interior upon which h assumes a value greater than that associated with a max-sum cone. This, of course, is a contradiction. □

The preceding theorem prompts another definition.

(3.2.5) Definition: Let $C(\pi)$ be pointed. $C(\pi)^+$ is a *hill* if and only if for any $k \in I$ such that $\langle \pi_k y_k \rangle$ is an isolated ray of $C(\pi)$ (i.e., is in the frame of $C(\pi)$) and $\pi_k = -1$, then there exists $j \in I$ such that $\langle -y_k \rangle = \langle y_j \rangle$.

Theorem (3.2.4) shows that every max-sum cone is a hill. The converse, however, does not hold. For example in Figure (3.1.5), while hills C and F are max-sum cones, hills A and D are not. The tree algorithm enumerates all of the hills and consequently, all of the max-sum cones. Fortunately, there are practically always far fewer hills than there are fully-dimensional $C(\pi)^+$ as will be indicated by the examples given in Chapter 9.

-- *more about hills* --

The nature of hills will be discussed from an intuitive standpoint after some of their properties have been summarized in formulas and after the $I^+(\pi)$ notation has been defined. Theorem-Definition (3.2.6) follows easily from material already developed.

(3.2.6) Theorem-Definition: Let $C(\pi)^+$ be a hill. Then:

(a) There exists $I^+(\pi) \subset I$ such that $0 \in I^+(\pi)$ and $\{\langle y_i \rangle, i \in I^+(\pi)\}$ is a frame for $C(\pi)$.

(b) $C(\pi)^+ = \{\tilde{x} \in \tilde{X}: [y_i, \tilde{x}] \geqslant 0 \text{ for } i \in I^+(\pi)\}$.

(c) $\text{int}\, C(\pi)^+ = \{\tilde{x} \in \tilde{X}: [y_i, \tilde{x}] > 0 \text{ for } i \in I^+(\pi) \text{ w.o. } 0\}$.

(d) Each $d-1$ dimensional boundary face of $C(\pi)^+$ is $F_{\{k\}}$ for some $k \in I^+(\pi)$ w.o. 0. where $F_{\{k\}} =$

$$\{\tilde{x} \in \tilde{X}: [y_i, \tilde{x}] > 0 \text{ for } i \in I^+(\pi) \text{ w.o. } \{0, k\}, [y_k, \tilde{x}] = 0\}$$

(e) For $k \in I^+(\pi)$ w.o. 0,

$$\{\tilde{x} \in \tilde{X}: [\pi_i y_i, \tilde{x}] > 0 \text{ for } i \in I \text{ w.o. } (I_0 \cup I_k(\pi)),$$
$$[\pi_i y_i, \tilde{x}] < 0 \text{ for } i \in I_k(\pi),$$
$$[y_i, \tilde{x}] = 0 \text{ for } i \in I_0\} \neq \emptyset.$$

Intuitively, $C(\pi)^+$ is a hill if and only if it is the d-dimensional intersection of closed positive halfspaces chosen from among those generated by the y_i. Consequently, if a point in the interior of $C(\pi)^+$ crosses over one of the $d-1$ dimensional boundary faces into the neighboring cone, then the criterion function h will lose the rewards σ_i associated with the y_i which generated the closed positive halfspace in question. Now even though h may pick up rewards associated with y_j where $\langle -y_j \rangle = \langle y_i \rangle$ and y_i generates the closed positive half-space just mentioned, the basic idea behind the hill concept is that when a point leaves a hill, h decreases. In other words, hills can be seen as being relative maxima.

-- *visualizing hills in the original space X* --

It is possible to visualize hills not only in the context of the dual space as done in Figure (3.1.5) but also in the context of the original space X. To accomplish this, associate each linear functional $\tilde{v} \in \tilde{X}$ with the open halfspace through the origin $\{x \in X: [x, \tilde{v}] > 0\}$.

For fixed $C(\pi)^+$, here is how to determine which open halfspaces in the original space correspond to the elements of the face

$$F_J(C(\pi)^+) = \{\tilde{x} \in \tilde{X}: [\pi_i y_i, \tilde{x}] > 0 \text{ for } i \notin J, [\pi_i y_i, \tilde{x}] = 0 \text{ for } i \in J\}:$$

Fix $\tilde{v} \in F_J(C(\pi)^+)$ and note that:

(a) for $i \notin J$, if $\pi_i = 1$ then $y_i \in \{x \in X: [x, \tilde{v}] > 0\}$,

(b) for $i \notin J$, if $\pi_i = -1$ then $y_i \in \{x \in X: [x, \tilde{v}] < 0\}$, and

(c) for $i \in J$, $y_i \in \{x \in X: [x, \tilde{v}] = 0\}$.

So, the open halfspace associated with any $\tilde{v} \in F_J(C(\pi)^+)$ contains only those y_i for which $i \notin J$ and $\pi_i = 1$ and contains on its boundary precisely those y_i for which $i \in J$.

Consider the set $\{y_i\}_0^7$ shown in Figure (3.2.7) where the open halfspace associated with each of $\tilde{v}_1$, $\tilde{v}_2$, and $\tilde{v}_3$ is indicated by dashed arrows pointing from its boundary into its interior. Observe that

$$\tilde{v}_1 \in F_\emptyset(\mathrm{C}\{y_0, -y_1, -y_2, -y_3, y_4, y_5, y_6, y_7\}^+),$$

$$\tilde{v}_2 \in F_\emptyset(\mathrm{C}\{y_0, -y_1, -y_2, y_3, y_4, y_5, y_6, y_7\}^+),$$

$$\text{and } \tilde{v}_3 \in F_{\{5\}}(\mathrm{C}\{y_0, -y_1, \ y_2, y_3, y_4, -y_5, -y_6 -y_7\}^+).$$

Note that as any $\tilde{v}$'s halfspace is moved back and forth in such a way that it does not cross over any y_i, then in such a way are generated all of the halfspaces associated with the elements in the $C(\pi)^+$ cone containing $\tilde{v}$.

In order to determine the frame of the dual of $C(\pi)^+$, recall that there is a one-to-one correspondence between the elements of the frame and the $d-1$ dimensional boundary faces of $C(\pi)^+$. Consequently, $\langle \pi_j y_j \rangle$ is in the frame of $C(\pi)$ if and only if as $\tilde{v}$ tries in every which way to leave $C(\pi)^+$, it will at some point run into the $d-1$ dimensional wall $\{\tilde{x}: [y_j, \tilde{x}] = 0\} \cap C(\pi)^+$ generated by y_j. Back in the original space, this corresponds to $\tilde{v}$'s halfspace being constrained by y_j as it wiggles about. The frames of the duals of $\tilde{v}_1$ and $\tilde{v}_2$'s cones in Figure (3.2.7) are $\{\langle y_0 \rangle, \langle -y_3 \rangle, \langle y_4 \rangle\}$ and $\{\langle y_0 \rangle, \langle y_3 \rangle, \langle y_7 \rangle\}$, respectively.

It is now clear that a hill can be seen in the original space as a set of open halfspaces whose boundaries pass through the origin, which correspond to the same cone in $\tilde{X}$, and each of which is constrained only by y_i contained in its interior or boundary. In Figure (3.2.7), $\tilde{v}_1$ is not in a hill and $\tilde{v}_2$ is. Note that $\tilde{v}_2$'s hill is also a max-sum cone. Also, it is immediately clear that $\tilde{v}_1$ is not in a max-sum cone because if $\tilde{v}_1$'s halfspace is moved counter-clockwise just past y_3, then a better halfspace results.

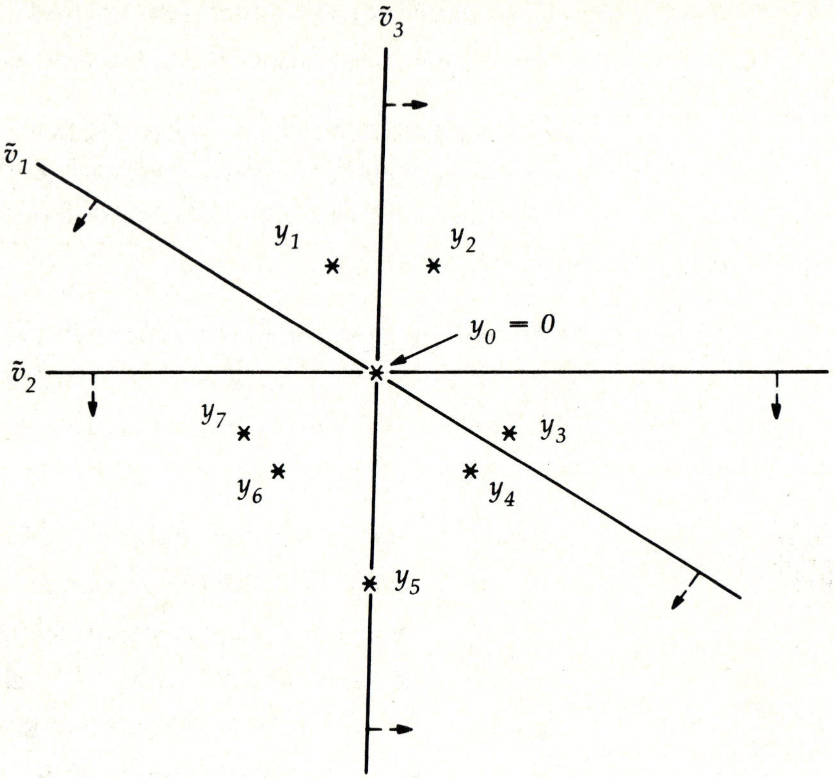

(3.2.7) Figure: The representations of three linear functionals in R^2 amidst $\{y_i\}_0^7$.

-- *relating hills and max-sum cones* --

As the reader may have guessed by now, the fundamental concept here is that of the hill. The max-sum cone concept is derived from the specific optimization problem defined by the WOH problem, i.e., that of maximizing over $\tilde{x}$, $\sum_1^n \sigma_i \; 1\{[y_i, \tilde{x}] > 0\}$. The hill concept, however, is central to solving problems of extremizing functions of systems of linear relations. For the rest of this chapter, the characteristics of both max-sum cones and hills will be discovered in parallel since both are of independent interest.

The next theorem shows that any hill can be made into a unique max-sum cone by the choice of suitable weights σ_i.

(3.2.8) Theorem: Given $\{y_i, i \in I\}$ where, recall, $I = \{0, \ldots, n\}$ for some n. For some $J \subset I$, let $\{\langle y_i \rangle, i \in J\}$ be the frame for the dual of a hill. Let $\sigma_i = 1$ for $i \notin J$ and $\sigma_i = n$ for $i \in J$. Then for the weights σ_i just defined, $\mathrm{C}\{y_i, i \in J\}^+$ is the unique max-sum cone for this problem.

Proof: Remembering that $0 \in J$, set $p = \#J - 1$ and note that h assumes a value greater than pn on $\mathrm{int}\ \mathrm{C}\{y_i, i \in J\}^+$. Now take any $\tilde{v} \notin \mathrm{int}\ \mathrm{C}\{y_i, i \in J\}^+$. By (3.2.6), there exists $k \in J$ such that $[y_k, \tilde{v}] \leqslant 0$. Observe that $h(\tilde{v}) \leqslant (p-1)n + n - p = p(n-1) < pn$. □

The next theorem contains one of the three main ideas behind the tree algorithm, namely, that if a vector $\tilde{v}$ is not in a particular hill (or max-sum cone), then that hill (or max-sum cone) will send a signal indicating this condition. The proof follows from (3.2.6).

(3.2.9) Theorem: Let $C(\pi)^+$ be a hill with the frame $\{\langle y_i \rangle, i \in I^+(\pi)\}$ for its dual. Let $\tilde{v} \notin C(\pi)^+$. Then there exists $j \in I^+(\pi)$ such that $[y_j, \tilde{v}] < 0$.

The next theorem yields a result which will be used in the tree algorithm to determine a sufficient condition for stopping.

(3.2.10) Theorem: Suppose $\tilde{v}$ is such that for all $i \in I$, $[y_i, \tilde{v}] \geqslant 0$. Then $\tilde{v}$ is in every hill (and hence every max-sum cone).

Proof: Suppose $C(\pi)^+$ is a hill such that $\tilde{v} \notin C(\pi)^+$. Then there exists j such that $[y_j, \tilde{v}] < 0$ which is a contradiction. □

-- *the essence of the tree algorithm* --

This is a good point to delineate in more detail what is to come. The tree algorithm has two phases. The objective of the first phase is to find for each (as yet unknown) hill or max-sum cone a vector which is contained in it. These vectors will usually be contained in the boundaries of cones and so the objective of the second phase is to displace these boundary vectors into the interiors of their respective hills (or max-sum cones). A notable characteristic of the second phase is that in some cases, in order to displace a given boundary vector into the interiors of cones, it is necessary for the entire procedure to be called recursively to solve certain lower dimensional versions of the original problem. Fortunately, the entire two phase procedure is guaranteed to obtain solution vectors in the interiors of cones before the procedure ever runs out of dimensions to recurse on.

-- *a particular WOH tree algorithm* --

In order to facilitate the reader's comprehension of later sections, the opportunity is taken now to present a tree algorithm for solving the WOH problem. Other variants of this algorithm will be discussed later. Since the assertion that this tree algorithm solves the WOH problem will only be discussed and validated in the following three sections, the reader should not expect to fully understand the algorithm at this point. The reader might want to refer back to this algorithm statement after reading each of the following sections in order to see how each of the subpieces of the algorithm fit back into the whole.

First, the algorithm's variables will be defined.

(3.2.11) Definitions: Recall that the WOH problem requests the identification of all those vectors $\tilde{x} \in \tilde{X}$ which maximize $h(\tilde{x}) := \sum_I \sigma_i \, 1\{[y_i, \tilde{x}] > 0\}$ for given $\{y_i, i \in I\} \subset X$ and $\sigma_i > 0$ for all $i \in I$.

For any nonempty subset J of I, let $R_J := \mathsf{L}\{y_i, i \in J\}$ and define $h_J: \tilde{R}_J \to [0, \infty)$ via

$$h_J(\tilde{r}) := \sum_J \sigma_i \, 1\{[y_i, \tilde{r}] > 0\}.$$

Note $h = h_I$ and by the assumption in (3.1.1), $\tilde{X} = \tilde{R}_I$.

Also, for $\emptyset \neq J \subset I$, for all $\tilde{r} \in \tilde{R}_J$, let

$$J_0 := \{i \in J : y_i = 0\},$$

$$N_J(\tilde{r}) := \{i \in J : [y_i, \tilde{r}] < 0\},$$

$$\text{and } Z_J(\tilde{r}) := \{i \in J \text{ w.o. } J_0 : [y_i, \tilde{r}] = 0\}.$$

y_{i_k} will be written as $y(i_k)$. Also, at times, "$\tilde{v}_k$" will be used to represent some $\tilde{v}_k(i_0, \ldots, i_{k-1})$ either generically or individually as the context will indicate.

From Chapter 2, recall that $\#A$ is the cardinality of the set A, $\underline{x}$ is the representation in R^d of the vector $x \in R_I$ according to some fixed arbitrary basis, and ψ_J is the vector space isomorphism mapping $S^{\perp}$ onto $S^{\perp}|_{R_J}$ where S is such that $R_J \oplus S = R_I$ (cf., (2.1.27)).

EXPLORE is the procedure which constructs and searches the boundary vector tree and periodically calls upon its subroutine UPDATE_B to update a set B_I which contains the most promising looking boundary vectors found so far. Once certain conditions regarding B_I have been satisfied, EXPLORE calls its subroutine DISPLACE to initiate the second phase of the tree algorithm where the boundary vectors in B_I are displaced and the resulting WOH solution vectors are saved in the set A_I (for "answer").

To help DISPLACE do its job, the subroutine COMP_DISP computes the α factor necessary to satisfactorily displace a given boundary vector $\tilde{v}_k(i_0, \ldots, i_{k-1})$ in the direction of a given $\tilde{z}$. The subroutine UPDATE_A updates A_I with candidate solution vectors as they are found.

The following tree algorithm is written in a hopefully self-explanatory hybrid of Fortran, BASIC, PL/I, and English. Succeeding sections will show that it solves the WOH problem.

(3.2.12) Algorithm: Obtain I, $\{y_i, i \in I\} \subset X$, and h_I where $\mathsf{L}\{y_i, i \in I\} = X$. If desired, modify the preceding to eliminate any $y_i = 0$ and to eliminate all ties among the $\langle y_i \rangle$. Regarding notation, in UPDATE_B, any sum involving $\{i_0, \ldots, i_{j-1}\} = \emptyset$ is to be ignored.

Obtain some nonzero $\tilde{v}_0 \in R_I = \tilde{X}$ and set $A_I = \emptyset$.

Call EXPLORE $(I, \{y_i, i \in I\}, h_I, \tilde{v}_0, A_I)$.

EXPLORE: Procedure $(I, \{y_i, i \in I\}, h_I, \tilde{v}_0, A_I)$;

Step 1: Set $B_I = \{\tilde{v}_0\}$.

If $\#N_I(-\tilde{v}_0) < \#N_I(\tilde{v}_0)$ then set $\tilde{v}_0 = -\tilde{v}_0$.

If $N_I(\tilde{v}_0) = \emptyset$ then do:

Call DISPLACE (h_I, B_I, A_I).

Return from EXPLORE.

end;

Step 2: For $k = 1, \ldots, d-2$, do:

For each $i_0 \in N_I(\tilde{v}_0)$,

$i_1 \in N_I(\tilde{v}_1(i_0)), \ldots, i_{k-1} \in N_I(\tilde{v}_{k-1}(i_0, \ldots, i_{k-2}))$, do:

Obtain $\tilde{x} \in y(i_0)^\perp \cap \cdots \cap y(i_{k-1})^\perp$ where $\tilde{x} \neq 0$.

If $\#N_I(-\tilde{x}) < \#N_I(\tilde{x})$ then set $\tilde{x} = -\tilde{x}$.

Set $\tilde{v}_k(i_0, \ldots, i_{k-1}) = \tilde{x}$.

If $N_I(\tilde{v}_k(i_0, \ldots, i_{k-1})) = \emptyset$ then do:

Set $B_I = \{\tilde{v}_k(i_0, \ldots, i_{k-1})\}$.

Call DISPLACE (h_I, B_I, A_I).

Return from EXPLORE.

end;

If $\#N_I(\tilde{v}_k) + \#Z_I(\tilde{v}_k) - k < \#N_I(\tilde{v}_0)$ then do:

Set $B_I = \{\tilde{v}_k(i_0, \ldots, i_{k-1})\}$.

Let g_I be such that $g_I(\tilde{x}) := \sum_I 1\{[y_i, \tilde{x}] > 0\}$.

Call DISPLACE (g_I, B_I, A_I).

Set $\tilde{v}_0$ equal to any element of A_I.

Go to "Step 1" of EXPLORE.

end;

Call UPDATE_B $(\tilde{v}_k(i_0, \ldots, i_{k-1}), \{i_0, \ldots, i_{k-1}\}, B_I)$.

next i_{k-1}; . . . ; next i_0;

next k;

Step 3: For each $i_0 \in N_I(\tilde{v}_0)$, . . . , $i_{d-2} \in N_I(\tilde{v}_{d-2}(i_0, \ldots, i_{d-3}))$, do:

Obtain $\tilde{x} \in y(i_0)^{\perp} \cap \cdots \cap y(i_{d-2})^{\perp}$ where $\tilde{x} \neq 0$.

If $\#N_I(-\tilde{x}) < \#N_I(\tilde{x})$ then set $\tilde{x} = -\tilde{x}$.

Set $\tilde{v}_k(i_0, \ldots, i_{k-1}) = \tilde{x}$.

If $N_I(\tilde{v}_k(i_0, \ldots, i_{k-1})) = \varnothing$ then do:

Set $B_I = \{\tilde{v}_k(i_0, \ldots, i_{k-1})\}$.

Call DISPLACE (h_I, B_I, A_I)

Return from EXPLORE.

end;

If $\#N_I(\tilde{v}_k) + \#Z_I(\tilde{v}_k) - k < \#N_I(\tilde{v}_0)$ then do:

Set $B_I = \{\tilde{v}_k(i_0, \ldots, i_{k-1})\}$.

Let g_I be such that $g_I(\tilde{x}) = \sum_I 1\{[y_i, \tilde{x}] > 0\}$.

Call DISPLACE (g_I, B_I, A_I).

Set $\tilde{v}_0$ equal to any element of A_I.

Go to "Step 1" of EXPLORE.

end;

If $h_I(\tilde{v}_k) > h_I(-\tilde{v}_k)$

then call UPDATE_B $(\tilde{v}_k, \{i_0, \ldots, i_{d-2}\}, B_I)$.

If $h_I(-\tilde{v}_k) > h_I(\tilde{v}_k)$

then call UPDATE_B $(-\tilde{v}_k, \{i_0, \ldots, i_{d-2}\}, B_I)$.

If $h_I(\tilde{v}_k) = h_I(-\tilde{v}_k)$ then do:

Call UPDATE_B $(\tilde{v}_k, \{i_0, \ldots, i_{d-2}\}, B_I)$.

Call UPDATE_B $(-\tilde{v}_k, \{i_0, \ldots, i_{d-2}\}, B_I)$.

end;

next i_{d-2}; $\cdots$; next i_0;

Step 4: Call DISPLACE (h_I, B_I, A_I).

Return from EXPLORE.

UPDATE_B: Procedure $(\tilde{x}, \{i_0, \ldots, i_{k-1}\}, B_I)$;

If $h_I(\tilde{x}) + \sum_{i \in Z_I(\tilde{x})} \sigma_i \geqslant$

$$\sup_{\tilde{v}_j(i_0, \ldots, i_{j-1}) \in B_I} h_I(\tilde{v}_j(i_0, \ldots, i_{j-1})) + \sum_{i \in \{i_0, \ldots, i_{j-1}\}} \sigma_i \quad \text{then do:}$$

For each $\tilde{v}_j \in B_I$ do:

If $h_I(\tilde{x}) + \sum_{i \in \{i_0, \ldots, i_{k-1}\}} \sigma_i > h_I(\tilde{v}_j) + \sum_{i \in Z_I(\tilde{v}_j)} \sigma_i$

then set $B_I = B_I$ w.o. $\{\tilde{v}_j\}$.

next $\tilde{v}_j$;

$B_I = B_I \cup \{\tilde{x}\}$.

end;

end UPDATE_B;

DISPLACE: Procedure (h_I, B_I, A_I);

Step 1: Set $A_I = \emptyset$.

Step 2: For each $\tilde{v}_k(i_0, \ldots, i_{k-1}) \in B_I$, do:

Set $J = \{i \in I: [y_i, \tilde{v}_k(i_0, \ldots, i_{k-1})] = 0\}$.

If dim $\mathsf{L}\{y_i, i \in J\} = 0$

then call UPDATE_A $(\tilde{v}_k(i_0, \ldots, i_{k-1}), A_I)$.

If dim $\mathsf{L}\{y_i, i \in J\} = 1$ then do:

Take $p \in J$ w.o. I_0.

Set $J_1 = \{i \in J: \langle y_i \rangle = \langle y_p \rangle\}$.

Set $J_2 = \{i \in J: \langle y_i \rangle = -\langle y_p \rangle\}$.

Set $\underline{\tilde{z}} = \underline{y}_p$.

Call COMP_DISP $(\tilde{v}_k(i_0, \ldots, i_{k-1}), J_1, \tilde{z}, \alpha)$.

Call UPDATE_A $((1-\alpha)\tilde{v}_k(i_0, \ldots, i_{k-1}) + \alpha\tilde{z}, A_I)$.

If $J_2 \neq \emptyset$ then do:

Set $\underline{\tilde{z}} = -\underline{y}_p$.

```
                Call COMP_DISP (ṽ_k(i_0, . . . ,i_{k-1}), J_2, z̃, α).
                Call UPDATE_A ((1−α)ṽ_k(i_0, . . . ,i_{k-1}) + αz̃, A_I).
            end;
        end;
        If dim L{y_i, i ∈ J} > 1 then do:
            Solve the linear program:
                maximize γ subject to γ ⩽ 1 and γ ⩽ y_i^T x for
                    i ∈ J w.o. J_0 and x ∈ R^d where d = dim R_I.
            If γ = 1 then do:
                Set z̃ = x.
                Call COMP_DISP (ṽ_k(i_0, . . . ,i_{k-1}), J, z̃, α).
                Call UPDATE_A ((1−α)ṽ_k(i_0, . . . ,i_{k-1}) + αz̃, A_I).
            end;
            If γ = 0 then do:
                Select some ṽ_0(J) ∈ R̃_J.
                Call EXPLORE (J, {y_i, i ∈ J}, h_J, ṽ_0(J), A_J);
                For each r̃_j ∈ A_J do:
                    Call COMP_DISP (ṽ_k(i_0, . . . ,i_{k-1}), J, ψ_J^{-1}(r̃_J), α).
                    Call UPDATE_A
                        ((1−α)ṽ_k(i_0, . . . ,i_{k-1}) + αψ_J^{-1}(r̃_J), A_I).
                end;
            end;
        end;
        next ṽ_k(i_0, . . . ,i_{k-1});
```

COMP_DISP: Procedure $(\tilde{x}, J, \tilde{z}, \alpha)$;

If $J \neq \{i \in I: [y_i, \tilde{x}] = 0\}$ then print "error in COMP_DISP".

Obtain for each $i \in I$ w.o. J, $\pi_i \in \{-1, 1\}$ such that $[\pi_i y_i, \tilde{x}] > 0$.

Set $K = \{i \in I \text{ w.o. } J: [\pi_i y_i, \tilde{z}] < 0\}$.

If $K = \varnothing$ then set $\alpha = \frac{1}{2}$.

If $K \neq \varnothing$ then set $\alpha = \frac{1}{2} \min_{i \in K} \left\{ \frac{[y_i, \tilde{x}]}{[y_i, \tilde{x}] - [y_i, \tilde{z}]} \right\}$.

end COMP_DISP;

UPDATE_A: Procedure $(\tilde{x}, A_I)$;

If $A_I = \varnothing$ then set $A_I = \{\tilde{x}\}$.

If $h_I(\tilde{x}) \geqslant \sup\{h_I(\tilde{w}): \tilde{w} \in A_I\}$ then do:

If $h_I(\tilde{x}) > \sup\{h_I(\tilde{w}): \tilde{w} \in A_I\}$ then set $A_I = \varnothing$.

Set $A_I = A_I \cup \{\tilde{x}\}$.

end;

end UPDATE_A;

end DISPLACE;

end EXPLORE;

As the reader will discover subsequently, Algorithm (3.2.12) does not incorporate the major improvements of trimming, depth-first searching, and the projection method of determining $\tilde{v}_k(i_0, \ldots, i_{k-1})$.

Summary For Section 3.2

In order to find all vectors $\tilde{v} \in \tilde{X}$ which maximize the criterion function h, where $h(\tilde{x}) = \sum_I \sigma_i \, 1\{[y_i, \tilde{x}] > 0\}$, first note that each $\tilde{v} \in \tilde{X}$ is in a face of some $C(\pi)^+$ where $C(\pi) = \mathrm{C}\{\pi_i y_i, i \in I\}$ and $\pi_i \in \{-1, 1\}$. Since h is constant on faces of $C(\pi)^+$, it suffices to enumerate the values of h on all nonempty faces of the $C(\pi)^+$ in order to determine which sets of vectors maximize the criterion function h.

The consideration of lower dimensional faces is eliminated upon discovering that the vectors maximizing h comprise the interiors of certain $C(\pi)^+$ called max-sum cones. Max-sum cones are elements of a larger class of cones called hills.

$C(\pi)^+$ is a hill if and only if it is the fully dimensional intersection of closed positive halfspaces chosen from among those generated by the y_i. Thus hills are seen to assume the character of relative maxima among the set of all faces. (It is of interest to note that while not every hill is a max-sum cone, it can be made into one through the choice of suitable σ_i.) As a result, the problem of maximizing h has been reduced to the problem of enumerating the values of h on the hills. This is what the tree algorithm does.

In the general systems of linear relations optimization problems which will be treated in Chapter 5, a hill is contained in every equivalence class of solution cones (this will be explained in more detail later) and so the general optimization problem is solved in the same way, i.e., by enumerating the values of the criterion function on the hills.

One of the primary concepts allowing the efficient enumeration of the hills is the fact that if $C(\pi)^+$ is a hill and $\tilde{v} \notin C(\pi)^+$, then there exists y_j such that $[y_j, \tilde{v}] < 0$ and $y_j^{\perp}$ contains a $d-1$ dimensional face of $C(\pi)^+$.

Section 3.3:

The Basic Algorithm For Obtaining Vectors On The Boundaries Of Hills

This section is concerned with developing the basic algorithm for obtaining at least one vector in, and usually on the boundary of, every hill. The proof that validates this algorithm is based on an induction on the dimensionality of the problem. The first thing to do is to establish how sets of points in lower dimensional subspaces are created from $\{y_i, i \in I\}$ and to understand how these lower dimensional sets relate back to the original set.

-- *projecting the problem* --

(3.3.1) Definition: Let $J \subset I$. Let $S := \mathsf{L}\{y_i, i \in J\}$. Suppose $1 \leqslant \dim S \leqslant d - 1$. Let R be any subspace such that $X = R \oplus S$. For $i \in I$, define $z_i := P[y_i \,|\, R, S]$. (Recall that $P[y_i \,|\, R, S]$ is the projection of y_i onto the subspace R along the subspace S. See Definition (2.1.25).) (Also, there is no need to notationally indicate the dependence on J in the expression $P[y_i \,|\, R, S]$ since J will be easily inferred from the context in what follows.).

(3.3.2) Theorem: The set $\{z_i, i \in I\} \subset R$ is a set of vectors which satisfies all of the assumptions listed for $\{y_i, i \in I\} \subset X$ in problem statement (3.1.1).

Proof: It is easy to see that $\dim R = d - \dim S \geqslant 1$. Observe that $z_0 = 0$. The last thing to show is that $\mathsf{L}\{z_i, i \in I\} = R$. To do this, begin by showing $\mathsf{L}\{z_i, i \in I\} \oplus S = X$. Clearly, $LHS \subset RHS$. For the other direction, recall that by assumption, $X = \mathsf{L}\{y_i, i \in I\}$ and observe that $\sum_I \alpha_i y_i = \sum_I \alpha_i z_i + s$ for some $s \in S$. Finally, $\mathsf{L}\{z_i, i \in I\} \cap S = \{0\}$ because $\mathsf{L}\{z_i, i \in I\} \subset R$. Since $\dim \mathsf{L}\{z_i, i \in I\} = d - \dim S = \dim R$, the conclusion is that $\mathsf{L}\{z_i, i \in I\} = R$. □

The reason for allowing zeros and "conical ties" in $\{y_i, i \in I\}$ can now be explained. Note that $\{\langle z_i \rangle, i \in I\}$ does not necessarily consist of distinct nonzero rays since $z_i = 0$ for $i \in J$ and because it may happen that for some $s \in S$, $\alpha > 0$, and $j \neq k$, that $y_j = \alpha y_k + s$ and so $\langle z_j \rangle = \langle z_k \rangle$. Consequently, if $\{\langle y_i \rangle, i \in I\}$ was required to consist of distinct nonzero rays, then $\{z_i, i \in I\}$ would not be a lower dimensional analog of $\{y_i, i \in I\}$.

-- *projecting cones* --

The next result shows that cones in the original space are projected down onto cones in any fixed lower dimensional space.

(3.3.3) Theorem: $P[C(\pi) \mid R, S] = \mathsf{C}\{\pi_i z_i, i \in I\}$.

Proof:

$$\begin{aligned} P[C(\pi) \mid R, S] &= \{P[\sum_I \lambda_i \pi_i y_i \mid R, S]: \lambda_i \geqslant 0\} \\ &= \{\sum_I \lambda_i \pi_i z_i : \lambda_i \geqslant 0\} \\ &= \mathsf{C}\{\pi_i z_i, i \in I\}. \end{aligned}$$

□

More notation is needed.

(3.3.4) Definitions: For z_i as in (3.3.1) and for $i \in I$, $\pi_i \in \{-1, 1\}$ with $\pi_i = 1$ for $i \in I_0$,

$$\dot{C}(\pi) := \mathsf{C}\{\pi_i z_i, i \in I\}.$$

For $j \in I$,

$$\dot{I}_j(\pi) := \{i \in I : \langle \pi_i z_i \rangle = \langle \pi_j z_j \rangle \}.$$

(3.3.5) Theorem: For some $I^*(\pi) \subset I$, let $\{ \langle \pi_i y_i \rangle, i \in I^*(\pi)\}$ be a frame for $C(\pi)$. Then $\{ \langle \pi_i z_i \rangle, i \in I^*(\pi)\}$ contains a frame for $\dot{C}(\pi)$.

Proof: $\dot{C}(\pi)$ is conically spanned by $\{\langle P[\pi_i y_i \mid R, S]\rangle : i \in I^*(\pi)\}$. □

-- the duals of projected cones --

(3.3.6) Theorem: For all $\tilde{v} \in S^{\perp}$, for all $i \in I$, $[y_i, \tilde{v}] = [z_i, \tilde{v}]$.

Recall that $\tilde{R} = S^{\perp}|_R$ by (2.1.27). This enables $\dot{C}(\pi)^+$ to be related to $C(\pi)^+$.

(3.3.7) Theorem: $\dot{C}(\pi)^+ = (C(\pi)^+ \cap S^{\perp})|_R$.

Proof:

$$\begin{aligned}
\dot{C}(\pi)^+ &= \{\tilde{r} \in \tilde{R} : [\pi_i z_i, \tilde{r}] \geqslant 0, i \in I\} \\
&= \{\tilde{u}|_R : [\pi_i z_i, \tilde{u}|_R] \geqslant 0, i \in I, \tilde{u} \in S^{\perp}\} \\
&= \{\tilde{u}|_R : [\pi_i z_i, \tilde{u}] \geqslant 0, i \in I, \tilde{u} \in S^{\perp}\} \\
&= \{\tilde{u}|_R : [\pi_i y_i, \tilde{u}] \geqslant 0, i \in I, \tilde{u} \in S^{\perp}\} \\
&= (C(\pi)^+ \cap S^{\perp})|_R .
\end{aligned}$$

□

-- some projected hills are hills --

One might hope that if $C(\pi)^+$ is a hill in the original space then $\dot{C}(\pi)^+$ would be a hill in the lower dimensional space. This is not generally true. Consider Example (2.3.10) and set $y_i = a_i$ for $i = 1, \ldots, 4$. So $y_1 = (1, 0, 1)$, $y_2 = (0, 1, 1)$, $y_3 = (-1, 0, 1)$, and $y_4 = (0, -1, 1)$. Let $S = \mathsf{L}\{y_1, y_3\}$ and $R = \mathsf{L}\{(0, 1, 0)\}$. Note that $(\mathsf{C}\{y_i\}_1^4)^+$ is a hill and $P[\mathsf{C}\{y_i\}_1^4 \mid R, S] = \mathsf{L}\{(0, 1, 0)\}$ which is not pointed. So, $\dot{C}(\pi)^+ = ((\mathsf{C}\{y_i\}_1^4)^+ \cap S^{\perp})|_R$ cannot be a hill in the lower dimensional space $\tilde{R}$ and is in fact $\langle 0 \rangle$.

There is a special case, however, when $\dot{C}(\pi)^+$ is necessarily a hill.

(3.3.8) Theorem: Suppose $\dim X \geqslant 2$. Let $C(\pi)^+$ be a hill and suppose $\langle \pi_k y_k \rangle \neq \langle 0 \rangle$ is an element of the frame of $C(\pi)$. Let $S = \mathsf{L}\{y_k\}$ and R be any subspace such that $R \oplus S = X$. Then:

(a) $\dot{C}(\pi)$ is pointed, i.e., there exists $\tilde{r} \in \tilde{R}$ such that $[\pi_i z_i, \tilde{r}] > 0$ for $i \in I$ w.o. $\dot{I}_0$.

(b) $\dot{C}(\pi)^+ = (C(\pi)^+ \cap y_k^{\perp})|_R$ is a hill in the lower dimensional problem.

Proof: For (a), first obtain from (2.4.11) a $\tilde{v} \in \tilde{X}$ such that $[\pi_i y_i, \tilde{v}] > 0$ for $i \in I$ w.o. $(I_0 \cup I_k(\pi))$ and $[y_i, \tilde{v}] = 0$ for $i \in I_0 \cup I_k(\pi)$. Note that since $\dim X \geqslant 2$, there are at least two distinct nonzero rays in $\{ \langle \pi_i y_i \rangle , i \in I\}$ and so there exists $i \in I$ w.o. $(I_0 \cup I_k(\pi))$. Clearly, $\dot{I}_0 \supset I_0 \cup I_k(\pi)$ and equality holds since $[\pi_i z_i, \tilde{v}] > 0$ for $i \in I$ w.o. $(I_0 \cup I_k(\pi))$. So $\tilde{v} \in y_k^{\perp}$ is such that $[\pi_i z_i, \tilde{v}] = [\pi_i y_i, \tilde{v}] > 0$ for $i \in I$ w.o. $\dot{I}_0$. Now set $\tilde{r} = \tilde{v}|_R$.

For (b), by (3.2.6) and (3.3.5), $\{\langle z_i \rangle, i \in I^+(\pi)\}$ contains every isolated ray of the pointed cone $\dot{C}(\pi)$ and so if $\langle \pi_j z_j \rangle$ is an isolated ray of $\dot{C}(\pi)$ then $\langle \pi_j z_j \rangle = \langle z_i \rangle$ for some $i \in I^+(\pi)$. □

(3.3.9) Example: Example (2.4.12) will help to illustrate these concepts if $y_0 = (0, 0, 0)$, $y_1 = (1, 0, 0)$, $y_2 = (0, 1, 0)$, $y_3 = (0, 0, 1)$, and $y_4 = (1, 1, 0)$. Let $S = \mathsf{L}\{y_3\}$ and $R = \mathsf{L}\{y_1, y_2\}$. Then $z_0 = y_0$, $z_1 = y_1$, $z_2 = y_2$, $z_3 = 0$, and $z_4 = y_4$. Setting $\pi_i = 1$ for all i, observe that $\dot{C}(\pi) = \mathsf{C}\{z_1, z_2\}$ is pointed and $\dot{C}(\pi)^+ = (C(\pi)^+ \cap y_3^{\perp})|_R$ is a hill which, more specifically, can be seen to be the positive x-y quadrant in the two dimensional dual space $\tilde{R}$.

-- *some projected max-sum cones are max-sum cones* --

The analog of the previous theorem holds for max-sum cones as well. To show this, it is necessary to define the maximization problem for the lower dimensional vectors and then to connect this up with a constrained maximization problem for the original vectors.

(3.3.10) Theorem-Definition: Define $\dot{h}: \tilde{R} \to \mathrm{R}$ via

$$\dot{h}(\tilde{r}) := \sum_I \sigma_i \, 1\{[z_i, \tilde{r}] > 0\}.$$

Then:

(a) Seeking to maximize $\dot{h}(\tilde{r})$ over $\tilde{r} \in \tilde{R}$ is a lower dimensional version of the original maximization problem for h and $\{y_i, i \in I\}$.

(b) Let Problem A be that of finding $\tilde{v}_0 \in S^\perp$ which maximizes $h(\tilde{x})$ over $\tilde{x} \in S^\perp$ and Problem B be that of finding $\tilde{r}_0 \in \tilde{R}$ which maximizes $\dot{h}(\tilde{r})$ over $\tilde{r} \in \tilde{R}$. Recall the vector space isomorphism $\psi : S^\perp \to \tilde{R}$ from (2.1.27). Then:

(i) If $\tilde{v}_0$ solves Problem A, then $\psi(\tilde{v}_0)$ solves Problem B.

(ii) If $\tilde{r}_0$ solves Problem B, then $\psi^{-1}(\tilde{r}_0)$ solves Problem A.

Proof: (a) follows directly from (3.3.2).

For (b), suppose for all $\tilde{v} \in S^\perp$,

$$\sum_I \sigma_i \, 1\{[y_i, \tilde{v}_0] > 0\} \geqslant \sum_I \sigma_i \, 1\{[y_i, \tilde{v}] > 0\}.$$

Then for all $\tilde{v} \in S^\perp$,

$$\sum_I \sigma_i \, 1\{[z_i, \tilde{v}_0] > 0\} \geqslant \sum_I \sigma_i \, 1\{[z_i, \tilde{v}] > 0\}.$$

And so, for all $\tilde{v} \in S^\perp$,

$$\sum_I \sigma_i \, 1\{[z_i, \tilde{v}_0|_R] > 0\} \geqslant \sum_I \sigma_i \, 1\{[z_i, \tilde{v}|_R] > 0\}.$$

And finally, for all $\tilde{r} \in \tilde{R}$, $\dot{h}(\tilde{v}_0|_R) \geqslant \dot{h}(\tilde{r})$.

Next suppose for all $\tilde{r} \in \tilde{R}$,

$$\sum_I \sigma_i \, 1\{[z_i, \tilde{r}_0] > 0\} \geqslant \sum_I \sigma_i \, 1\{[z_i, \tilde{r}] > 0\}.$$

Let $\tilde{v}_0 = \psi^{-1}(\tilde{r}_0) \in S^\perp$. Then for all $\tilde{v} \in S^\perp$,

$$\sum_I \sigma_i \, 1\{[z_i, \tilde{v}_0|_R] > 0\} \geqslant \sum_I \sigma_i \, 1\{[z_i, \tilde{v}|_R] > 0\}.$$

And then for all $\tilde{v} \in S^\perp$,

$$\sum_I \sigma_i \, 1\{[y_i, \tilde{v}_0] > 0\} \geqslant \sum_I \sigma_i \, 1\{[y_i, \tilde{v}] > 0\}.$$

□

Now for the analog.

(3.3.11) Theorem: Let $C(\pi)^+$ be a max-sum cone when $\dim X \geqslant 2$. Let $\langle \pi_k y_k \rangle \neq \langle 0 \rangle$ be an isolated ray of $C(\pi)$, $S = \mathrm{L}\{y_k\}$, and R be any subspace such that $R \oplus S = X$. Then $\dot{C}(\pi)^+$ is a max-sum cone in the lower dimensional problem.

Proof: By (3.3.8), $\dot{C}(\pi)^+$ is a hill in the lower dimensional problem since $C(\pi)^+$ is a hill in the original problem. Take $\tilde{r}_0 \in \dot{C}(\pi)^+$ such that $[\pi_i z_i, \tilde{r}_0] > 0$ for $i \in I$ w.o. $(I_0 \cup I_k(\pi))$. It is necessary to show that for all $\tilde{r} \in \tilde{R}$,

$$\sum_I \sigma_i \, 1\{[z_i, \tilde{r}_0] > 0\} \geqslant \sum_I \sigma_i 1\{[z_i, \tilde{r}] > 0\}.$$

By (3.3.10), setting $\tilde{v}_0 = \psi^{-1}(\tilde{r}_0) \in y_k^\perp$, it suffices to show for all $\tilde{v} \in y_k^\perp$,

$$\sum_I \sigma_i \, 1\{[y_i, \tilde{v}_0] > 0\} \geqslant \sum_I \sigma_i \, 1\{[y_i, \tilde{v}] > 0\}.$$

Now, $[\pi_i y_i, \tilde{v}_0] > 0$ for all $i \in I$ w.o. $(I_0 \cup I_k(\pi))$ and $[\pi_i y_i, \tilde{v}_0] = 0$ for $i \in I_0 \cup I_k(\pi)$. Furthermore, there exists $\tilde{u}_0 \in \text{int } C(\pi)^+$ where $[\pi_i y_i, \tilde{u}_0] > 0$ for all $i \in I$ w.o. I_0. Let $J_1 = \{i : [y_i, \tilde{u}_0] > 0\}$. Clearly,

$$h(\tilde{v}_0) = \sum_{J_1} \sigma_i - \sum_{\substack{i \in I_k(\pi) \\ \pi_i = 1}} \sigma_i > 0$$

where the latter strict inequality holds because $C(\pi)$ contains at least two distinct nonzero rays and $C(\pi)^+$ is a hill.

Suppose then that there exists $\tilde{v}_1 \in y_k^\perp$ such that $h(\tilde{v}_1) > h(\tilde{v}_0)$. A contradiction will be obtained by constructing a vector which is better than $\tilde{u}_0$. Note that this is accomplished if $\alpha > 0$ can be chosen such that

$$h(\tilde{v}_1 + \alpha\tilde{u}_0) \geqslant h(\tilde{v}_1) + \sum_{\substack{i \in I_k(\pi) \\ \pi_i = 1}} \sigma_i > h(\tilde{u}_0).$$

So, let $K_1 = \{i : [y_i, \tilde{v}_1] > 0\}$. Observe that $\alpha > 0$ can be chosen such that for all $i \in K_1$, $[y_i, \tilde{v}_1 + \alpha\tilde{u}_0] = [y_i, \tilde{v}_1] + \alpha[y_i, \tilde{u}_0] > 0$ and for all $i \in I_k(\pi)$ such that $\pi_i = 1$, $[y_i, \tilde{v}_1 + \alpha\tilde{u}_0] > 0$. □

-- *hill boundary vector collection algorithm* --

The algorithm stated next is designed to obtain a vector in every hill; it is thought of as a hill boundary vector collection algorithm since all but at most one of these vectors will be on the boundaries of hills, not in their interiors. The proof follows a discussion of how one might and might not interpret the algorithm.

(3.3.12) Definitions: $N(\tilde{x}) := \{i \in I : [y_i, \tilde{x}] < 0\}$.
$y(i_k) := y_{i_k}$.

The following algorithm is the first phase of the basic or generic tree algorithm.

(3.3.13) Algorithm:

For $k = 0, \ldots, d-2$, do: Set $V_k = \varnothing$. next k;

Set $V_{d-1,1} = V_{d-1,2} = \varnothing$.

Step 1: Obtain $\tilde{v}_0 \neq 0$. Set $V_0 = \{\tilde{v}_0\}$. If $N(\tilde{v}_0) = \varnothing$ then exit.

Step 2: For $k = 1, \ldots, d-2$, do:

For each $i_0 \in N(\tilde{v}_0)$, $i_1 \in N(\tilde{v}_1(i_0)), \ldots,$

$i_{k-1} \in N(\tilde{v}_{k-1}(i_0, \ldots, i_{k-2}))$, do:

Obtain $\tilde{x} \in y(i_0)^\perp \cap \cdots \cap y(i_{k-1})^\perp$ where $\tilde{x} \neq 0$.

Set $\tilde{v}_k(i_0, \ldots, i_{k-1}) = \tilde{x}$.

Set $V_k = V_k \cup \{\tilde{v}_k(i_0, \ldots, i_{k-1})\}$.

If $N(\tilde{v}_k(i_0, \ldots, i_{k-1})) = \varnothing$ then exit.

next i_{k-1}; $\cdots$; next i_0;

next k;

Step 3: For each $i_0 \in N(\tilde{v}_0), \ldots, i_{d-2} \in N(\tilde{v}_{d-2}(i_0, \ldots, i_{d-3}))$, do:

Obtain $\tilde{x} \in y(i_0)^\perp \cap \cdots \cap y(i_{d-2})^\perp$ where $\tilde{x} \neq 0$.

Set $\tilde{v}_{d-1,1}(i_0, \ldots, i_{d-2}) = \tilde{x}$ and $\tilde{v}_{d-1,2}(i_0, \ldots, i_{d-2}) = -\tilde{x}$.

Set $V_{d-1,1} = V_{d-1,1} \cup \{\tilde{v}_{d-1,1}(i_0, \ldots, i_{d-2})\}$.

Set $V_{d-1,2} = V_{d-1,2} \cup \{\tilde{v}_{d-1,2}(i_0, \ldots, i_{d-2})\}$.

next i_{d-2}; $\cdots$; next i_0;

Subsequently, "$\tilde{v}_{d-1}(i_0, \ldots, i_{d-2})$" will be used to ambiguously represent one of $\tilde{v}_{d-1,1}(i_0, \ldots, i_{d-2})$ and $\tilde{v}_{d-1,2}(i_0, \ldots, i_{d-2})$.

The reader may wish to observe how a modified and expanded form of Algorithm (3.3.13) lies at the heart of EXPLORE in Algorithm (3.2.12).

-- *the algorithm finds a vector from each hill* --

(3.3.14) Theorem: Let X be a d-dimensional vector space over R with $d \geqslant 2$. Let $I = \{0, \ldots, n\}$ for some n. Let $\{y_i, i \in I\} \subset X$ where $y_0 := 0$. Suppose $\mathsf{L}\{y_i, i \in I\} = X$ and suppose the V_k are created as in Algorithm (3.3.13).

Then:

(a) For $k = 0, \ldots, d-2$, for $i_0 \in N(\tilde{v}_0), \ldots, i_k \in N(\tilde{v}_k(i_0, \ldots, i_{k-1}))$, $\{y(i_0), \ldots, y(i_k)\}$ is linearly independent.

(b) Algorithm (3.3.13) is well-defined.

(c) If $C(\pi)^+$ is a hill, then there exists $\tilde{v} \in \bigcup_0^{d-2} V_k \cup V_{d-1,1} \cup V_{d-1,2}$ such that $\tilde{v} \in C(\pi)^+$. Since every max-sum cone is a hill, there is a vector from every max-sum cone in $\bigcup_0^{d-2} V_k \cup V_{d-1,1} \cup V_{d-1,2}$ as well.

(3.3.15) Discussion: The output $\bigcup_0^{d-2} V_k \cup V_{d-1,1} \cup V_{d-1,2}$ of Algorithm (3.3.13) can be conveniently organized into a data structure known as a tree. (Definitions of trees and related concepts may be found in Knuth (1973a)). Figure (3.3.16) shows how this may be done for an example in three dimensions. $\tilde{v}_0$ is in the root node of the tree. $N(\tilde{v}_0) = \{1, 4, 5\}$ in this example and so $\tilde{v}_0$'s node has three children named by the numerals stationed along the three paths leading out of this node. $\tilde{v}_1(5)$'s node has four children, two for each element of $N(\tilde{v}_1(5)) = \{1, 4\}$. In general, the number of children generated by a node containing $\tilde{v}_k(i_0, \ldots, i_{k-1})$ is $\#N(\tilde{v}_k(i_0, \ldots, i_{k-1}))$ if $k \leqslant d-3$ and $2 \cdot \#N(\tilde{v}_k(i_0, \ldots, i_{k-1}))$ if $k = d-2$.

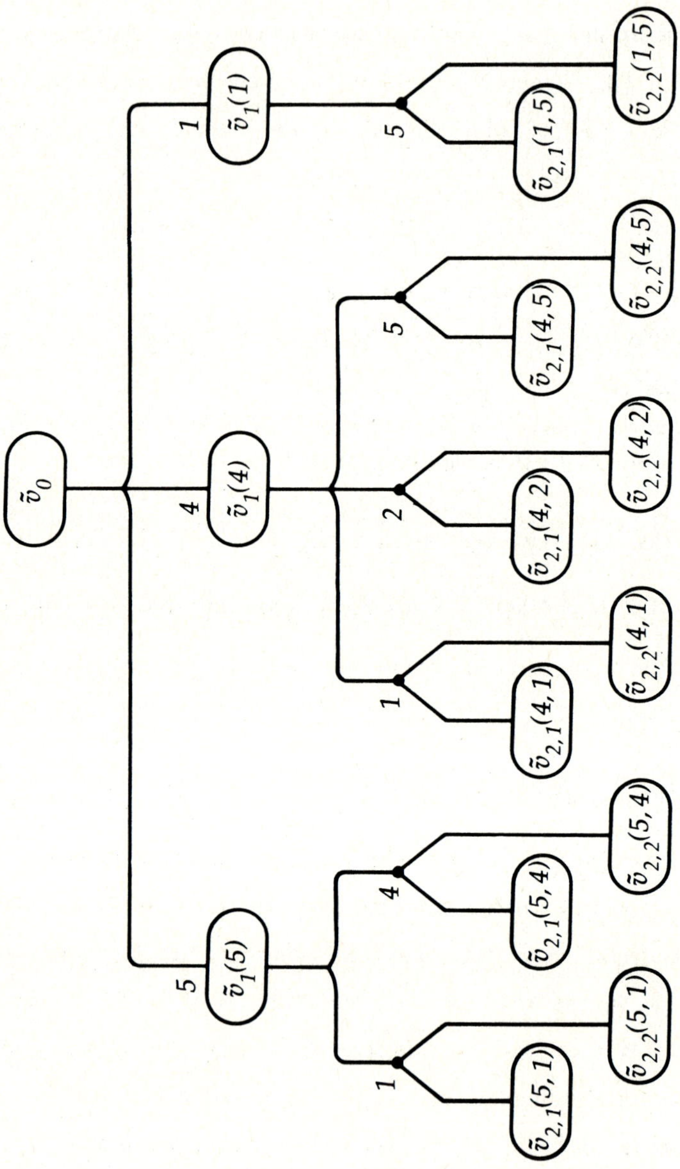

(3.3.16) Figure: A sample tree when $d = 3$.

-- *examples of boundary vector collection* --

Note that Step 2 of the algorithm is omitted when $d = 2$. The only way then to observe the behavior of the complete algorithm is to work with examples with $d \geqslant 3$. The situation in two dimensions is important to understand however and so a two dimensional example will be given followed by a three dimensional example.

(3.3.17) Example: Consider Example (3.1.2). Suppose for the choice of $\tilde{v}_0$, a vector is selected in the interior of cone D which is one of the four hills in this example. $N(\tilde{v}_0) = \{1, 2, 5\}$. Step 3 of the algorithm selects six vectors, two conically spanning each of $y_1^\perp$, $y_2^\perp$, and $y_5^\perp$. Since $y_1^\perp = y_2^\perp$, some effort has been wasted here. Note that for each of the four hills (A, C, D, F), the algorithm has selected a vector which is contained in that hill. Note, in particular, in order to obtain vectors in the hills A and F, it was necessary for the algorithm to include in $V_{d-1,1} \cup V_{d-1,2}$ both $\tilde{x}$ and $-\tilde{x}$ for some nonzero $\tilde{x} \in y_5^\perp$. The reader may wish to follow the algorithm through with different $\tilde{v}_0$. Theorem (3.3.14) guarantees that no matter what $\tilde{v}_0$ is selected, the algorithm will produce vectors in each of the hills.

(3.3.18) Example: Consider the following example in R^3. Let $y_0 = (0, 0, 0)$, $y_1 = (1, 0, 0)$, $y_2 = (0, 1, 0)$, $y_3 = (-1, -1, 0)$, $y_4 = (1, -1, 0)$, and $y_5 = (0, 0, 1)$. Suppose $\sigma_i = 1$ for all i. All but one of the vectors lie in the x-y plane and these vectors are pictured in Figure (3.3.19).

There are 16 $C(\pi)^+$, all shaped like wedges. The intersection of these wedges with the $x-y$ plane is shown in Figure (3.3.20). Note that there are three hills. The two max-sum cones are $\mathsf{C}\{y_1, y_2, -y_3, y_4, y_5\}^+$ and $\mathsf{C}\{y_1, -y_2, y_3, y_4, y_5\}^+$.

Now, since the standard basis is being used in the original space and so also its dual in the dual space, the vectors and their representations look the same. This enables $[y, \tilde{v}]$ to be computed as the Euclidean inner product is

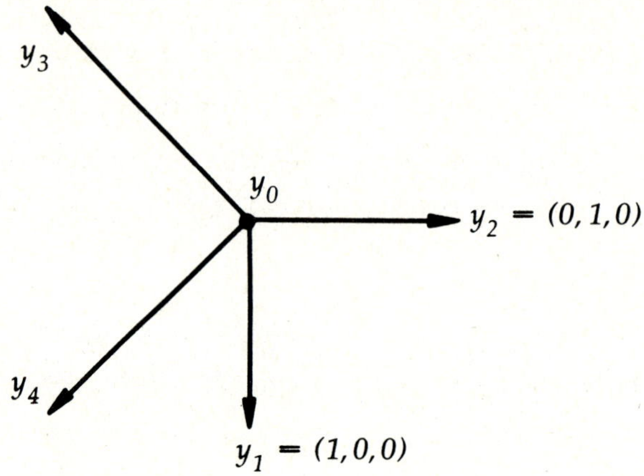

(3.3.19) Figure: The y_i from Example (3.3.18) which lie in the $x-y$ plane.

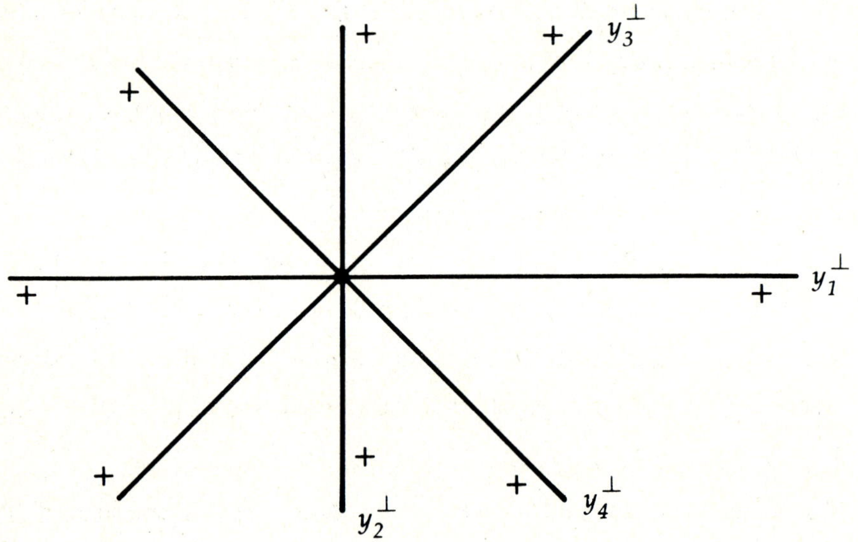

(3.3.20) Figure: The intersection of the cones in the dual space with the $x-y$ plane for Example (3.3.18).

calculated, i.e., by $\underline{\tilde{v}}^T \underline{y}$.

Here is a sample tree for this example. Let $\tilde{v}_0 = (-1, 0, -1)$. Then $N(\tilde{v}_0) = \{1, 4, 5\}$. Let $\tilde{v}_1(1) = (0, 0, -1) \in y_1^\perp$. Then $N(\tilde{v}_1(1)) = \{5\}$ and, for this, select $\tilde{v}_{2,1}(1,5) = (0, 1, 0) = -\tilde{v}_{2,2}(1,5)$. Let $\tilde{v}_1(4) = (-\frac{1}{2}, -\frac{1}{2}, -1) \in y_4^\perp$. Then $N(\tilde{v}_1(4)) = \{1, 2, 5\}$ for which $\tilde{v}_{2,1}(4,1) = \tilde{v}_{2,1}(4,2) = (0, 0, 1)$ and $\tilde{v}_{2,1}(4,5) = (1, 1, 0)$ are selected. Let $\tilde{v}_1(5) = (-1, 0, 0)$. Then $N(\tilde{v}_1(5)) = \{1, 4\}$ for which $\tilde{v}_{2,1}(5,1) = (0, 1, 0)$ and $\tilde{v}_{2,1}(5,4) = (1, 1, 0)$ are selected. The tree corresponding to this information is shown in Figure (3.3.16).

Note that the vector $(0, 0, 1)$ is in every hill and is in the tree. Also $(1, 1, 0)$ is in a max-sum cone and is in the tree.

While it may seem like an anomaly to have $[y_2, \tilde{v}_0] = 0$ or $[y_2, \tilde{v}_1(5)] = 0$ when neither is required, Algorithm (3.3.13) is insensitive to it.

One of the curious features about this algorithm is the enormous amount of freedom one has in the selection of $\tilde{v}_k(i_0, \ldots, i_{k-1})$ for $k = 0, \ldots, d-2$. In the next section, various ways are suggested which capitalize on this freedom and produce more efficient algorithms.

-- *how not to prove the validity of the algorithm* --

An intuitive explanation of why this algorithm works is not as simple as one might hope for. In fact, what might be considered the obvious and natural approach turns out to be wrong. More specifically, the following argument is incorrect. (The false assertion made in this argument is also made in the analogous Warmack and Gonzalez' argument. This error is what invalidates their main algorithm proof.)

(3.3.21) Fallacious Argument: "Let the tree be constructed as in (3.3.13) and let $C(\pi)^+$ be a hill. We show that the tree contains a nonzero vector in $C(\pi)^+$. If $\tilde{v}_0 \in C(\pi)^+$, then we are done. If $\tilde{v}_0 \notin C(\pi)^+$, then by Theorem (3.2.9), there exists $\langle y(j_0) \rangle$ in the frame of $C(\pi)$ such that

$j_0 \in N(\tilde{v}_0)$. $y(j_0)^\perp$ is a bounding hyperspace of $C(\pi)^+$ and so contains nonzero vectors in $C(\pi)^+$. We now search this subspace. We have $\tilde{v}_1(j_0) \in y(j_0)^\perp$ in the tree. If $\tilde{v}_1(j_0) \in C(\pi)^+$, we are done. If $\tilde{v}_1(j_0) \notin C(\pi)^+$, then by Theorem (3.2.9), there exists $j_1 \in N(\tilde{v}_1(j_0))$ such that $y(j_1)^\perp$ is a bounding hyperspace of $C(\pi)^+$. $y(j_0)^\perp \cap y(j_1)^\perp$ contains a nonzero vector in $C(\pi)^+$ and we now search this subspace. $\tilde{v}_2(j_0, j_1) \in y(j_0)^\perp \cap y(j_1)^\perp$ is in the tree. If $\tilde{v}_2(j_0, j_1) \in C(\pi)^+$, then we are done. If not, then by the preceding argument, we obtain $\langle y(j_2)\rangle$ in the frame of $C(\pi)$ and start the search of $y(j_0)^\perp \cap y(j_1)^\perp \cap y(j_2)^\perp$ by examining $\tilde{v}_3(j_0, j_1, j_2)$ which is in the tree. Suppose we consistently fail to find a nonzero vector in $C(\pi)^+$ and work our way down to $\tilde{v}_{d-2}(j_0, \ldots, j_{d-3}) \notin C(\pi)^+$. Then there exists $j_{d-2} \in N(\tilde{v}_{d-2}(j_0, \ldots, j_{d-3}))$ such that $y(j_0)^\perp \cap \cdots \cap y(j_{d-2})^\perp$ contains a nonzero vector of $C(\pi)^+$. But this subspace is one-dimensional and the algorithm at this point selects out a conical spanning set for this subspace, which must therefore include a nonzero vector in $C(\pi)^+$. In short, the tree has been shown to contain a nonzero vector in $C(\pi)^+$."

This argument rests upon the false assertion that if $\langle y(j_0)\rangle, \ldots, \langle y(j_k)\rangle$ are elements of the frame of $C(\pi)$, then

$$y(j_0)^\perp \cap \cdots \cap y(j_k)^\perp \cap C(\pi)^+ \neq \langle 0\rangle.$$

As a counter-example, consider the cone of Example (2.3.10) for which $\underline{y}_1 = (1, 0, 1)$, $\underline{y}_2 = (0, 1, 1)$, $\underline{y}_3 = (-1, 0, 1)$, and $\underline{y}_4 = (0, -1, 1)$. All of the $\langle y_i\rangle$ are isolated rays of $\mathrm{C}\{y_i\}_1^4$. Note that $y_1^\perp \cap y_3^\perp = \mathrm{L}\{\tilde{u}\}$ where $\underline{\tilde{u}} = (0,1,0)$. Observe $y_1^\perp \cap y_3^\perp \cap (\mathrm{C}\{y_i\}_1^4)^+ = \langle 0\rangle$ since $[y_2, -\tilde{u}] < 0$ and $[y_4, \tilde{u}] < 0$.

-- *proving the validity of the boundary vector algorithm* --

The proof of (3.3.14), which will appear shortly, is based on the following idea. First, it is shown that the algorithm works for the case $\dim X = 2$. For $\dim X > 2$, the proof proceeds by induction on $\dim X$. If, for some fixed hill

$C(\pi)^+$, $\tilde{v}_0 \notin C(\pi)^+$, then by Theorem (3.2.9), there exists $j_0 \in N(\tilde{v}_0)$ such that $\langle y(j_0) \rangle$ is an element of the frame of $C(\pi)$. A lower dimensional problem is then formed by projecting the y_i onto $z_i = P[y_i \,|\, R, \mathrm{L}\{y(j_0)\}]$ for any suitable R. $\dot{C}(\pi)^+$ is a hill in this lower dimensional problem by (3.3.8). Next, it is shown that if each linear functional in the subtree of the original tree with root node $\tilde{v}_1(j_0)$ is restricted to have domain R, then the resulting tree is in fact a tree which satisfies the algorithm's requirements for this lower dimensional problem. The induction hypothesis says that there is a linear functional in this lower dimensional tree which is in $\dot{C}(\pi)^+$. This, in turn, yields a vector in $C(\pi)^+$ and the proof is complete. In more detail:

(3.3.22) Proof Of (3.3.14): First, (a) is shown and this is done by induction on k. The assertion is true for $k = 0$ because for all $i_0 \in N(\tilde{v}_0)$, $y(i_0) \neq 0$.

Assume (a) holds for $k = r$. To prove that it holds for $k = r + 1 \leqslant d - 2$, suppose that one is given

$$i_0 \in N(\tilde{v}_0), \ldots, i_r \in N(\tilde{v}_r(i_0, \ldots, i_{r-1})),\; i_{r+1} \in N(\tilde{v}_{r+1}(i_0, \ldots, i_r))$$

and $\alpha(i_j)$ such that $\sum_0^{r+1} \alpha(i_j) y(i_j) = 0$. Then

$$0 = \Big[\sum_0^{r+1} \alpha(i_j) y(i_j),\, \tilde{v}_{r+1}(i_0, \ldots, i_r)\Big] = \alpha(i_{r+1})\,[y(i_{r+1}),\, \tilde{v}_{r+1}(i_0, \ldots, i_r)]$$

and so $\alpha(i_{r+1}) = 0$. Hence, $\sum_0^{r} \alpha(i_j) y(i_j) = 0$ and all of the rest of the $\alpha(i_j)$ are 0 by the induction hypothesis.

To show (b), namely that the algorithm is well-defined, it must shown that it is actually possible to obtain $\tilde{x}$ at each step prior to an exit, if any. This is the case since for $k = 0, \ldots, d - 2$, $i_0 \in N(\tilde{v}_0), \ldots,$ $i_k \in N(\tilde{v}_k(i_0, \ldots, i_{k-1}))$, $\{y(i_0), \ldots, y(i_k)\}$ is linearly independent and so

$$\dim y(i_0)^\perp \cap \cdots \cap y(i_k)^\perp = d - (k + 1) \geqslant 1.$$

As mentioned before, the proof that the algorithm is valid proceeds by an induction on $\dim X$. Let $\dim X = 2$. Suppose for some $\tilde{v} \in V_0 \cup V_{1,1} \cup V_{1,2}$, $N(\tilde{v}) = \emptyset$. Then by Theorem (3.2.10), this $\tilde{v}$ is in every hill and the proof is complete in this event.

Suppose for all $\tilde{v} \in V_0 \cup V_{1,1} \cup V_{1,2}$, $N(\tilde{v}) \neq \emptyset$. Let $C(\pi)^+$ be a hill and suppose for all $\tilde{v} \in V_0 \cup V_{1,1} \cup V_{1,2}$, $\tilde{v} \notin C(\pi)^+$. Since $\tilde{v}_0 \notin C(\pi)^+$, by (3.2.9), there exists $p \in N(\tilde{v}_0)$ such that $\langle y_p \rangle$ is an element of the frame of $C(\pi)$. Hence by (2.4.11),

$$F_p := \{\tilde{x} \in \tilde{X}: [\pi_i y_i, \tilde{x}] > 0 \text{ for } i \in I \text{ w.o. } (I_0 \cup I_p(\pi)), [y_p, \tilde{x}] = 0\} \neq \emptyset.$$

Now, $F_p \subset C(\pi)^+ \cap y_p^\perp$. Hence for any nonzero $\tilde{x} \in y_p^\perp$, $F_p = \langle \tilde{x} \rangle$ or $\langle -\tilde{x} \rangle$ but each of these rays, one of which must be in $C(\pi)^+$, is in $V_{1,1} \cup V_{1,2}$, a contradiction.

The next step is to show that the algorithm is valid when $\dim X = d \geqslant 3$, assuming that it is valid for lower dimensional problems. Suppose there exists $\tilde{v} \in \bigcup_0^{d-2} V_k \cup V_{d-1,1} \cup V_{d-1,2}$ such that $N(\tilde{v}) = \emptyset$. Then by (3.2.10), this $\tilde{v}$ is in every hill and the proof is complete.

So assume for all $\tilde{v} \in \bigcup_0^{d-2} V_k \cup V_{d-1,1} \cup V_{d-1,2}$, $N(\tilde{v}) \neq \emptyset$. Let $C(\pi)^+$ be a hill and suppose for all $\tilde{v} \in \bigcup_0^{d-2} V_k \cup V_{d-1,1} \cup V_{d-1,2}$, $\tilde{v} \notin C(\pi)^+$. Since $\tilde{v}_0 \notin C(\pi)^+$, there exists $p \in N(\tilde{v}_0)$ such that $\langle y_p \rangle$ is an element of the frame of $C(\pi)^+$. By (3.3.8), $\dot{C}(\pi)^+$ is a hill in the lower dimensional z_i problem. The next step is to show how to construct a tree for the lower dimensional problem out of the original tree.

For $\tilde{r} \in \tilde{R}$, define $\dot{N}(\tilde{r}) := \{i \in I: [z_i, \tilde{r}] < 0\}$. Since for all $i \in I$ and $\tilde{v} \in y_p^\perp$, $[z_i, \tilde{v}] = [y_i, \tilde{v}]$, it is clear that $\dot{N}(\tilde{v}|_R) = N(\tilde{v})$ for all $\tilde{v} \in y_p^\perp$. Here is the procedure:

For $k = 1, \ldots, d-3$, do: Set $\dot{V}_k = \varnothing$. next k;

Set $\dot{V}_{d-2,1} = \dot{V}_{d-2,2} = \varnothing$.

Step 1: Set $\dot{\tilde{v}}_0 = \tilde{v}_1(p)\big|_R$.

Set $\dot{V}_0 = \{\dot{\tilde{v}}_0\}$.

Step 2: For $k = 1, \ldots, d-3$, do:

For $i_0 \in \dot{N}(\dot{\tilde{v}}_0)$, $i_1 \in \dot{N}(\dot{\tilde{v}}_1(i_0)), \ldots,$

$i_{k-1} \in \dot{N}(\dot{\tilde{v}}_{k-1}(i_0, \ldots, i_{k-2}))$, do:

Set $\dot{\tilde{v}}_k(i_0, \ldots, i_{k-1}) = \tilde{v}_{k+1}(p, i_0, \ldots, i_{k-1})\big|_R$.

next i_{k-1}; $\cdots$; next i_0;

Set $\dot{V}_k = \dot{V}_k \cup \{\dot{\tilde{v}}_k(i_0, \ldots, i_{k-1})\}$.

next k;

Step 3: For each $i_0 \in \dot{N}(\dot{\tilde{v}}_0), \ldots, i_{d-3} \in \dot{N}(\dot{\tilde{v}}_{d-3}(i_0, \ldots, i_{d-4}))$, do:

Set $\dot{\tilde{v}}_{d-2,1}(i_0, \ldots, i_{d-3}) = \tilde{v}_{d-1,1}(p, i_0, \ldots, i_{d-3})\big|_R$ and

$\dot{\tilde{v}}_{d-2,2}(i_0, \ldots, i_{d-3}) = \tilde{v}_{d-1,2}(p, i_0, \ldots, i_{d-3})\big|_R$.

next i_{d-3}; $\cdots$; next i_0;

Set $\dot{V}_{d-2,1} = \dot{V}_{d-2,1} \cup \{\dot{\tilde{v}}_{d-2,1}(i_0, \ldots, i_{d-3})\}$.

Set $\dot{V}_{d-2,2} = \dot{V}_{d-2,2} \cup \{\dot{\tilde{v}}_{d-2,2}(i_0, \ldots, i_{d-3})\}$.

It must be shown that not only can this procedure find the necessary vectors in the original tree but also that it constructs a tree which satisfies all of the requirements for a tree constructed by Algorithm (3.3.13) for this lower dimensional problem.

As far as Step 1 goes, $\tilde{v}_1(p)$ is certainly in the original tree and since $\tilde{v}_1(p) \in y_p^{\perp}$, $\tilde{v}_1(p)\big|_R$ is a valid choice for $\dot{\tilde{v}}_0$.

For Step 2, the proof proceeds by induction on k. When $k = 1$, $\dot{N}(\dot{\tilde{v}}_0) = N(\tilde{v}_1(p)) \neq \varnothing$ and so for any $i_0 \in \dot{N}(\dot{\tilde{v}}_0)$, $\tilde{v}_2(p, i_0) \neq 0$ is in the original tree. Now $\tilde{v}_2(p, i_0) \in y(p)^{\perp} \cap y(i_0)^{\perp}$ so that $\tilde{v}_2(p, i_0)\big|_R \in \tilde{R}$ and

$$[z(i_0), \tilde{v}_2(p, i_0)|_R] = [z(i_0), \tilde{v}_2(p, i_0)] = [y(i_0), \tilde{v}_2(p, i_0)] = 0.$$

Consequently, $\dot{\tilde{v}}_1(i_0) = \tilde{v}_2(p, i_0)|_R \in z(i_0)^{\perp}$.

Assume that the procedure works as claimed for $k = 1, \ldots, q$. To see that it works for the $k = q + 1 \leqslant d - 3$ level of the tree, take

$$i_0 \in \dot{N}(\dot{\tilde{v}}_0) = N(\tilde{v}_1(p)) \neq \varnothing,\ i_1 \in \dot{N}(\dot{\tilde{v}}_1(i_0)) = N(\tilde{v}_2(p, i_0)) \neq \varnothing, \ldots,$$
$$i_q \in \dot{N}(\dot{\tilde{v}}_q(i_0, \ldots, i_{q-1})) = N(\tilde{v}_{q+1}(p, i_0, \ldots, i_{q-1})) \neq \varnothing.$$

Then $\tilde{v}_{q+2}(p, i_0, \ldots, i_q) \neq 0$ exists in the original tree. Note that $\tilde{v}_{q+2}(p, i_0, \ldots, i_q)|_R \in \tilde{R}$ and for $j = 0, \ldots, q$, $[z(i_j), \tilde{v}_{q+2}(p, i_0, \ldots, i_q)|_R] = 0$ so that

$$\dot{\tilde{v}}_{q+1}(i_0, \ldots, i_q) \in z(i_0)^{\perp} \cap \cdots \cap z(i_q)^{\perp}.$$

Step 3 of the construction procedure is validated in a similar fashion.

Now since $\dot{C}(\pi)^+$ is a hill in the lower dimensional problem and since the tree constructed for this problem is a valid one, by the induction hypothesis, there exists $\tilde{v} \in \bigcup_0^{d-2} V_k \cup V_{d-1,1} \cup V_{d-1,2}$ such that $\tilde{v} \in y_p^{\perp}$, $\tilde{v}|_R \in \bigcup_0^{d-3} \dot{V}_k \cup \dot{V}_{d-2,1} \cup \dot{V}_{d-2,2}$, and $\tilde{v}|_R \in \dot{C}(\pi)^+ = (C(\pi)^+ \cap y_p^{\perp})|_R$. So for some $\tilde{u} \in C(\pi)^+ \cap y_p^{\perp}$, $\psi(\tilde{v}) = \psi(\tilde{u})$ and so $\tilde{v} = \tilde{u} \in C(\pi)^+$ and $\tilde{v} \neq 0$ is in the original tree. This is a contradiction. □

-- *efficiencies available when searching for a max-sum cone* --

When searching for a max-sum cone, Step 3 of Algorithm (3.3.13) can be made more efficient.

(3.3.23) Theorem: Suppose $\dim X \geqslant 2$. If Step 3 of Algorithm (3.3.13) is replaced by the following set of instructions, then the resulting algorithm will place a nonzero vector in $\bigcup_0^{d-2} V_k \cup V_{d-1,1} \cup V_{d-1,2}$ for every max-sum cone.

Max-sum Cone Step 3:

For each $i_0 \in N(\tilde{v}_0), \ldots, i_{d-2} \in N(\tilde{v}_{d-2}(i_0, \ldots, i_{d-3}))$,
obtain $\tilde{x} \in y(i_0)^\perp \cap \cdots \cap y(i_{d-2})^\perp$ where $\tilde{x} \neq 0$.
If $h(\tilde{x}) \geqslant h(-\tilde{x})$ then set $\tilde{v}_{d-1,1}(i_0, \ldots, i_{d-2}) = \tilde{x}$.
If $h(\tilde{x}) < h(-\tilde{x})$ then set $\tilde{v}_{d-1,1}(i_0, \ldots, i_{d-2}) = -\tilde{x}$.
Set $V_{d-1,1} = V_{d-1,1} \cup \{\tilde{v}_{d-1,1}(i_0, \ldots, i_{d-2})\}$.
If $h(\tilde{x}) = h(-\tilde{x})$ then set $\tilde{v}_{d-1,2}(i_0, \ldots, i_{d-2}) = -\tilde{x}$
and set $V_{d-1,2} = V_{d-1,2} \cup \{\tilde{v}_{d-1,2}(i_0, \ldots, i_{d-2})\}$.

Proof: The proof proceeds as that of (3.3.14). The first thing to notice is that since every max-sum cone is a hill and since by (3.3.11), $\dot{C}(\pi)^+$ is a max-sum cone if $C(\pi)^+$ is a max-sum cone, a valid induction step is obtained for this theorem if only the word "hill" is replaced with "max-sum cone" in (3.3.22)'s induction step. This being done it is necessary only to show that this theorem is true when $\dim X = 2$.

Suppose $\dim X = 2$. If there exists $\tilde{v} \in V_0 \cup V_{1,1} \cup V_{1,2}$ such that $N(\tilde{v}) = \varnothing$, then $\tilde{v}$ is in every max-sum cone and the proof is complete.

So, suppose for all $\tilde{v} \in V_0 \cup V_{1,1} \cup V_{1,2}$ that $N(\tilde{v}) \neq \varnothing$ and fix a max-sum cone $C(\pi)^+$. Suppose further that for all $\tilde{v} \in V_0 \cup V_{1,1}\ V_{1,2}$, $\tilde{v} \notin C(\pi)^+$. Since $\tilde{v}_0 \notin C(\pi)^+$, there exists $p \in N(\tilde{v}_0)$ such that $\langle y_p \rangle$ is in the frame of $C(\pi)$. By (3.3.11), $\dot{C}(\pi)^+$ is a max-sum cone in the associated lower dimensional problem. By (3.3.10), there is nonzero $\tilde{u}_0 \in y_p^\perp$ such that for all $\tilde{t} \in y_p^\perp$, $h(\tilde{u}_0) \geqslant h(\tilde{t})$, $\tilde{u}_0|_R \in \dot{C}(\pi)^+$, and $\tilde{u}_0 \in C(\pi)^+$.

Let $\tilde{x} \neq 0$ be the vector which is selected by the modified algorithm. Now, $y_p^\perp = \langle -\tilde{x} \rangle \cup \langle 0 \rangle \cup \langle \tilde{x} \rangle$. Suppose $h(\tilde{x}) > h(-\tilde{x})$. Then it must be the case that $\langle \tilde{x} \rangle = \langle \tilde{u}_0 \rangle$ for if $\langle \tilde{x} \rangle = \langle -\tilde{u}_0 \rangle$ then $h(\tilde{x}) = h(-\tilde{u}_0) \leqslant h(\tilde{u}_0) = h(-\tilde{x})$. Since, in this case, $\tilde{x}$ is saved into $V_{1,1}$, a nonzero vector in $C(\pi)^+$ has been saved by the algorithm and a contradiction is obtained. A similar argument holds when $h(\tilde{x}) < h(-\tilde{x})$. When $h(\tilde{x}) = h(-\tilde{x})$, then the algorithm saves $\tilde{x}$ and $-\tilde{x}$, yielding another contradiction. □

(3.3.24) Example: The condition above that both $\tilde{x}$ and $-\tilde{x}$ be saved when $h(\tilde{x}) = h(-\tilde{x})$ is in fact necessary for the algorithm to obtain a vector in every max-sum cone as the following example shows. Referring back to Figures (3.1.3) to (3.1.5), let $\tilde{v}_0$ be in the interior of cone A. $N(\tilde{v}_0) = \{3, 6\}$. When it comes time to select $\tilde{v}_{1,1}(3)$, the max-sum cone modified algorithm will select a nonzero vector in cone C and not select any $\tilde{v}_{1,2}(3)$. Now let nonzero $\tilde{x} \in y_6^{\perp}$ be in cone C. If the algorithm saved only this vector, then the max-sum cone F would be missed. However, the algorithm saves both $\tilde{x}$ and $-\tilde{x}$ since $h(\tilde{x}) = h(-\tilde{x}) = 4$.

Summary For Section 3.3

This section presented the first phase of the basic tree algorithm which finds vectors in the interiors of all of the hills and consequently in all of the max-sum cones. The first phase finds at least one vector in every hill where, with at most one exception, all of the vectors produced by the first phase lie on the boundaries of cones in the dual space, not in their interiors. This provides the raison d'être for the second phase which is designed to displace desired boundary vectors into neighboring interiors.

The first or boundary vector collection phase of the tree algorithm works in the following way. An initial hyperspace $\tilde{v}_0^{\perp}$ is chosen arbitrarily. ($\tilde{v}_0$ is the only vector the first phase produces which may lie in the interior of a cone). The set $N(\tilde{v}_0) = \{i \in I: [y_i, \tilde{v}_0] < 0\}$ is computed. If $N(\tilde{v}_0) = \varnothing$, then $\tilde{v}_0$ is in every hill and the first phase ends. If not, then for each $i_0 \in N(\tilde{v}_0)$, a hyperspace $\tilde{v}_1(i_0)^{\perp}$ is chosen arbitrarily which contains $y(i_0)$. This finishes the construction of the second level of a tree where $\tilde{v}_0$ is the root and the $\tilde{v}_1(i_0)$ comprise the children of $\tilde{v}_0$.

The process now becomes recursive. Each $\tilde{v}_1(i_0)$ is processed like $\tilde{v}_0$ with the added proviso that each new hyperspace be constrained additionally to pass through the same y_i constraining its predecessor. For each $i_0 \in N(\tilde{v}_0)$, $N(\tilde{v}_1(i_0))$ is computed. If $N(\tilde{v}_1(i_0))$ is empty, then $\tilde{v}_1(i_0)$ is in every hill. If not, then for each $i_1 \in N(\tilde{v}_1(i_0))$, a hyperspace $\tilde{v}_2(i_0, i_1)^{\perp}$ is chosen arbitrarily which contains both $y(i_0)$ and $y(i_1)$. For fixed i_0, the children of $\tilde{v}_1(i_0)$ are $\tilde{v}_2(i_0, i_1)$ for $i_1 \in N(\tilde{v}_1(i_0))$. Once all of the $\tilde{v}_2(i_0, i_1)$ have been selected, the third level of the tree has been completed.

The process is then carried on recursively for each of the $\tilde{v}$ in the current level until the $(d - 1)$st level is reached at which point one is asked to find hyperspaces constrained to pass through linearly independent sets of $d - 1$ points. An $\tilde{x} \neq 0$ is found which generates each desired hyperspace and both

$\tilde{x}$ and $-\tilde{x}$ become children in the tree. After this level is completed, the algorithm stops.

The associated theorem states that for every hill, there is a $\tilde{v}$ in this tree which is in that hill. In order to prove this theorem, one shows directly that the algorithm is valid for $\dim X = 2$ and then proceeds by induction. The idea behind the induction step is that if $\tilde{v}_0$ is not in a specified hill $C(\pi)^+$, then there is a certain lower dimensional analog of the original problem which is solvable by virtue of the induction hypothesis and whose solutions provide a vector in $C(\pi)^+$.

A certain property of max-sum cones enables one to modify the first phase of the tree algorithm for hills into a somewhat more efficient algorithm designed to find a vector in every max-sum cone. More specifically, when an $\tilde{x}^{\perp}$ is obtained passing through $d - 1$ y_i in the construction of the last level of the tree, then $\tilde{x}$ alone is added to the tree if $h(\tilde{x}) > h(-\tilde{x})$, $-\tilde{x}$ alone is added to the tree if $h(\tilde{x}) < h(-\tilde{x})$, and both $\tilde{x}$ and $-\tilde{x}$ are added to the tree if $h(\tilde{x}) = h(-\tilde{x})$.

Section 3.4:

Improvements To The Boundary Vector Collection Phase Of The Basic Tree Algorithm

As remarked earlier, the first phase of the basic tree algorithm (3.3.13) gives the user a great deal of freedom in constructing the tree of boundary vectors. This section is concerned with presenting a number of ways to restrict that freedom in the direction of improving the efficiency of the algorithm. Different tree algorithms will be obtained depending upon which of the following improvements are incorporated.

-- *eliminating zeros and conical ties* --

(3.4.1) Improvement: Naturally, the smaller the set $\{y_i, i \in I\}$, the smaller the tree created by Algorithm (3.3.13). The first thing to do in this regard is to delete all $y_i = 0$ since 0 can never be in the interior of a halfspace through the origin. The next thing to do is to consolidate each group of "conical ties" into a single representative vector. A "conical tie" is said to occur if $\langle y_i \rangle = \langle y_j \rangle$ for $i \neq j$. When a conical tie occurs and all but one of the tied vectors is deleted, then precisely the same polyhedral cones are present after the deletions as before; the cones just have different names/labels attached to them.

To find all conical ties, it suffices to first remove the $y_i = 0$ and replace the set $\{y_i, i \in I\}$ with $\{y_i / \| y_i \|, i \in I\}$ where $\| \cdot \|$ is some convenient norm such as the sup norm. Next, sort lexicographically the resulting set of unit vectors and eliminate ties as they are encountered.

In most cases, the formula defining the criterion function h will have to be modified in some obvious way when zeros are eliminated and conical ties are consolidated. When $h(\tilde{x}) = \sum_1^n \sigma_i 1\{[y_i, \tilde{x}] > 0\}$ and conical ties are present, then the obvious modification is to add together the weights σ_i for each fixed group of "tied" vectors and let that be the weight for the remaining representative vector.

Note that all tie consolidation and zero elimination is to be done before the tree algorithm starts to work.

-- *discarding hopeless vectors* --

(3.4.2) Improvement: In most cases, the objective is not to obtain a vector in the interior of every hill. Instead, the objective is to find all of those hills for which a certain criterion function achieves its maximum value (as is the case when seeking to find all max-sum cones). One soon discovers, in this context, that there are many boundary vectors in the tree which when, displaced into the interiors of any of their neighboring cones, do not yield vectors with the maximum criterion function value. This gives rise to the idea of exploring the tree vector by vector, saving only those vectors which have the highest potentially realizable criterion function values found so far and ignoring those which don't. After the entire tree has been explored, all of the saved vectors can be turned over to the second phase of the algorithm for further processing. (Since the tree algorithm is recursive in nature, it must be shown that in order to find all max-sum cones, it is sufficient to find all max-sum cones at any and all levels of recursion. This is done in Section 3.5).

The selective saving of promising boundary vectors is precisely what UPDATE_B does in Algorithm (3.2.12). Examination of this routine reveals that essentially two types of quantities are being compared. The quantity of the form $h_I(\tilde{x}) + \sum_{i \in Z_I(\tilde{x})} \sigma_i$ is the largest possible h_I value that a vector legally displaced from $\tilde{x}$ could achieve, namely, that which is achieved when $\tilde{x}$ is displaced in such a way as to have the displaced vector on the positive side of all of the $d-1$ dimensional hyperspaces $y_i^{\perp}$ that contain $\tilde{x}$. The quantity of the form $h_I(\tilde{v}_j(i_0, \ldots, i_{j-1})) + \sum_{i \in \{i_0, \ldots, i_{j-1}\}} \sigma_i$ represents the smallest possible h_I value for a vector legally displaced from $\tilde{v}_j(i_0, \ldots, i_{j-1})$. This value is correct since $\{y(i_0), \ldots, y(i_{j-1})\}$ is linearly independent and hence generates a pointed cone with the concomitant implication that a legally displaced vector can always be obtained on the positive side of $y(i_0)^{\perp}, \ldots,$ and $y(i_{j-1})^{\perp}$. (This entire technical discussion here may mean more to the reader after he or she has read the displacement section 3.5.)

The logic then behind UPDATE_B is that if the best displaced vector that a given boundary vector can produce has an h_I value which is less than the best guaranteed minimum h_I value for displaced vectors produced from the boundary vectors in the set B_I, then the given boundary vector should be ignored. On the other hand, if there is a possibility that the given boundary vector could produce through displacement one or more solution vectors, then it should be saved into B_I and any boundary vector in B_I whose best possible displaced h_I value is less than the guaranteed minimum displaced h_I value for the newly added boundary vector should be discarded.

-- *searching depth first* --

(3.4.3) Improvement: The tree constructed in (3.3.13) and, for that matter, in (3.2.12) is constructed in a breadth-first manner, i.e., one entire level at a time. In order to generate the next level of nodes, it is necessary to have available information on all of the nodes at the current level. As one goes deeper in the tree, the number of nodes at each level grows geometrically. From the standpoint of creating a computer program to implement (3.3.13), it is simply not feasible to store all of the information necessary to do a breadth-first search for problems of reasonable size.

So it is better to have the computer program explore the tree via a depth-first search (see Knuth (1973a) for the precise mathematical definition). The following algorithm will accomplish a depth-first search of the tree constructed in (3.3.13). It is well-known that a depth-first search is equivalent to a breadth-first search.

(3.4.4) Algorithm:

Compute $\tilde{v}_0$. If $N(\tilde{v}_0) = \varnothing$ then exit.

For each $i_0 \in N(\tilde{v}_0)$ do:

  Compute $\tilde{v}_1(i_0)$. If $N(\tilde{v}_1(i_0)) = \varnothing$ then exit.

  For each $i_1 \in N(\tilde{v}_1(i_0))$ do:

    Compute $\tilde{v}_2(i_0, i_1)$. If $N(\tilde{v}_2(i_0, i_1)) = \varnothing$ then exit.

    For each $i_2 \in N(\tilde{v}_2(i_0, i_1))$ do:

    ⋮

      For each $i_{d-2} \in N(\tilde{v}_{d-2}(i_0, \ldots, i_{d-3}))$ do:

        Compute $\tilde{v}_{d-1,1}(i_0, \ldots, i_{d-2})$ and $\tilde{v}_{d-1,2}(i_0, \ldots, i_{d-2})$.

      next i_{d-2};

    next i_{d-3};

    ⋮

  next i_1;

next i_0;

The depth-first search algorithm is more economical than the breadth-first search algorithm because it only requires storing the children of $d - 1$ nodes instead of all of the nodes in the next-to-the-last level of the tree.

-- *looking for instant termination* --

(3.4.5) Improvement: As mentioned before, any vector $\tilde{v}$ for which $N(\tilde{v}) = \varnothing$ is in every hill and so if the tree searching algorithm should find such a vector, then it should save it and stop immediately. Consequently, in the innermost loop of Algorithm (3.4.4), it may be worthwhile to insert an instruction which causes termination of the algorithm if either $N(\tilde{v}_{d-1,1}(i_0, \ldots, i_{d-2})) = \varnothing$ or $N(\tilde{v}_{d-1,2}(i_0, \ldots, i_{d-2})) = \varnothing$. This situation is very unlikely if the set of inequalities is inconsistent. On the other hand, if the set of inequalities is known to be consistent, then this improvement should be included.

-- flipping vectors when beneficial --

(3.4.6) Improvement: Step 2 of Algorithm (3.3.13) allows considerable freedom in the choice of $\tilde{x} \neq 0$. The efficiency of the algorithm is highly dependent on the $\tilde{x}$ chosen. Since the number of children of $\tilde{v}_k(i_0, \ldots, i_{k-1}) = \tilde{x}$ is $N(\tilde{x})$, an obvious way to improve matters is to replace $\tilde{x}$ with $-\tilde{x}$ if $\#N(-\tilde{x}) < \#N(\tilde{x})$ (cf., Algorithm (3.2.12)).

-- using heuristically good vectors --

(3.4.7) Improvement: Continuing in this vein, consider now an ad hoc way of obtaining with minimal effort what should be "good" $\tilde{v}$ in each level of the tree. The first $\tilde{v}$ to consider is $\tilde{v}_0$. Since the objective at this point is to modify (3.3.13) so that it generates as small a tree as possible, clearly a best choice for $\tilde{v}_0$ is one which achieves the smallest $N(\tilde{v}_0)$ possible.

This is equivalent to seeking to maximize over $\tilde{v}_0$, $\sum_I 1\{[y_i, \tilde{v}_0] > 0\}$ which is a special case of the problem being solved. In order to escape the circularity of trying to improve a solution method by using a solution method, it is necessary to resort to heuristics. As it turns out then, in the forthcoming paragraphs, inner products and norms will be used in order to obtain heuristically good $\tilde{v}_k$. Not only does this not contradict the author's position that, conceptually speaking, norms and inner products are artificial constructs for these problems, it supports it: the heuristically generated $\tilde{v}_k$ are not the best possible in general and the argument supporting their generation is flawed with arbitrary assumptions. It is adopted however because it is necessary to have some computationally stable way of generating reasonably good $\tilde{v}_k$ and this one at least is computationally stable and, furthermore, makes a certain amount of sense.

To recapitulate, it is desired to have $\tilde{v}_0$ such that $[y_i, \tilde{v}_0]$ is positive for as many y_i as possible. The following procedure appears to be a sensible and economical way to approximate a best $\tilde{v}_0$. Let $w = \frac{1}{n} \sum_1^n y_i$ be the centroid of the $\{y_i, i \in I\}$. It would be good to have $[w, \tilde{v}_0] > 0$ and in some sense as large as possible.

In order to make this criterion more precise and reasonable, recall a few standard definitions. The Euclidean inner product $(\cdot, \cdot)$ on $\mathsf{R}^d = \{\underline{x} : x \in X\}$ is defined via $(\underline{x}, \underline{z}) := \sum_1^d \xi_i \zeta_i = \underline{x}^T \underline{z}$. The associated norm is $\| \underline{x} \| := \sqrt{(\underline{x}, \underline{x})}$. The distance between $\underline{x}$ and $\underline{z}$ is said to be $\| \underline{x} - \underline{z} \|$.

Now, $[w, \tilde{v}_0] = \underline{\tilde{v}}_0^T \underline{w}$ where $\underline{w}$ is the vector representing w with respect to some basis of X and $\underline{\tilde{v}}_0$ is the vector representing $\tilde{v}_0$ with respect to the dual basis in $\tilde{X}$. To find $\tilde{v}_0$, it suffices to find $\underline{\tilde{v}}_0 \in \mathsf{R}^d$. The first thing to note is that $\{i \in I : (\alpha \underline{\tilde{v}}_0)^T \underline{y}_i > 0\}$ is the same for all $\alpha > 0$ while for any i in the above set, $(\alpha \underline{\tilde{v}}_0)^T \underline{y}_i$ increases to infinity as α increases to infinity. Consequently, in seeking to maximize $\underline{\tilde{v}}_0^T \underline{w}$, it would be good to work with representative elements of each $\langle \underline{\tilde{v}}_0 \rangle$ which are all equal in size. A convenient way to do this is to maximize $\underline{\tilde{v}}_0^T \underline{w}$ subject to the condition that $\| \underline{\tilde{v}}_0 \| = 1$. This can be done by finding any nonzero $\underline{\tilde{v}}_0$ which maximizes $\underline{\tilde{v}}_0^T \underline{w} / \| \underline{\tilde{v}}_0 \|$ and then normalizing to unit length.

Along these same lines, any particular $\underline{y}_i$ is prevented from having disproportionate influence on $\underline{w}$ by normalizing all $\underline{y}_i$ to unit length before computing $\underline{w}$.

To continue, the following definitions are needed.

(3.4.8) Definition: The *orthogonal complement* of a set $A \subset \mathsf{R}^d$ is $\{\underline{x} : (\underline{a}, \underline{x}) = 0\}$ and is denoted by $A^\perp$. (The context will infer whether $A^\perp$ is the annihilator of A or the orthogonal complement of A.) Let S be a subspace of R^d. Then $\mathsf{R}^d = S \oplus S^\perp$ and the projector on S along $S^\perp$ is called the *orthogonal projector on* S and is denoted by $P[\cdot | S]$ (cf., $P[\cdot | R, S]$ in (2.1.25)).

(3.4.9) Theorem: Consider $\mathsf{R}^d = \{\underline{x} : x \in X\}$.

(a) Let $S = \mathsf{L}\{\underline{a}\} \neq \langle 0 \rangle$. Then $P[\underline{x}\,|\,S] = (\frac{\underline{a}^T\underline{x}}{\|\,\underline{a}\,\|^2})\underline{a}$.

(b) Let S be a subspace of R^d and $\underline{a} \in \mathsf{R}^d$. Then the distance between $\underline{a}$ and S, $\inf_{\underline{s} \in S} \|\,\underline{a} - \underline{s}\,\|$, is achieved by $\underline{s} = P[\underline{a}\,|\,S]$ and is $\|\,P[\underline{a}\,|\,S^\perp]\,\|$.

(c) Let $S \neq \langle 0 \rangle$ be a subspace of R^d and $\underline{a} \in \mathsf{R}^d$.

Then $\sup_{\substack{\underline{s} \in S \\ \underline{s} \neq 0}} \frac{\underline{s}^T\underline{a}}{\|\,\underline{s}\,\|} = \sup_{\substack{\underline{s} \in S \\ \underline{s} \neq 0}} \frac{|\underline{s}^T\underline{a}|}{\|\,\underline{s}\,\|}$ is achieved by $\underline{s} = \alpha P[\underline{a}\,|\,S]$ for any $\alpha > 0$.

Proof:

(a) $\underline{x} = (\underline{a}^T\underline{x}/\,\|\,\underline{a}\,\|^2)\underline{a} + (\underline{x} - (\underline{a}^T\underline{x}/\,\|\,\underline{a}\,\|^2)\,\underline{a}) \in S \oplus S^\perp$.

(b) $\|\,\underline{s} - \underline{a}\,\|$ is minimized when

$$\begin{aligned}\|\,\underline{s} - \underline{a}\,\|^2 &= \|\,\underline{s} - P[\underline{a}\,|\,S] - P[\underline{a}\,|\,S^\perp]\,\|^2 \\ &= \|\,\underline{s} - P[\underline{a}\,|\,S]\,\|^2 + \|\,P[\underline{a}\,|\,S^\perp]\,\|^2\end{aligned}$$

is minimized.

(c) By the Cauchy-Schwarz inequality, $\frac{|\underline{s}^T\underline{a}|}{\|\,\underline{s}\,\|} = \frac{|\underline{s}^TP[\underline{a}\,|\,S]|}{\|\,\underline{s}\,\|}$ is maximized for $\underline{s} \in S$ when $\underline{s} = \alpha\,P[\underline{a}\,|\,S]$ for $\alpha \neq 0$. The two supremums are identical since if $\underline{s} \in S$ then $-\underline{s} \in S$. Finally, note that since

$$\mathsf{sgn}\,(\alpha P[\underline{a}\,|\,S\,]^T\underline{a}) = \mathsf{sgn}\,(\alpha\,P[\underline{a}\,|\,S\,]^T\,P[\underline{a}\,|\,S\,]) = \mathsf{sgn}\,(\alpha),$$

α must be chosen > 0. □

Applying (c) of (3.4.9) with $S = \mathsf{R}^d$, it is clear that $\underline{\tilde{v}}_0^T\underline{w}\,/\,\|\,\underline{\tilde{v}}_0\,\|$ is maximized by $\underline{\tilde{v}}_0 = \underline{w}$.

Consider next the problem of finding good $\tilde{v}_k(i_0, \ldots, i_{k-1})$ for $k = 1, \ldots, d-2$. The same heuristics as before indicate the desirability of finding $\underline{\tilde{v}} \in S^{\perp} = \mathrm{L}\{\underline{y}(i_0), \ldots, \underline{y}(i_{k-1})\}^{\perp}$ such that $\dfrac{\underline{\tilde{v}}^T \underline{w}}{\| \underline{\tilde{v}} \|}$ is maximized. By (c) of (3.4.9), this occurs when $\underline{\tilde{v}} = P[\underline{w} \,|\, S^{\perp}]$.

Note that when the standard basis is used for $X = \mathrm{R}^d$, then the vectors and their representations are the same. Consequently, (3.4.9) shows that the distance from $\underline{w}$ to the hyperspace $\underline{\tilde{v}}^{\perp}$ is $\| P[\underline{w} \,|\, \mathrm{L}\{\underline{\tilde{v}}\}] \| = \dfrac{| \underline{\tilde{v}}^T \underline{w} |}{\| \underline{\tilde{v}} \|}$ which is maximized subject to the constraint that $\underline{\tilde{v}} \in S^{\perp}$ by $P[\underline{w} \,|\, S^{\perp}]$. This is further justification for this procedure.

-- *using Modified Gram-Schmidt to compute* $\underline{\tilde{v}}_k$ --

This is an economical process because in order to compute $\underline{\tilde{v}}_k(i_0, \ldots, i_{k-1}) = P[\underline{w} \,|\, \mathrm{L}\{\underline{y}(i_0), \ldots, \underline{y}(i_{k-1})\}^{\perp}]$, it turns out to be sufficient to perform a few modifications on $\underline{\tilde{v}}_{k-1}(i_0, \ldots, i_{k-2})$ via a variant of the Modified Gram-Schmidt procedure which will be stated just after (3.5.12) since its details are not needed at the moment. To be more specific about the modifications performed on $\underline{\tilde{v}}_{k-1}(i_0, \ldots, i_{k-2})$, the computer program written to implement the tree algorithm conducts a depth-first search of the tree in such a way that when it comes time to compute $\underline{\tilde{v}}_k(i_0, \ldots, i_{k-1})$ for some k, both $\underline{\tilde{v}}_{k-1}(i_0, \ldots, i_{k-2})$ and an orthonormal basis for $\mathrm{L}\{\underline{y}(i_0), \ldots, \underline{y}(i_{k-2})\}$ are available. Using the Modified Gram-Schmidt procedure on $\underline{y}(i_{k-1})$, a unit vector $\underline{e}(i_{k-1})$ orthogonal to the existing orthonormal basis is obtained such that adjoining $\{\underline{e}(i_{k-1})\}$ to this basis yields an orthonormal basis for $\mathrm{L}\{\underline{y}(i_0), \ldots, \underline{y}(i_{k-1})\}$. Observe that

$$\begin{aligned}\underline{\tilde{v}}_k &= P\left[\underline{w} \mid \mathsf{L}\{\underline{y}(i_0), \ldots, \underline{y}(i_{k-1})\}^{\perp}\right] \\ &= P\left[P[\underline{w} \mid \mathsf{L}\{\underline{y}(i_0), \ldots, \underline{y}(i_{k-2})\}^{\perp}] \mid \mathsf{L}\{\underline{y}(i_0), \ldots, \underline{y}(i_{k-1})\}^{\perp}\right] \\ &= P\left[\underline{\tilde{v}}_{k-1}(i_0, \ldots, i_{k-2}) \mid \mathsf{L}\{\underline{y}(i_0), \ldots, \underline{y}(i_{k-1})\}^{\perp}\right] \\ &= \underline{\tilde{v}}_{k-1}(i_0, \ldots, i_{k-2}) \\ &\qquad - P\left[\underline{\tilde{v}}_{k-1}(i_0, \ldots, i_{k-2}) \mid \mathsf{L}\{\underline{y}(i_0), \ldots, \underline{y}(i_{k-2})\} + \mathsf{L}\{\underline{e}(i_{k-1})\}\right] \\ &= \underline{\tilde{v}}_{k-1}(i_0, \ldots, i_{k-2}) - \left(\underline{e}(i_{k-1})^T\, \underline{\tilde{v}}_{k-1}(i_0, \ldots, i_{k-2})\right)\underline{e}(i_{k-1})\end{aligned}$$

-- *an example* --

As an example of how the tree algorithm works with this improvement, consider $\{y_i, i \in I\} \subset \mathsf{R}^3$ of Example (3.3.18). Here $\underline{w} = \frac{1}{5}(1, 1 - \sqrt{2}, 1)$ and so $\underline{\tilde{v}}_0$ is taken to be $(1, 1 - \sqrt{2}, 1)$. $N(\tilde{v}_0) = \{2, 3\}$.

$$\underline{\tilde{v}}_1(2) = P[\underline{\tilde{v}}_0 \mid \underline{y}_2^{\perp}] = (1, 1 - \sqrt{2}, 1) - (1 - \sqrt{2})(0, 1, 0) = (1, 0, 1)$$

which is an element of a max-sum cone. $N(\tilde{v}_1(2)) = \{3\}$ and thus one obtains $\underline{\tilde{v}}_{2,1}(2, 3) = (0, 0, 1)$ which is in every hill. Similarly, $P[\underline{\tilde{v}}_0 \mid \underline{y}_3^{\perp}] = \frac{1}{2}(\sqrt{2}, -\sqrt{2}, 2)$ which is also in a max-sum cone. $N(\tilde{v}_1(3)) = \{2\}$ and thus $\underline{\tilde{v}}_{2,1}(3, 2) = (0, 0, 1)$ is obtained. The associated tree contains 7 nodes.

The $\tilde{v}_1(i_0)$ in Example (3.3.18) were also obtained by projection with the sole exception that the $\tilde{v}_0$ used there was not the one suggested by the above procedure. The tree in (3.3.18) contains 16 nodes.

The procedure for approximating the best $\tilde{v}_0$ clearly does just that. The best choice for $\tilde{v}_0$ here is (0, 0, 1) which leads to a tree consisting of one node since $N((0, 0, 1)) = \emptyset$.

-- *starting over* --

(3.4.10) Improvement: It is so important to have a good $\tilde{v}_0$ that it is actually worthwhile to start over again when, during the course of exploring the tree, a $\tilde{v}$ is discovered which is sure to have a smaller $N(\tilde{v})$ than $\tilde{v}_0$ once $\tilde{v}$ is displaced into an appropriate interior. So, when such a $\tilde{v}$ is discovered, the algorithm should be restarted with the displaced $\tilde{v}$ as the new $\tilde{v}_0$.

This improvement appears in Algorithm (3.2.12). The expression $\#N_I(\tilde{v}_k) + \#Z_I(\tilde{v}_k) - k$ is the largest possible number of elements in the N_I set associated with a legally displaced $\tilde{x}$. If this number is strictly smaller than the number of children $\tilde{v}_0$ now has, then $\tilde{x}$ is displaced in the best manner possible and the algorithm is started over. (To give further explanation for the "$-k$" above, it will be shown in the displacement section 3.5 that it is always possible to properly displace $\tilde{v}_k(i_0, \ldots, i_{k-1})$ so that the displaced vector is on the positive side of $y(i_0)^{\perp}, \ldots, y(i_{k-1})^{\perp}$.)

-- *tree trimming* --

(3.4.11) Improvement: The efficiency of the depth-first tree searching algorithm would naturally be increased if it were unnecessary to search the entire tree. Fortunately, there is a method whereby certain subtrees of the original tree may be safely left unexplored by the depth-first search algorithm. Such trimming, as described here, can only occur when $d \geqslant 3$.

The use of trimming substantially improves the efficiency of the algorithm. In a sample six-dimensional problem with $\#I = 61$ and with the objective of finding all max-sum cones, the final (after several restarts) trimmed tree consisted of 8778 nodes after an estimated 63,000 nodes had been trimmed by the method described subsequently and another estimated 1000 nodes trimmed by another method which won't be described.

Warmack and Gonzalez (1973) suggest the following way to trim the tree but only offer heuristic justification for it. (In fact, Warmack and Gonzalez trimming is probably not valid for the Warmack and Gonzalez tree algorithm because this tree algorithm arbitrarily constrains all $\tilde{v}_k$ to be in precisely $d - 1$ hyperspaces and as will be seen shortly, Example (3.4.16) shows that the assumption that each $\tilde{v}_k(i_0, \ldots, i_{k-1})^{\perp}$ not contain any more y_i than is required is a critical assumption in this monograph's proof that the trimming algorithm is valid. This, of course, does not directly prove that trimming should not be used with the Warmack and Gonzalez tree algorithm but it is a strong indication in that direction.)

(3.4.12) Algorithm: In this algorithm for searching the tree, the children of every parent node except that containing $\tilde{v}_0$ may be reduced in number by the effects of trimming. For $k = 1, \ldots, d - 2$ and $i_0, \ldots, i_{k-1} \in I$, if $\tilde{v}_k(i_0, \ldots, i_{k-1})$ is in the trimmed tree then, by definition, $N_t(\tilde{v}_k(i_0, \ldots, i_{k-1}))$ is the set of indices indicating the children of $\tilde{v}_k(i_0, \ldots, i_{k-1})$ that are in the trimmed tree. As a special example, $N_t(\tilde{v}_0) = N(\tilde{v}_0)$.

It is necessary in this algorithm to order the elements of each $N_t(\tilde{v})$. This may be done in any desired way by associating the integers $1, \ldots, \#N_t(\tilde{v})$ in a one-to-one fashion with the elements of $N_t(\tilde{v})$. So, for example, $i_0(p_0)$ is the $p_0{}^{th}$ element of $N_t(\tilde{v}_0)$, for fixed $i_0(p_0)$, $i_1(p_1)$ is the $p_1{}^{th}$ element of $N_t\left[\tilde{v}_1(i_0(p_0))\right]$, and in general, for fixed $i_0(p_0), i_1(p_1), \ldots, i_{k-1}(p_{k-1}), i_k(p_k)$ is the $p_k{}^{th}$ element of $N_t\left[\tilde{v}_k(i_0(p_0), \ldots, i_{k-1}(p_{k-1}))\right]$. Here is the algorithm. Examples follow.

Compute $\tilde{v}_0$.

If $N(\tilde{v}_0) = \varnothing$ then exit.

For $p_0 = 1$ to $\#N(\tilde{v}_0)$ do:

Compute $\tilde{v}_1(i_0(p_0))$.

If $N\left[\tilde{v}_1(i_0(p_0))\right] = \varnothing$ then exit.

Set $N_t\left[\tilde{v}_1(i_0(p_0))\right] =$

$$N\left[\tilde{v}_1(i_0(p_0))\right] \text{ w.o. } \{i_0(q):\ q = p_0 + 1, \ldots, \#N(\tilde{v}_0)\}$$

If $N_t\left[\tilde{v}_1(i_0(p_0))\right] = \varnothing$ then next p_0.

For $p_1 = 1$ to $\#N_t\left[\tilde{v}_1(i_0(p_0))\right]$ do:

Compute $\tilde{v}_2(i_0(p_0), i_1(p_1))$.

If $N\left[\tilde{v}_2(i_0(p_0), i_1(p_1))\right] = \varnothing$ then exit.

Set $N_t\left[\tilde{v}_2(i_0(p_0), i_1(p_1))\right] = N\left[\tilde{v}_2(i_0(p_0), i_1(p_1))\right]$

$$\text{w.o. } \bigcup_{j=0}^{1}\{i_j(q): q = p_j + 1, \ldots, \#N_t\left[\tilde{v}_j(i_0(p_0), \ldots, i_{j-1}(p_{j-1}))\right]\}.$$

If $N_t\left[\tilde{v}_2(i_0(p_0), i_1(p_1))\right] = \varnothing$ then next p_1.

⋮

For $p_{k-1} = 1$ to $\#N_t\left[\tilde{v}_{k-1}(i_0(p_0), \ldots, i_{k-2}(p_{k-2}))\right]$ do:

Compute $\tilde{v}_k(i_0(p_0), \ldots, i_{k-1}(p_{k-1}))$.

If $N(\tilde{v}_k(i_0(p_0), \ldots, i_{k-1}(p_{k-1})) = \varnothing$ then exit.

Set $N_t\left[\tilde{v}_k(i_0(p_0), \ldots, i_{k-1}(p_{k-1}))\right] = N\left[\tilde{v}_k(i_0(p_0), \ldots, i_{k-1}(p_{k-1}))\right]$

$$\text{w.o. } \bigcup_{j=0}^{k-1}\{i_j(q): q = p_j + 1, \ldots, \#N_t\left[\tilde{v}_j(i_0(p_0), \ldots, i_{j-1}(p_{j-1}))\right]\}.$$

If $N_t\left[\tilde{v}_k(i_0(p_0), \ldots, i_{k-1}(p_{k-1}))\right] = \varnothing$ then next p_{k-1}.

⋮

For $p_{d-3} = 1$ to $\#N_t\left[\tilde{v}_{d-3}(i_0(p_0), \ldots, i_{d-4}(p_{d-4}))\right]$ do:

Compute $\tilde{v}_{d-2}(i_0(p_0), \ldots, i_{d-3}(p_{d-3}))$. If $N(\tilde{v}_{d-2}(i_0(p_0), \ldots, i_{d-3}(p_{d-3}))) = \varnothing$ then exit.

Set $N_t\left[\tilde{v}_{d-2}(i_0(p_0), \ldots, i_{d-3}(p_{d-3}))\right] = N\left[\tilde{v}_{d-2}(i_0(p_0), \ldots, i_{d-3}(p_{d-3}))\right]$

$$\text{w.o.} \bigcup_{j=0}^{d-3} \{i_j(q) : q = p_j + 1, \ldots, \#N_t \left[\tilde{v}_j(i_0(p_0), \ldots, i_{j-1}(p_{j-1}))\right]\}.$$

If $N_t \left[\tilde{v}_{d-2}(i_0(p_0), \ldots, i_{d-3}(p_{d-3}))\right] = \emptyset$ then next p_{d-3}.

For $p_{d-2} = 1$ to $\#N_t \left[\tilde{v}_{d-2}(i_0(p_0), \ldots, i_{d-3}(p_{d-3}))\right]$ do:

Compute $\tilde{v}_{d-1,1}(i_0(p_0), \ldots, i_{d-2}(p_{d-2}))$ and $\tilde{v}_{d-1,2}(i_0(p_0), \ldots, i_{d-2}(p_{d-2}))$.

next p_{d-2};

next p_{d-3};

⋮

next p_1;

next p_0;

It is understood in the above algorithm that the $\tilde{v}_k$ are computed and stored away as in (3.3.13).

This algorithm is actually easier to understand than it may appear at first glance. As an example of how it works, consider the tree of Figure (3.3.16). After computing $\tilde{v}_0$, compute $N_t(\tilde{v}_0) = N(\tilde{v}_0)$. Define the ordering of $N(\tilde{v}_0)$ to be the left-to-right ordering of the associated paths on the page. In this case, $i_0(1) = 5$, $i_0(2) = 4$, and $i_0(3) = 1$. Setting $p_0 = 1$, compute $\tilde{v}_1(i_0(p_0))$ and $N\left[\tilde{v}_1(i_0(p_0))\right] = \{1, 4\}$ and discover that $N_t\left[\tilde{v}_1(i_0(p_0))\right] = \emptyset$. Next, set $p_0 = 2$ and calculate $\tilde{v}_1(i_0(2))$ and $N\left[\tilde{v}_1(i_0(2))\right] = \{1, 2, 5\}$. Since $i_0(3) = 1$, $N_t\left[\tilde{v}_1(i_0(2))\right] = \{2, 5\}$ which is then ordered by the left-to-right method as before. So, for $i_1(1) = 2$ and $i_1(2) = 5$, compute $\tilde{v}_{2,1}(i_0(2), i_1(1))$, $\tilde{v}_{2,2}(4, 2)$, $\tilde{v}_{2,1}(i_0(2), i_1(2))$, and $\tilde{v}_{2,2}(4, 5)$. Set $p_0 = 3$ and compute $\tilde{v}_1(i_0(3))$ and $N(\tilde{v}_1(1)) = \{5\}$. This last tree cannot be trimmed and so set $i_1(1) = 5$ and compute $\tilde{v}_{2,1}(i_0(3), i_1(1))$ and $\tilde{v}_{2,2}(1, 5)$. The trimming algorithm has reduced the tree from 16 nodes to 10 nodes. Note if the original ordering of $N(\tilde{v}_0)$ was $i_0(1) = 1$, $i_0(2) = 5$, and $i_0(3) = 4$, then four nodes instead of six would be trimmed.

For a more general example, consider Figure (3.4.13) where a subsection of an untrimmed tree is portrayed. As before, consider each $N_t(\tilde{v})$ to be ordered by the left-to-right ordering of the appropriate indices on the page. Running through the trimming algorithm for the $\tilde{v}$ which are circled, first obtain as required $N_t(\tilde{v}_0) = N(\tilde{v}_0)$. Setting $p_0 = 3$, obtain $i_0(p_0) = 4$ and $N\left[\tilde{v}_1(i_0(3))\right] = \{5, 10, 12, 8\}$. Since $i_0(4) = 8$, $N_t\left[\tilde{v}_1(i_0(3))\right] = \{5, 10, 12\}$. For $p_1 = 1$ and $i_1(p_1) = 5$, $N_t\left[\tilde{v}_2(i_0(3), i_1(1))\right] = \{17, 6, 7, 9\}$ since $i_0(5) = 3$. Note that 7 is not removed even though $i_0(2) = 7$. For $p_2 = 3$ and $i_2(p_2) = 7$, $N_t\left[\tilde{v}_3(i_0(3), i_1(1), i_2(3))\right] = \{16\}$ since $i_0(5) = 3$, $i_1(2) = 10$, and $i_2(4) = 9$.

-- *a required lemma* --

A lemma is needed in order to show that the trimmed tree still contains a vector in every hill. First, more notation is needed.

(3.4.14) Definition: For some $k = 0, \ldots, d-2$ and $i_0, \ldots, i_{k-1} \in I$, let $\tilde{v}_k(i_0, \ldots, i_{k-1})$ be a node of the tree constructed by (3.3.13). Then $\mathrm{Tr}(\tilde{v}_k(i_0, \ldots, i_{k-1}))$ is the subtree of the untrimmed tree which has root node containing the vector $\tilde{v}_k(i_0, \ldots, i_{k-1})$ and whose position in the untrimmed tree is uniquely determined by $i_0, \ldots, i_{k-1}$.

The next theorem shows that if the untrimmed tree is constructed in such a way that no $\tilde{v}_k^{\perp}$ contains more y_i than is necessary, then for any hill $C(\pi)^+$ and node $\tilde{v}_k(i_0, \ldots, i_{k-1})$, if $y(i_0)^{\perp} \cap \cdots \cap y(i_{k-1})^{\perp}$ contains a nonzero vector of $C(\pi)^+$, then so does $\mathrm{Tr}(\tilde{v}_k(i_0, \ldots, i_{k-1}))$.

(3.4.15) Theorem: Let $d = \dim X \geqslant 3$. Suppose the tree constructed by (3.3.13) has been constructed in such a way that for all $k = 1, \ldots, d-2$, for all $i_0 \in N(\tilde{v}_0)$, $i_1 \in N(\tilde{v}_1(i_0)), \ldots, i_{k-1} \in N(\tilde{v}_{k-1}(i_0, \ldots, i_{k-2}))$,

$$\{i \in I : [y_i, \tilde{v}_k(i_0, \ldots, i_{k-1})] = 0 \text{ and } y_i \notin \mathrm{L}\{y(i_0), \ldots, y(i_{k-1})\}\} = \varnothing.$$

Let $C(\pi)^+$ be a hill. Suppose for some fixed $k = 1, \ldots, d-2$ and

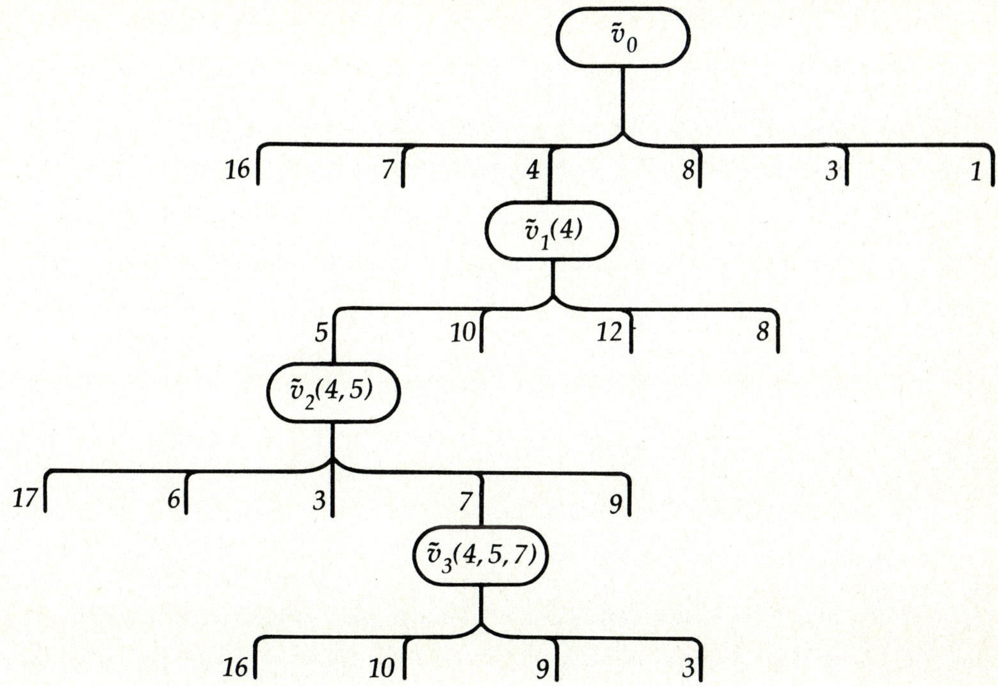

(3.4.13) Figure: A subsection of an untrimmed tree.

$i_0, \ldots, i_{k-1} \in I$ that $\tilde{v}_k(i_0, \ldots, i_{k-1})$ is in the untrimmed tree and $y(i_0)^\perp \cap \cdots \cap y(i_{k-1})^\perp \cap C(\pi)^+ \neq \langle 0 \rangle$. Then $\mathrm{Tr}(\tilde{v}_k(i_0, \ldots, i_{k-1}))$ contains a vector of $C(\pi)^+$.

Proof: The basic idea behind this proof is that if it can't be immediately shown that $\mathrm{Tr}(\tilde{v}_k(i_0, \ldots, i_{k-1}))$ contains a vector of $C(\pi)^+$, then it is shown that the hypothesis of the theorem holds for a subtree of $\mathrm{Tr}(\tilde{v}_k(i_0, \ldots, i_{k-1}))$. Fortunately, if one is ever forced to consider $\mathrm{Tr}(\tilde{v}_{d-2}(i_0, \ldots, i_{d-3}))$, then the result follows directly.

Let $S = \mathrm{L}\{y(i_0), \ldots, y(i_{k-1})\}$. Let R be any subspace such that $X = R \oplus S$. In the context of the lower dimensional problem as defined in (3.3.1), there are three cases:

Case 1: $\dot{C}(\pi)$ is pointed.

Since $\dot{C}(\pi)$ is pointed and since $\{\langle z_i \rangle : i \in I^+(\pi)\}$ contains a frame for $\dot{C}(\pi)$, $\dot{C}(\pi)^+$ is a hill in the lower dimensional problem. Following the same argument as given in (3.3.22), if each linear functional in $\text{Tr}(\tilde{v}_k(i_0, \ldots, i_{k-1}))$ is restricted to have domain R, then the resulting tree is a tree which satisfies the hypotheses of (3.3.14) for this lower dimensional problem. Consequently, the "restricted" tree contains a vector in $\dot{C}(\pi)^+$ and so, following the final argument in (3.3.22), $\text{Tr}(\tilde{v}_k(i_0, \ldots, i_{k-1}))$ contains a vector in $C(\pi)^+$.

Case 2: $\dot{C}(\pi)$ is not pointed and $k = d-2$.

By the hypothesis, there exists nonzero $\tilde{x} \in S^\perp \cap C(\pi)^+$. $\tilde{x}\big|_R \neq 0$ and is an element of $\tilde{R}$. Since $\dot{C}(\pi)^+ = (C(\pi)^+ \cap S^\perp)\big|_R$, $\tilde{x}\big|_R \in \dot{C}(\pi)^+$.

Since $\dim R = 2$ and $\dot{C}(\pi)$ is not pointed, then either $\dim \text{Lin}\, \dot{C}(\pi) = 1$ or $\dot{C}(\pi) = R$. If $\dot{C}(\pi) = R$, then $\dot{C}(\pi)^+ = \langle 0 \rangle$, which is a contradiction. So $\text{Lin}\, \dot{C}(\pi)$ is a one-dimensional subspace.

Since $\{\langle z_i \rangle, i \in I^+(\pi)\}$ contains a frame for $\dot{C}(\pi)$,

$$\text{Lin}\, \dot{C}(\pi) = \text{C}\{z_i, i \in I^+(\pi) \text{ and } z_i \in \text{Lin}\, \dot{C}(\pi)\}.$$

So there exist $p, q \in I^+(\pi)$ such that $\langle z_p \rangle = \langle -z_q \rangle \neq \langle 0 \rangle$, $\text{Lin}\, \dot{C}(\pi) = \langle z_p \rangle \cup \langle 0 \rangle \cup \langle z_q \rangle$, and $\text{L}(\dot{C}(\pi)^+) = z_p^\perp$. Now if $[z_p, \tilde{v}_{d-2}(i_0, \ldots, i_{d-3})\big|_R] = 0$, then $[y_p, \tilde{v}_{d-2}(i_0, \ldots, i_{d-3})] = 0$. But this cannot happen by the hypothesis since $z_p \neq 0$ implies $y_p \notin \text{L}\{y(i_0), \ldots, y(i_{d-3})\}$. As a result, either

$$[z_p, \tilde{v}_{d-2}(i_0, \ldots, i_{d-3})\big|_R] < 0 \text{ or } [z_q, \tilde{v}_{d-2}(i_0, \ldots, i_{d-3})\big|_R] < 0.$$

Suppose without loss of generality, that $[y_p, \tilde{v}_{d-2}(i_0, \ldots, i_{d-3})] < 0$. Algorithm (3.3.13) will then obtain a conical spanning set for the one-dimensional subspace $y(i_0)^\perp \cap \cdots \cap y(i_{d-3})^\perp \cap y_p^\perp$. But inasmuch as $\tilde{x}\big|_R \in \dot{C}(\pi)^+ \subset z_p^\perp$, it is clear that $\tilde{x} \in y_p^\perp$ and so the algorithm will find $\langle \tilde{x} \rangle$.

Case 3: $\dot{C}(\pi)$ is not pointed and $k < d-2$.

Let $J = \{i : i \in I^+(\pi)$ and $z_i \in \text{Lin } \dot{C}(\pi)\}$. The claim is that there exists $p \in J$ w.o. $\dot{I}_0(\pi) \neq \emptyset$ such that $[z_p, \tilde{v}_k(i_0, \ldots, i_{k-1})|_R] < 0$. Suppose not, i.e., for all $j \in J$ w.o. $\dot{I}_0(\pi)$, $[z_j, \tilde{v}_k(i_0, \ldots, i_{k-1})|_R] \geqslant 0$. Observe though that for all $j \in J$ w.o. $\dot{I}_0(\pi)$, $z_j \neq 0$ which implies $y_j \notin S$ and then $[z_j, \tilde{v}_k(i_0, \ldots, i_{k-1})|_R] > 0$. This implies that $\mathsf{C}\{z_i, i \in J\}$ is pointed which is impossible for a non-trivial subspace.

So there is a p such that $[y_p, \tilde{v}_k(i_0, \ldots, i_{k-1})] < 0$ and consequently, $\tilde{v}_{k+1}(i_0, \ldots, i_{k-1}, p)$ is in the tree. There is also a nonzero $\tilde{x} \in S^\perp \cap C(\pi)^+$ which yields $\tilde{x}|_R \in \dot{C}(\pi)^+$. Since $z_p \in \text{Lin } \dot{C}(\pi)$, $\tilde{x}|_R \in z_p^\perp$ and so

$$\tilde{x} \in y(i_0)^\perp \cap \cdots \cap y(i_{k-1})^\perp \cap y(p)^\perp \cap C(\pi)^+.$$

Now refer to Case 1, 2, or 3 as appropriate. Since d is finite and the cases are exhaustive, this process will stop after a while and the result will follow. □

-- *illustrating the necessity of an hypothesis* --

(3.4.16) Example: The hypothesis of the last theorem states that for each $\tilde{v}_k(i_0, \ldots, i_{k-1})$ in the tree, the only y_i allowed to be in $\tilde{v}_k(i_0, \ldots, i_{k-1})^\perp$ are the ones that are supposed to be in $\tilde{v}_k(i_0, \ldots, i_{k-1})^\perp$, namely, the y_i in $\mathsf{L}\{y(i_0), \ldots, y(i_{k-1})\}$. To give an example where this assumption is violated, where $\dot{C}(\pi)$ is not pointed, and where the theorem does not hold, consider Example (3.3.18) in R^3. Note that $\{y_0, \ldots, y_4\} \subset \tilde{v}_1(1)^\perp$ whereas only y_0 and y_1 are required to be in $\tilde{v}_1(1)^\perp$. Setting $S = \mathsf{L}\{\underline{y}_1\}$ and letting R be $\underline{y}_1^\perp$, the set of vectors in R^3 whose inner products with $\underline{y}_1$ are 0, one obtains $z_0 = (0, 0, 0)$, $z_1 = (0, 0, 0)$, $z_2 = (0, 1, 0)$, $z_3 = (0, -1, 0)$, $z_4 = (0, -1, 0)$, and $z_5 = (0, 0, 1)$. Consider the max-sum cone $C(\pi)^+ = \mathsf{C}\{y_1, y_2, -y_3, y_4, y_5\}^+$. Note that $\dot{C}(\pi)$ is a halfspace in R and $\text{Lin } \dot{C}(\pi) = \mathsf{L}\{(0, 1, 0)\}$. $C(\pi)^+ \cap y_1^\perp = \langle(0, 0, 1)\rangle$ which is not the two dimensional cone that would result if $\langle y_1\rangle$ were in the frame of $C(\pi)$. Anyway, $C(\pi)^+ \cap y_1^\perp \neq \langle 0\rangle$ and yet $\mathsf{Tr}(\tilde{v}_1(1))$ does not contain any vector in

$C(\pi)^+$.

-- *the trimmed tree is sufficient* --

The next theorem states that if the untrimmed tree is constructed in such a way that no $\tilde{v}_k^\perp$ contains more y_i than is necessary, then the trimmed tree contains a vector in every hill.

(3.4.17) Theorem: Let $d = \dim X \geqslant 3$. Suppose the tree constructed by (3.3.13) has been constructed in such a way that for all $k = 1, \ldots, d-2$, for all $i_0 \in N(\tilde{v}_0)$, $i_1 \in N(\tilde{v}_1(i_0)), \ldots,$ $i_{k-1} \in N(\tilde{v}_{k-1}(i_0, \ldots, i_{k-2}))$, $\{i \in I: [y_i, \tilde{v}_k(i_0, \ldots, i_{k-1})] = 0$ and $y_i \notin \mathrm{L}\{y(i_0), \ldots, y(i_{k-1})\}\} = \emptyset$. Let $C(\pi)^+$ be a hill. Then there is a vector in $C(\pi)^+$ which is in the trimmed tree constructed by (3.4.12).

Proof: The basic idea behind this proof is that if for any hill, a vector in that hill is in a subtree that is trimmed away, then the index which was responsible for the trimming generates a subtree which contains a vector in the hill.

Fix a method for ordering the $N_t(\tilde{v})$. Let $C(\pi)^+$ be a hill. If $\tilde{v}_0$ or $\tilde{v}_1(i_0)$ for some $i_0 \in N(\tilde{v}_0)$ are in $C(\pi)^+$ then the proof is complete since these vectors are in the trimmed tree. So, suppose that neither $\tilde{v}_0$ nor $\tilde{v}_1(i_0)$ for any $i_0 \in N(\tilde{v}_0)$ are in $C(\pi)^+$.

In what follows, $\mathrm{Tr}(\tilde{v})$ will continue as in Definition (3.4.14) to indicate the subtree of the untrimmed tree of (3.3.13) which has root node containing $\tilde{v}$.

The next few paragraphs will witness the following abuse of notation: "$\tilde{v}_{d-1}(i_0, \ldots, i_{d-2})$" will be used to ambiguously represent one of $\tilde{v}_{d-1,1}(i_0, \ldots, i_{d-2})$ and $\tilde{v}_{d-1,2}(i_0, \ldots, i_{d-2})$. There is no danger in doing this since if one of these two vectors is trimmed then both are.

Another definition is needed: $\tilde{v}_k(i_0(q_0), \ldots, i_{k-1}(q_{k-1}))$ is said to be *further along in the search sequence in the trimmed tree* than is $\tilde{v}_j(i_0(p_0), \ldots, i_{j-1}(p_{j-1}))$ if and only if one of the following cases hold:

(i) $p_0 < q_0$

(ii) there exists ℓ with $1 \leqslant \ell \leqslant \min(j-1,k-1)$ such that $p_0 = q_0, \ldots, p_{\ell-1} = q_{\ell-1}$, and $p_\ell < q_\ell$.

(iii) $k > j$ and $p_0 = q_0, \ldots, p_{j-1} = q_{j-1}$.

If this definition is considered to define a relation on the $\tilde{v}$ in the trimmed tree, then it is easily shown that this relation is transitive and not reflexive.

Consider the following claim:

Claim #: Suppose $\tilde{v}_j(i_0(p_0), \ldots, i_{j-1}(p_{j-1}))$ is in the trimmed tree for some $1 \leqslant j \leqslant d-2$. Suppose $\tilde{v}_k(i_0(p_0), \ldots, i_{j-1}(p_{j-1}), i_j, \ldots, i_{k-1})$ for $j+1 \leqslant k \leqslant d-1$ is an element of the hill $C(\pi)^+$, is in $\mathrm{Tr}(\tilde{v}_j(i_0(p_0), \ldots, i_{j-1}(p_{j-1})))$, and is not in the trimmed tree. Then there exists ℓ with $0 \leqslant \ell \leqslant k-2$ and suitable q_i such that $\tilde{v}_{\ell+1}(i_0(q_0), \ldots, i_\ell(q_\ell))$ is in the trimmed tree, is further along in the search sequence than $\tilde{v}_j(i_0(p_0), \ldots, i_{j-1}(p_{j-1}))$ and is such that $\mathrm{Tr}(\tilde{v}_{\ell+1}(i_0(q_0), \ldots, i_\ell(q_\ell)))$ contains a vector of $C(\pi)^+$.

In order to show this, begin by considering the obvious condition sufficient for any $\tilde{v}_k(j_0, \ldots, j_{k-1})$ to be in the trimmed tree, namely, if $\tilde{v}_k(j_0, \ldots, j_{k-1})$ is in the untrimmed tree and if for each $m = 1, \ldots, k-1$, there exists p_m such that $j_m = i_m(p_m)$ and $i_m(p_m) \in N_t\left[\tilde{v}_m(i_0(p_0), \ldots, i_{m-1}(p_{m-1}))\right]$, then $\tilde{v}_k(j_0, \ldots, j_{k-1})$ is in the trimmed tree as well. So, since $\tilde{v}_k(i_0(p_0), \ldots, i_{j-1}(p_{j-1}), i_j, \ldots, i_{k-1})$ is in $\mathrm{Tr}\left[\tilde{v}_j(i_0(p_0), \ldots, i_{j-1}(p_{j-1}))\right]$ and is not in the trimmed tree, there exists m where $j \leqslant m \leqslant k-1$ such that if $m = j$ then $i_m \notin N_t\left[\tilde{v}_j(i_0(p_0), \ldots, i_{j-1}(p_{j-1}))\right]$ whereas if $m > j$ then there exists suitable p_i such that $i_j = i_j(p_j), \ldots, i_{m-1} = i_{m-1}(p_{m-1})$ and $i_m \notin N_t\left[\tilde{v}_m(i_0(p_0), \ldots, i_{m-1}(p_{m-1}))\right]$. So, in short, there exists m where $j \leqslant m \leqslant k-1$ and suitable p_i such that $\tilde{v}_m(i_0(p_0), \ldots, i_{m-1}(p_{m-1}))$ is either equal to or is further along in the search sequence than $\tilde{v}_j(i_0(p_0), \ldots, i_{j-1}(p_{j-1}))$, $\tilde{v}_k(i_0(p_0), \ldots, i_{j-1}(p_{j-1}), i_j, \ldots, i_{k-1})$ is in $\mathrm{Tr}\left[\tilde{v}_m(i_0(p_0), \ldots, i_{m-1}(p_{m-1}))\right]$, and $i_m \notin N_t\left[\tilde{v}_m(i_0(p_0), \ldots, i_{m-1}(p_{m-1}))\right]$.

So, there exists ℓ and q with $0 \leqslant \ell \leqslant m-1$ and $p_\ell + 1 \leqslant q \leqslant \#N_t\left[\tilde{v}_\ell(i_0(p_0), \ldots, i_{\ell-1}(p_{\ell-1}))\right]$ such that $i_\ell(q) = i_m$. Note that $\tilde{v}_{\ell+1}(i_0(p_0), \ldots, i_{\ell-1}(p_{\ell-1}), i_\ell(q))$ is in the trimmed tree. Since $\tilde{v}_k(i_0(p_0), \ldots, i_{m-1}(p_{m-1}), i_m, \ldots, i_k) \in C(\pi)^+$,

$$C(\pi)^+ \cap y(i_0(p_0))^\perp \cap \cdots \cap y(i_{m-1}(p_{m-1}))^\perp \cap y(i_m)^\perp \neq \langle 0 \rangle.$$

Consequently,

$$C(\pi)^+ \cap y(i_0(p_0))^\perp \cap \cdots \cap y(i_{\ell-1}(p_{\ell-1}))^\perp \cap y(i_\ell(q))^\perp \neq \langle 0 \rangle$$

and so by (3.4.15), $\mathrm{Tr}\left[\tilde{v}_{\ell+1}(i_0(p_0), \ldots, i_{\ell-1}(p_{\ell-1}), i_\ell(q))\right]$ contains a vector in $C(\pi)^+$. Note that $\tilde{v}_{\ell+1}(\cdots)$ is further along than $\tilde{v}_m(\cdots)$ which is either equal to or further along than $\tilde{v}_j(\cdots)$ in the search sequence. The claim now holds.

To wrap up the proof, proceed as follows. By (3.3.14) and by the assumption at the very start of this proof, there exists $i_0 \in N(\tilde{v}_0)$ such that $\tilde{v}_1(i_0)$ is in the trimmed tree and $\mathrm{Tr}(\tilde{v}_1(i_0))$ contains a vector of $C(\pi)^+$. If this vector is not trimmed out, then the result follows. If it is then, by claim #, there exists a $\tilde{v}_j$ in the trimmed tree which is further along in the search sequence than $\tilde{v}_1(i_0)$ and whose tree contains a vector of $C(\pi)^+$. If this vector is not trimmed, then the result follows. If it is trimmed away, then, by claim #, there exists still another $\tilde{v}_k$ in the trimmed tree which is further along in the search sequence than $\tilde{v}_j$ and whose tree contains a vector of $C(\pi)^+$. Continue on in this fashion.

One of two things will happen. Either a finite or an infinite sequence of $\tilde{v}$ in the trimmed tree will be generated. If an infinite sequence is generated, then since there are only a finite number of vectors in the trimmed tree, at least one of them must be repeated in the sequence. But since each vector in the sequence is further along than its predecessors and since the "further along" relation is transitive, the repeated vector is necessarily found to be further along than itself which is impossible.

So, the sequence is finite and stops with some $\tilde{v}_j$ in the trimmed tree for which $\mathsf{Tr}(\tilde{v}_j)$, contains a vector in $C(\pi)^+$ which is not trimmed away by the algorithm. □

(3.4.18) Remarks: Note that the last theorem (3.4.17) remains valid if "hill" is replaced by "max-sum cone". In fact, if the p_{d-2} loop of Algorithm (3.4.12) is modified to compute $\tilde{v}_{d-1,1}(i_0(p_0), \ldots, i_{d-2}(p_{d-2}))$ and $\tilde{v}_{d-1,2}(i_0(p_0), \ldots, i_{d-2}(p_{d-2}))$ as was done in (3.3.23), then the resulting trimmed tree still contains a vector for each max-sum cone. This is easy to see because this tree is precisely the trimmed tree produced by (3.4.12) with possibly some of its bottom leaves removed since these leaves are known to be worse than other leaves remaining in the tree.

The hypothesis of (3.4.17), stating that for all but one $\tilde{v}$ in the tree, $\tilde{v}^{\perp}$ cannot contain y_i that needn't be there, is something of a nuisance since it is intuitively unlikely to happen with $\tilde{v}_0$ computed as in (3.4.7). Yet, as Example (3.4.16) shows, one does not want this hypothesis to be violated. There are at least two ways to deal with this situation.

The first way is to check the hypothesis each time a $\tilde{v}_k$ in the trimmed tree is computed. (Note that it is safe to assume that any trimmed $\tilde{v}_k$ would have been computed satisfactorily if the opportunity had existed.) If

$$J := \{i \in I \text{ w.o. } I_0\colon\ [y_i, \tilde{v}_k(i_0, \ldots, i_{k-1})] = 0 \text{ and } y_i \notin \mathsf{L}\{y(i_0), \ldots, y(i_{k-1})\}\} = \varnothing,$$

then all is well. If J is not empty, then it is necessary to compute another $\tilde{v}_k$. Clearly, there are a large number to choose from even when working in the context of the fixed precision of computer numbers. Chances are then that a reasonable perturbation of $\tilde{v}_k$ will work.

If $\tilde{v}_k$ is selected by projecting $\tilde{v}_0$ down, then the following procedure is suggested. Select $\tilde{v}_0$ such that $[y_i, \tilde{v}_0] \neq 0$ for any $i \in I$. (If $[y_i, \tilde{v}_0] = 0$ for some initial $\tilde{v}_0$ then add $\alpha \underline{y}_i$ to $\underline{\tilde{v}}_0$ for small $\alpha > 0$ in order to get a new $\underline{\tilde{v}}_0$. This procedure converges quickly in practice). Suppose $\tilde{v}_0 \in C(\pi)^+$ for some $\{\pi_i, i \in I\}$. Now suppose that for some $\tilde{v}_k$, the hypothesis fails. Let $\underline{\tilde{x}} = \sum_J \underline{y}_i$. Choose $\alpha > 0$ such that $\pi_i \underline{y}_i^T(\underline{\tilde{v}}_0 + \alpha\underline{\tilde{x}}) > 0$ for $i \in I$. Let the new $\underline{\tilde{v}}_0$ be $\underline{\tilde{v}}_0 + \alpha\underline{\tilde{x}}$ and project this down to obtain $\underline{\tilde{v}}_k$. If this $\tilde{v}_k$ satisfies the hypothesis, then the process stops. If not, then iterate.

It is of course time consuming to do all of this checking. In practice, the next stratagem will probably only rarely fail. Since $\{\underline{y}_i, i \in I\}$ is usually given with the components of each vector being recorded to less than four decimal places and since most computers can work with upwards of 16 decimal places, compute each component of $\underline{\tilde{v}}_0$ to say four places and then fill out the rest of the number with random digits. This should cause the hypothesis to hold and obviate the need for checking.

The next two suggestions for improvement have not been tested yet but they appear promising.

-- *ordering the trimmed children* --

(3.4.19) Improvement: Another idea for obtaining a smaller tree at the expense of either additional computation or storage is to order each $N_t\left[\tilde{v}_k(i_0(p_0), \ldots, i_{k-1}(p_{k-1}))\right]$ for $0 \leqslant k \leqslant d-2$ according to decreasing $\#N\left[\tilde{v}_{k+1}(i_0(p_0), \ldots, i_{k-1}(p_{k-1}), q)\right]$ for each $q \in N_t\left[\tilde{v}_k(i_0(p_0), \ldots, i_{k-1}(p_{k-1}))\right]$. This way those indices which are likely to generate relatively small trees are placed at the end of the sequence and thus have maximum opportunity to trim the probably larger trees generated by the preceding indices.

There is reason to suspect that a good deal of trimming will go on. These $\tilde{v}$ associated with indices early in the sequence are likely to be in a region of $\tilde{X}$ "far from" those hills in $\tilde{X}$ whose interiors contain vectors with the smallest $N(\tilde{v})$. The opposite would tend to be true for those $\tilde{v}$ associated with indices later in the sequence. The trees of indices early in the sequence would then tend to contain indices corresponding to $d-1$ dimensional boundary hyperspaces for the hills associated with the smallest $N(\tilde{v})$. These indices are precisely the indices which would tend to constrain the $\tilde{v}$ corresponding to indices later in the sequence and thus there is increased likelihood of trimming.

-- *best-first depth-first search* --

(3.4.20) Improvement: In some situations, it may be desired to find a good vector $\tilde{v}$ (in the sense of having a large $h(\tilde{v})$ value) without exploring the whole tree. Of course, if a search stops before examining the whole tree, then there is no guarantee that the best $\tilde{v}$ found so far is in fact one of the optimal ones. The hope here is that a search procedure can be developed which will find an optimal $\tilde{v}$ (although it won't be known as such) early in the search sequence. One possible procedure would be to do a best-first depth-first search without trimming. In this case, $N(\tilde{v}_k(i_0, \ldots, i_{k-1}))$ would be ordered by decreasing values of the best guaranteed h value associated with $\tilde{v}_{k+1}(i_0, \ldots, i_{k-1}, q)$ for each $q \in N(\tilde{v}_k(i_0, \ldots, i_{k-1}))$ so that what look like "better" subtrees will be searched first. (The best guaranteed h value associated with $\tilde{v}_{k+1}(i_0, \ldots, i_{k-1}, q)$ is the maximum of $h(\tilde{v}_k(i_0, \ldots, i_{k-1}))$ and that number for which it is certain that a displaced vector with a larger h value can be found (cf., (3.4.2)).) Trimming would probably not be advantageous in this situation since the better vectors are typically located near the bottom of the tree and one does not want to risk trimming them away.

This best-first depth-first search procedure has been implemented and has been found to work quite well in practice. The procedure used for controlling and terminating the program in this instance is to first define a number $\#s(k)$

for $k = 0, \ldots, d-2$ which represents the maximum number of best children that will be examined for each $\tilde{v}_k(i_0, \ldots, i_{k-1})$ that is examined at the kth level. So for example, the process starts with $\tilde{v}_0$ and considers that number not exceeding $\#s(0)$ of $\tilde{v}_0$'s children with the best guaranteed h values. Then for each of these selected $\tilde{v}_1$'s, the best guaranteed h-values for its children are calculated, at most $\#s(1)$ (since there may be fewer than $\#s(1)$) of these children with the largest best guaranteed h values are identified, and then these children are visited. The process continues on in this fashion, always restarting whenever possible. Once all of the selected best $\tilde{v}_k$ have been explored, the process stops.

For instance, one of the examples in Chapter 9 shows for a given data set that when the two best children alone were visited at each level (i.e., $\#s(k) = 2$ for all k) and when this algorithm was restarted whenever possible, this fast approximate algorithm terminated with a vector with h value 55 after computing the h values of 323 nodes (i.e., $\tilde{v}_k$) whereas the full standard tree algorithm with improvements even starting with this good vector as $\tilde{v}_0$ nonetheless had to examine 8778 nodes in order to prove that the best possible h value was 56.

As a final note in this regard, when deliberately using this fast approximate (i.e., best-first depth-first) tree algorithm to provide a good $\tilde{v}_0$ for a run of the standard tree algorithm to find the best possible h value and all associated vectors, it would be better to use the criterion function $\sum_I 1\{[y_i, \tilde{x}] > 0\}$ instead of h in the fast algorithm because it is best to give the standard algorithm a $\tilde{v}_0$ with small cardinality $N(\tilde{v}_0)$, not necessarily with a large $h(\tilde{v}_0) = \sum_I \sigma_i \, 1\{[y_i, \tilde{v}_0] > 0\}$.

Summary For Section 3.4

Several improvements to the first phase of the basic tree algorithm (3.3.13) were presented in this section.

The tree created by (3.3.13) is made smaller if the set $\{y_i, i \in I\}$ is made smaller. Consequently, all $y_i = 0$ should be dropped, all conical ties should be consolidated, and the criterion function h should be modified accordingly.

A depth-first search of the tree created by (3.3.13) which saves only those $\tilde{v}$ which have the highest potentially realizable criterion function values is a more efficient way to explore the tree than is Algorithm (3.3.13).

When choosing $\tilde{v}_k(i_0, \ldots, i_{k-1})$ for $k = 0, \ldots, d-2$, it is wise to ensure that $\#N(\tilde{v}_k(i_0, \ldots, i_{k-1})) < \#N(-\tilde{v}_k(i_0, \ldots, i_{k-1}))$.

Another way to obtain $\tilde{v}_k(i_0, \ldots, i_{k-1})$ with small numbers of children is to let $\underline{w} = \sum_I \underline{y}_i$ for $\underline{y}_i$ of unit length and then to seek to maximize $\underline{\tilde{v}}^T \underline{w} \,/\, \| \underline{\tilde{v}} \|$ subject to $\underline{\tilde{v}} \in \{\underline{y}(i_0), \ldots, \underline{y}(i_{k-1})\}^{\perp}$. This is accomplished when $\underline{\tilde{v}} = P[\underline{w} \,|\, \mathrm{L}\{\underline{y}(i_0), \ldots, \underline{y}(i_{k-1})\}^{\perp}]$.

It is so important to have a $\tilde{v}_0$ with as few children as possible that it is actually worthwhile to start the entire algorithm over again when a $\tilde{v}_k$ is found in the tree which is guaranteed to be better than $\tilde{v}_0$.

There is a method whereby certain subtrees of the original tree may be safely left unexplored by the depth-first search algorithm. The basic idea behind this method is that it is OK to trim away a given subtree if another subtree of a certain type can be found elsewhere in the tree and explored fully.

Examples show that by visiting the children of nodes in the tree according to decreasing order of their h-values (i.e., best-first depth-first), very good if not optimal vectors are encountered very early in the search sequence.

Section 3.5:

Displacing Boundary Vectors Into The Interiors Of Cones

The first phase of the tree algorithm constructs a tree of vectors, all of which, with the possible exception of $\tilde{v}_0$, lie in the boundary faces of cones in the dual space. This section describes the second phase of this algorithm, namely, the procedure whereby boundary vectors are displaced into the interiors of appropriate cones in such a way as to produce an interior vector for every hill (or max-sum cone, as desired). Interestingly enough, the second phase of the tree algorithm will on certain occasions ask for the entire tree algorithm (i.e., both phases) to solve certain lower dimensional problems. Note that the displacement operation is necessary since no boundary vector is ever a solution vector to Problem (3.1.1).

-- *how to displace boundary vectors* --

Consider first then the mechanics of displacing a boundary vector into the interior of a cone. Suppose for some index set $J \neq I_0$ and known $\pi_i \in \{-1, 1\}$ for $i \in I$ w.o. J, there is some boundary vector $\tilde{v}$ where $[\pi_i y_i, \tilde{v}] > 0$ for $i \in I$ w.o. J and $[y_i, \tilde{v}] = 0$ for $i \in J$. Suppose also that one is given a $\tilde{z}$ and some $\theta_i \in \{-1, 1\}$ for $i \in J$ w.o. I_0 such that $[\theta_i y_i, \tilde{z}] > 0$ for $i \in J$ w.o. I_0. The next theorem shows that there exists $\alpha \in (0, 1)$ such that

$$(1 - \alpha)\tilde{v} + \alpha\tilde{z} \in \text{int}\, \mathrm{C}\{\pi_i y_i, i \in I \text{ w.o. } J, \theta_i y_i, i \in J \text{ w.o. } I_0\}^+.$$

An example will follow.

(3.5.1) Theorem: Suppose for $I_0 \neq J \subset I$, $\pi_i \in \{-1, 1\}$ for $i \in I$ w.o. J, and some $\tilde{v} \in \tilde{X}$, $[\pi_i y_i, \tilde{v}] > 0$ for $i \in I$ w.o. J and

$[y_i, \tilde{v}] = 0$ for $i \in J$. Let $\tilde{z} \in \tilde{X}$ and $\theta_i \in \{-1, 1\}$ for $i \in J$ w.o. I_0 be such that $[\theta_i y_i, \tilde{z}] > 0$ for $i \in J$ w.o. I_0. Let $K = \{i \in I \text{ w.o. } J : [\pi_i y_i, \tilde{z}] < 0\}$. If $K = \emptyset$, then set $\alpha = ½$. If $K \neq \emptyset$, then choose α such that $0 < \alpha < \min_{i \in K} \dfrac{[y_i, \tilde{v}]}{[y_i, \tilde{v}] - [y_i, \tilde{z}]} < 1$. Then $[\pi_i y_i, (1 - \alpha)\tilde{v} + \alpha\tilde{z}] > 0$ for $i \in I$ w.o. J and $[\theta_i y_i, (1-\alpha)\tilde{v} + \alpha\tilde{z}] > 0$ for $i \in J$ w.o. I_0.

Proof: When $i \in J$ w.o. I_0 or when $i \in I$ w.o. J and $[\pi_i y_i, \tilde{z}] \geqslant 0$, then it suffices to have $\alpha \in (0, 1)$. If $K \neq \emptyset$, then it is necessary to have for all $i \in K$, $\alpha < \dfrac{[\pi_i y_i, \tilde{v}]}{[\pi_i y_i, \tilde{v}] - [\pi_i y_i, \tilde{z}]}$. □

An equivalent way of visualizing the displacement operation is to think of finding $\lambda > 0$ such that $\tilde{v} + \lambda\tilde{z}$ is in the interior of the cone. (3.5.1), however, is the procedure used in the author's computer program implementing the tree algorithm.

-- *the identification of hills by displacement* --

Consider now the tree created by the first phase of the tree algorithm. With the possible exception of $\tilde{v}_0$, each $\tilde{v}_k$ in the tree is a boundary vector for which there exists an index set $J \neq I_0$ and $\pi_i \in \{-1, 1\}$ for $i \in I$ w.o. J such that $[\pi_i y_i, \tilde{v}_k] > 0$ for $i \in I$ w.o. J and $[y_i, \tilde{v}_k] = 0$ for $i \in J$. The object is to first find which choices for π_i, $i \in J$ will recover $C(\pi)^+$ for all hills $C(\pi)^+$ which contain $\tilde{v}_k$ as a boundary vector (if any) and then to obtain via (3.5.1) an interior vector for each such hill.

The following example gives an indication of the complexity of this problem.

(3.5.2) Example: Consider Example (3.3.18). The vector $\tilde{v}_{2,1}(4, 1) = (0, 0, 1)$ is a boundary vector in each of the three hills. The first observation to make is that one should not think solely of displacing a boundary vector into the interior of a unique hill for, as seen here, the vector in question may be on the boundary of several hills.

To be specific, $C(\pi)^+$ is a hill in this example if and only if $(\pi_0, \ldots, \pi_5)$ is equal to one of $(1, 1, 1, -1, 1, 1)$, $(1, 1, -1, 1, 1, 1)$, and $(1, -1, 1, 1, -1, 1)$. Observe that, for each of these three sets of π_i,

$$\tilde{v}_{2,1}(4, 1) \in \{\tilde{x} \in \tilde{X} : [\pi_5 y_5, \tilde{x}] > 0, [\pi_i y_i, \tilde{x}] = 0 \text{ for } i = 0, \ldots, 4\}$$
$$= F_{\{0, \ldots, 4\}}(C(\pi)^+).$$

Naturally, when the first phase of the algorithm finishes and $\tilde{v}_{2,1}(4,1)$ is produced, while it is known that $\pi_5 = 1$, it is certainly not known what choices of $\pi_1, \ldots, \pi_4$ will make $C(\pi)^+$ into a hill with $\tilde{v}_{2,1}(4, 1)$ in a boundary face. This section will develop a way to find this out.

To show how (3.5.1) works in this example, suppose one is given $\tilde{z} = (1, \frac{1}{2}, 0)$. Note that for $(\theta_1, \theta_2, \theta_3, \theta_4) = (1, 1, -1, 1)$, $[\theta_i y_i, \tilde{z}] > 0$ for $i = 1, \ldots, 4$. Since $[y_5, \tilde{z}] = 0$, $K = \emptyset$ and so for any $\alpha \in (0, 1)$, $(1-\alpha)\tilde{v}_{2,1}(4, 1) + \alpha\tilde{z} \in \operatorname{int} C\{y_0, y_1, y_2, -y_3, y_4, y_5\}^+$.

Now, add $y_6 = (-1, -1, 1)$ to $\{y_i\}_0^5$. Then $C\{y_0, y_1, y_2, -y_3, y_4, y_5, y_6\}^+$ is a hill where $\{\langle y_0\rangle, \langle y_2\rangle, \langle y_4\rangle, \langle y_6\rangle\}$ is the frame of its dual. This hill is the subset of the previous hill $C\{y_0, y_1, y_2, -y_3, y_4, y_5\}^+$ formed by intersecting this cone with $\{\tilde{x} \in \tilde{X} : [y_6, \tilde{x}] \geqslant 0\}$. Since $[y_6, \tilde{z}] < 0$, $K = \{6\}$ and so for any $0 < \alpha < \frac{2}{5}$, $(1-\alpha)\tilde{v}_{2,1}(4, 1) + \alpha\tilde{z} \in \operatorname{int} C\{y_0, y_1, y_2, -y_3, y_4, y_5, y_6\}^+$.

-- *displacing when the annihilating y_i generate a pointed cone* --

There is one important special case when, for $\tilde{v}_k \in F_J(C(\pi)^+)$, the π_i, $i \in J$, can be identified immediately if $C(\pi)^+$ is a hill.

(3.5.3) Theorem: Suppose $C(\pi)^+$ is a hill and

$$\tilde{v} \in \{\tilde{x} : [\pi_i y_i, \tilde{x}] > 0 \text{ for } i \in I \text{ w.o. } J, [\pi_i y_i, \tilde{x}] = 0 \text{ for } i \in J\}.$$

Suppose there exists $\tilde{z}$ such that $[y_i, \tilde{z}] > 0$ for all $i \in J$ w.o. I_0. Then

$\pi_j = 1$ for all $j \in J$.

Proof: Let $\{\langle y_i \rangle, i \in I^+(\pi)\}$ be the frame for $C(\pi)$. Take $j \in J$ w.o. I_0. Then there exist $\lambda_i \geqslant 0$, not all 0, such that $\pi_j y_j = \sum_{I^+(\pi)} \lambda_i y_i$. In particular, there must be some $i \in I^+(\pi)$ w.o. I_0 such that $\lambda_i > 0$. Consequently, $0 = [\pi_j y_j, \tilde{v}] = \sum_{I^+(\pi)} \lambda_i [y_i, \tilde{v}] = \sum_{I^+(\pi) \text{ w.o. } J} \lambda_i [y_i, \tilde{v}]$ and therefore $\lambda_i = 0$ for $i \in I^+(\pi)$ w.o. J. Since $[\pi_j y_j, \tilde{z}] = \sum_{I^+(\pi) \cap J} \lambda_i [y_i, \tilde{z}] > 0$, $\pi_j = 1$. □

-- *using Theorem (3.5.3) to displace* --

This theorem is used in the following way. Suppose $\tilde{v}_k(i_0, \ldots, i_{k-1})$ is in the tree and is in $F_J(C(\pi)^+)$ where $J \neq I_0$ and $C(\pi)^+$ may or may not be a hill. Use the linear program described in (2.3.33) to determine whether or not $\mathrm{C}\{y_i, i \in J\}$ is pointed. If it is pointed, then the LP provides $\tilde{z}$ such that $[y_i, \tilde{z}] > 0$ for $i \in J$ w.o. I_0. Use this $\tilde{z}$ and (3.5.1) to displace $\tilde{v}_k$ into the interior of $\mathrm{C}\{\pi_i y_i, i \in I \text{ w.o. } J, y_i, i \in J\}^+$ which, by (3.5.3), is $C(\pi)^+$ if $C(\pi)^+$ is a hill. If $\mathrm{C}\{y_i, i \in J\}$ is not pointed, then other techniques will have to be used.

If $\{y_i, i \in I \text{ w.o. } I_0\}$ is in general position (cf., (2.1.19)), then the above procedure can always be used to displace $\tilde{v}_k(i_0, \ldots, i_{k-1})$ for $0 \leqslant k \leqslant d-1$. To show this, let $J = \{i \in I \text{ w.o. } I_0: [y_i, \tilde{v}_k(i_0, \ldots, i_{k-1})] = 0\}$. To see that $\#J \leqslant d-1$, suppose that $\#J \geqslant d$. Then there is a subset of $\{y_i, i \in I \text{ w.o. } I_0\}$ of size d contained in a $d-1$ dimensional subspace which is a contradiction. Since $\#J \leqslant d-1$, $\{y_j, j \in J\}$ is linearly independent. Consequently $\mathrm{C}\{y_j, j \in J\}$ is pointed and so the above procedure can be used to displace $\tilde{v}_k(i_0, \ldots, i_{k-1})$. Note that the general position assumption, while sufficient for the above procedure to successfully displace $\tilde{v}_k(i_0, \ldots, i_{k-1})$, is not necessary.

A weaker sufficient condition for ensuring that it will always be possible to displace boundary vectors using the linear programming method just described (and consequently, never having to resort to recursion) is that of requiring $\{y_i, i \in I\}$ to be in pointed position. Recall (cf., (2.3.34)) that $\{y_i, i \in I\}$ is said to be in *pointed position* if and only if for all nonempty subsets J of I, if $\{y_j, j \in J\}^\perp \neq \langle 0 \rangle$, then $\mathsf{C}\{y_j, j \in J\}$ is pointed.

-- displacing when the annihilating y_i have dimension 1 --

The next theorem describes a situation where if $\tilde{v}_k \in F_J(C(\pi)^+)$ for any dual cone $C(\pi)^+$, then there are *at most* two choices for $\{\pi_i, i \in J\}$.

(3.5.4) Theorem: Let $C(\pi)^+$ be a hill. Suppose $\tilde{v} \in F_J(C(\pi)^+)$ where $\dim \mathsf{L}\{y_i, i \in J\} = 1$. Let $p \in J$ w.o. I_0, $J_1 = \{i \in J : \langle y_i \rangle = \langle y_p \rangle\}$, and $J_2 = \{i \in J : \langle y_i \rangle = -\langle y_p \rangle\}$. So, $J = J_1 \cup J_2 \cup I_0$ and $J_1 \neq \varnothing$. If $J_2 = \varnothing$, then $\pi_i = 1$ for $i \in J_1$. If $J_2 \neq \varnothing$, then either $\pi_i = 1$ for $i \in J_1$ and $\pi_i = -1$ for $i \in J_2$ or $\pi_i = 1$ for $i \in J_2$ and $\pi_i = -1$ for $i \in J_1$.

Proof: If $J_2 = \varnothing$, then $\mathsf{C}\{y_i, i \in J\}$ is pointed and (3.5.3) applies.

Suppose $J_2 \neq \varnothing$. Let $\tilde{u} \in \text{int } C(\pi)^+$. Now either $\pi_p = 1$ or $\pi_p = -1$. Suppose $\pi_p = 1$. Take any $j \in J_1$. Then there exists $\alpha > 0$ such that $y_j = \alpha y_p$ and since $0 < [\pi_j y_j, \tilde{u}] = \alpha \pi_j [y_p, \tilde{u}]$, $\pi_j = 1$. Now take $j \in J_2$. Then $y_j = -\alpha y_p$ for some $\alpha > 0$ and since $0 < [\pi_j y_j, \tilde{u}] = -\alpha \pi_j [y_p, \tilde{u}]$, $\pi_j = -1$. The other case is handled similarly. □

-- using Theorem (3.5.4) to displace --

This theorem is used as follows. Suppose $\tilde{v}_k(i_0, \ldots, i_{k-1})$ is in the tree and is in $F_J(C(\pi)^+)$ where $J \neq I_0$ and $C(\pi)^+$ may or may not be a hill. Suppose further that $\dim \mathsf{L}\{y_i, i \in J\} = 1$. Consequently, $k = 0$ or 1. Form J_1 and J_2 as in (3.5.4).

If $J_2 = \varnothing$, then let $\underline{z} = \underline{y}_p$ and displace $\tilde{v}_k(i_0, \ldots, i_{k-1})$ using (3.5.1) to obtain a vector in int $C(\pi)^+$ if $C(\pi)^+$ is a hill.

Suppose $J_2 \neq \varnothing$. First, let $\tilde{z}$ be such that $\underline{z} = \underline{y}_p$ and use (3.5.1) to displace $\tilde{v}_k(i_0, \ldots, i_{k-1})$ into the interior of the cone $\mathsf{C}\{\pi_i y_i, i \in I \text{ w.o. } J,$

$\theta_i y_i,\ i \in J$ w.o. $I_0\}$ where the θ_i are chosen such that $\theta_i = 1$ for $i \in J_1$ and $\theta_i = -1$ for $i \in J_2$. Second, let $\tilde{z}$ be such that $\underline{\tilde{z}} = -\underline{y}_p$ and use (3.5.1) to displace $\tilde{v}_k(i_0, \ldots, i_{k-1})$ into the interior of the cone $\mathrm{C}\{\pi_i y_i,\ i \in I$ w.o. J, $\theta_i y_i,\ i \in J$ w.o. $I_0\}$ where the θ_i are chosen such that $\theta_i = 1$ for $i \in J_2$ and $\theta_i = -1$ for $i \in J_1$.

-- *the remaining problem and its solution* --

The problem is now reduced to that of determining how to displace $\tilde{v}_k(i_0, \ldots, i_{k-1}) \in F_J(C(\pi)^+)$ into the interior of all neighboring hills (if any) when $\dim \mathrm{L}\{y_i,\ i \in J\} > 1$ and $\mathrm{C}\{y_i,\ i \in J\}$ is not pointed. To suggest what might be done in this situation, refer back to Example (3.3.18).

$\tilde{v}_{2,1}(4, 1) = (0, 0, 1)$ is a vector on the boundary of three hills and $J = \{i : [y_i, \tilde{v}_{2,1}(4, 1)] = 0\} = \{0, 1, 2, 3, 4\}$. Since

$$\dim \mathrm{L}\{y_0, \ldots, y_4\} = 2 = d - 1 < d,$$

consider $\{y_0, \ldots, y_4\}$ in the context of its own lower dimensional dimensionality space. The two-dimensional hills generated by $\{y_0, \ldots, y_4\}$ can be seen in Figure (3.3.20). Now, imagine adding a third dimension to this flat drawing by introducing the z-axis. In this three-dimensional context, the three-dimensional hills of the original problem can be seen sitting on the two-dimensional hills of the lower dimensional problem.

If it were possible to obtain a vector (here, a linear functional with a two dimensional domain) in each of these two-dimensional hills and then somehow generate its geometrically obvious counterpart with a three-dimensional domain, then a positive multiple of each counterpart could be added to $\tilde{v}_{2,1}(4, 1)$ to obtain a vector in each three-dimensional hill.

-- hills generating lower-dimensional hills --

Before developing this any further, it is necessary to establish a correspondence between hills in the original problem and hills in a lower dimensional problem. It is important to note that this lower dimensional problem is not the same as the one defined in section 3.3.

(3.5.5) Theorem: Suppose $C(\pi)^+$ is a hill. Let $\{\langle y_i \rangle : i \in I^+(\pi)\}$ be the frame of $C(\pi)$. Suppose there exists

$$\tilde{v} \in F_J(C(\pi)^+) = \{\tilde{x} : [\pi_i y_i, \tilde{x}] > 0 \text{ for } i \in I \text{ w.o. } J, [\pi_i y_i, \tilde{x}] = 0 \text{ for } i \in J\}$$

where $J \neq I_0$. Let $R = \mathrm{L}\{y_i, i \in J\}$ and let S be such that $R \oplus S = X$. Then $\{\tilde{r} \in \tilde{R} : [\pi_i y_i, \tilde{r}] \geqslant 0, i \in J\}$ is a hill relative to $\{y_i, i \in J\}$.

Proof: Since $C(\pi)^+$ is a hill, there exists $\tilde{x} \in \operatorname{int} C(\pi)^+$ such that $[\pi_i y_i, \tilde{x}] > 0$ for $i \in I$ w.o. I_0. Since $[\pi_i y_i, \tilde{x}|_R] > 0$ for $i \in J$ w.o. I_0, $\mathrm{C}\{\pi_i y_i, i \in J\}$ is pointed.

The next thing to show is that if $\langle \pi_j y_j \rangle$ is an isolated ray of $\mathrm{C}\{\pi_i y_i, i \in J\}$, then $\langle \pi_j y_j \rangle$ is an isolated ray of $C(\pi)$. Assume there exist $\lambda_i \geqslant 0$, not all 0, such that $\pi_j y_j = \sum_{I \text{ w.o. } J} \lambda_i \pi_i y_i + \sum_{J \text{ w.o. } I_j(\pi)} \lambda_i \pi_i y_i$. Now

$0 = [\pi_j y_j, \tilde{v}] = \sum_{I \text{ w.o. } J} \lambda_i [\pi_i y_i, \tilde{v}]$ and so $\lambda_i = 0$ for $i \in I$ w.o. J. Consequently, $\pi_j y_j = \sum_{J \text{ w.o. } I_j(\pi)} \lambda_i \pi_i y_i$ for $\lambda_i \geqslant 0$, not all 0, which is a contradiction.

The final step is easy. Take $\langle \pi_j y_j \rangle$ in the frame of $\mathrm{C}\{\pi_i y_i, i \in J\}$. Then $\langle \pi_j y_j \rangle$ is an isolated ray of $C(\pi)$ and so there exists $i \in I^+(\pi)$ such that $\langle \pi_j y_j \rangle = \langle y_i \rangle$. It must be the case that $i \in J$ or otherwise for some $\alpha > 0$, $0 < [y_i, \tilde{v}] = \alpha[\pi_j y_j, \tilde{v}] = 0$. □

This theorem shows that each boundary vector of a hill establishes a certain lower dimensional problem which contains a hill generated in a natural way from the original hill.

It can now be observed that in fact (3.5.3) and (3.5.4) follow from (3.5.5). With regard to (3.5.3), if there exists $\tilde{z}$ such that $[y_i, \tilde{z}] > 0$ for $i \in J$ w.o. I_0, then of course $[y_i, \tilde{z}|_R] > 0$ for $i \in J$ w.o. I_0 and $\mathrm{C}\{y_i, i \in J\}^+$ is a hill in the lower dimensional problem. In fact, by (3.2.10), since $\tilde{z}|_R \in \mathrm{int}\ \mathrm{C}\{y_i, i \in J\}^+$, $\mathrm{C}\{y_i, i \in J\}^+$ is the only hill in the lower dimensional problem and so (3.5.5) forces $\pi_i = 1$ for all $i \in J$. With regard to (3.5.4), if $C(\pi)^+$ is a hill and $J_2 \neq \emptyset$, then the conclusion of (3.5.4) must hold by (3.5.5) since the only two hills in the lower dimensional problem occur when $\pi_i = 1$ for $i \in J_1$ and $\pi_i = -1$ for $i \in J_2$ or vice-versa.

-- *the converse of Theorem (3.5.5) doesn't hold* --

It is interesting to note that the converse of (3.5.5) does not hold in general. It is not generally true that if $C(\pi)^+$ is a hill in the original problem and $\mathrm{C}\{\theta_i y_i, i \in J\}^+$ is a hill in the lower dimensional problem then $\mathrm{C}\{\pi_i y_i, i \in I \text{ w.o. } J, \theta_i y_i, i \in J\}^+$ is necessarily a hill in the original problem. (3.5.3) gives an exception. The following is a counter-example.

(3.5.6) Counter-Example: Let $y_0 = (0, 0, 0)$, $y_1 = (-1, 1, 0)$, $y_2 = (0, -1, 0)$, $y_3 = (1, 1, 0)$, $y_4 = (0, -1, -1)$, and $y_5 = (0, 0, 1)$ in R^3. The first claim is that $\mathrm{C}\{y_1, -y_2, y_3, -y_4, y_5\}^+$ is a hill whose dual cone has frame $\{\langle y_0\rangle, \langle y_1\rangle, \langle y_3\rangle, \langle y_5\rangle\}$. $\mathrm{C}\{y_1, -y_2, y_3, -y_4, y_5\}$ was shown to be pointed in (2.3.30). Note that $-y_2 = \frac{1}{2}y_1 + \frac{1}{2}y_3$ and $-y_4 = -y_2 + y_5$. $\langle y_1\rangle$ is isolated since if $y_1 = (-1, 1, 0) = \lambda_1(1, 1, 0) + \lambda_2(0, 0, 1)$ for $\lambda_i \geqslant 0$, then λ_1 is not a real number. Similarly, $\langle y_3\rangle$ and $\langle y_5\rangle$ are isolated.

Letting $\tilde{v} = (0, 0, 1)$, observe that if $J = \{0, 1, 2, 3\}$, then $[y_i, \tilde{v}] = 0$ for $i \in J$ and $[y_i, \tilde{v}] \neq 0$ for $i \notin J$. Using the same techniques as above, it can be seen that $\mathrm{C}\{-y_1, y_2, y_3\}^+$ is a hill in the lower dimensional problem for $\{y_1, y_2, y_3\}$. Yet $\mathrm{C}\{-y_1, y_2, y_3, -y_4, y_5\}^+$ is not a hill in the original problem since $\langle -y_4\rangle$ is isolated and no $\langle y_i\rangle = \langle -y_4\rangle$.

-- using Theorem (3.5.5) to displace --

(3.5.5) is used in the following way. Suppose $C(\pi)^+$ is a hill and $\tilde{v}_k(i_0, \ldots, i_{k-1})$ for some $k \geqslant 0$ is in the tree and is an element of $F_J(C(\pi)^+) = \{\tilde{x}: [\pi_i y_i, \tilde{x}] > 0$ for $i \in I$ w.o. J, $[\pi_i y_i, \tilde{x}] = 0$ for $i \in J\}$. Suppose $\dim \mathrm{L}\{y_i, i \in J\} > 1$ and $\mathrm{C}\{y_i, i \in J\}$ is not pointed. Consider the hills (or hill) of the lower dimensional problem for $\{y_i, i \in J\}$. One of these must be $\mathrm{C}\{\pi_i y_i, i \in J\}^+$ by (3.5.5).

The next step is to apply the entire tree algorithm in a recursive fashion on the lower dimensional problem for the set $\{y_i, i \in J\}$ in order to obtain a vector in each of the lower dimensional hills. It has not yet been shown that this can be accomplished so the reader is asked to accept this on faith for the moment. One of the vectors that will be obtained is an $\tilde{r}_0 \in \tilde{R}$ such that $[\pi_i y_i, \tilde{r}_0] > 0$ for $i \in J$ w.o. I_0. It would be nice to use (3.5.1) and add a positive multiple of $\tilde{r}_0$ to $\tilde{v}_k(i_0, \ldots, i_{k-1})$ in order to obtain an interior vector of $C(\pi)^+$. This is patently impossible, of course, since $\tilde{r}_0$ and $\tilde{v}_k(i_0, \ldots, i_{k-1})$ are in different dual spaces. However, using (2.1.27), observe that $\psi^{-1}(\tilde{r}_0)$ and $\tilde{v}_k(i_0, \ldots, i_{k-1})$ are both in $\tilde{X}$ and $0 < [\pi_i y_i, \tilde{r}_0] = [\pi_i y_i, \psi^{-1}(\tilde{r}_0)]$ for $i \in J$ w.o. I_0. So, the idea is to compute $\psi^{-1}(\tilde{r}_0)$ and use it, $\tilde{v}_k(i_0, \ldots, i_{k-1})$, and (3.5.1) to compute an interior vector of $C(\pi)^+$.

Naturally, it isn't known which lower dimensional interior hill vector will displace $\tilde{v}_k(i_0, \ldots, i_{k-1})$ into int $C(\pi)^+$. So, it is necessary to run through the displacing procedure for each solution to the lower dimensional problem. The end result is the desired one, namely, a collection of interior vectors containing an interior vector for each hill containing $\tilde{v}_k(i_0, \ldots, i_{k-1})$.

-- using a computer to implement this displacing --

Next, it will be shown how a computer algorithm would employ this procedure on Example (3.3.18). Note the strong emphasis in what follows on the use of representations of vectors. This is because it is generally easier for computers to work with the one-dimensional arrays representing vectors with respect to some fixed basis than it is for them to work with the vectors themselves: for example, how would a computer work directly with elements of the vector space consisting of all polynomial functions on R^6 of degree $\leqslant 3$?

Let $\{y_1, y_2, y_5\}$ be a basis for $R = \mathsf{L}\{y_0, \ldots, y_5\}$ in Example (3.3.18). Note that the representations of the y_i with respect to this basis look the same as the vectors themselves. Furthermore, let $\{y_1, y_2\}$ be a basis for $\mathsf{L}\{y_0, \ldots, y_4\}$ and define $\underline{u}_i$ to be the representation of y_i, $i = 0, \ldots, 4$ with respect to this basis. Each linear functional $\tilde{r}$ on $\mathsf{L}\{y_0, \ldots, y_4\}$ has a two-dimensional vector representation $\underline{\tilde{r}}$ with respect to the dual basis such that $[y_i, \tilde{r}] = \underline{\tilde{r}}^T \underline{u}_i$ for $i = 0, \ldots, 4$.

An as yet unspecified algorithm will be used to find representations $\underline{\tilde{r}}$ for linear functionals in the interiors of hills in the $\mathsf{L}\{\underline{u}_0, \ldots, \underline{u}_4\}$ lower dimensional problem. In this example, one might obtain $\underline{\tilde{r}}_0 = (1, \frac{1}{2})$. Observe then that $[\theta_i y_i, \tilde{r}_0] = \theta_i \underline{u}_i^T \underline{\tilde{r}}_0 > 0$ for $(\theta_1, \ldots, \theta_4) = (1, 1, -1, 1)$. It is now necessary to compute $\underline{\psi^{-1}(\tilde{r}_0)}$.

Let $S = \mathsf{L}\{(0, 1, 1)\}$ so that $R \oplus S = \mathsf{R}^3$ and $\tilde{R} = S^{\perp}|_R$. All linear functionals in $S^{\perp}$ have representations with respect to the dual basis in $\tilde{X}$ of the form $(\alpha, \beta, -\beta)$ for some $\alpha, \beta \in \mathsf{R}$. Each $\underline{\tilde{r}}$ is of course of the form (α, β) for some $\alpha, \beta \in \mathsf{R}$. Fix $\underline{\tilde{r}} = (\alpha, \beta)$. It will now be shown how to compute $\underline{\psi^{-1}(\tilde{r})}$. $\underline{\psi^{-1}(\tilde{r})} = (\gamma_1, \gamma_2, \gamma_3)$ for suitable γ_i. Since $\psi^{-1}(\tilde{r}) \in S^{\perp}$, $\gamma_3 = -\gamma_2$. Since $\alpha = \underline{\tilde{r}}^T \underline{u}_1 = [y_1, \tilde{r}] = [y_1, \psi^{-1}(\tilde{r})] = \gamma_1$, it must be that $\gamma_1 = \alpha$. Similarly, $\gamma_2 = \beta$. So, in this very special case, if $\underline{\tilde{r}} = (\alpha, \beta)$, then $\underline{\psi^{-1}(\tilde{r})} = (\alpha, \beta, -\beta)$. In general, it is not this simple. Theorem (3.5.13) will elaborate on this.

At any rate, $\underline{\psi^{-1}(\tilde{r}_0)} = (1, \frac{1}{2}, -\frac{1}{2})$ and a multiple of it can be added to $\underline{\tilde{v}_{2,1}(4, 1)} = (0, 0, 1)$ to obtain an interior vector of a hill in the three-dimensional problem.

It is of course painful to keep in mind two vector spaces and their dual spaces, bases, dual bases, and various representations. All of this is necessary, however, because the theory is best couched in a coordinate-free context whereas the computer works best with the representations of vectors, not with the vectors themselves. The general procedures which the computer uses to go back and forth between lower dimensional and original problems will be presented after the complete procedure for the second phase of the tree algorithm has been stated and validated.

-- *the displacing algorithm* --

To summarize, the function of the second phase of the hill-finding tree algorithm is to take the tree of vectors produced by the first phase and to displace all of the boundary vectors in the tree into the interiors of appropriate adjacent cones in such a way as to obtain an interior vector for every hill. Here is the procedure followed by the second or displacement phase for each boundary vector.

(3.5.7) Algorithm: Given $\tilde{v}_k(i_0, \ldots, i_{k-1}) \in F_J(C(\pi)^+)$ for some $C(\pi)^+$ and $k \geqslant 0$. Select the appropriate case:

Case 1: $\dim \mathsf{L}\{y_i, i \in J\} = 0$.

Here $k = 0$, $\tilde{v}_0 \in \operatorname{int} C(\pi)^+$, and no displacing is needed.

Case 2: $\dim \mathsf{L}\{y_i, i \in J\} = 1$.

Here $k = 0$ or 1. Let $p \in J$ w.o. I_0, $J_1 = \{i \in J: \langle y_i \rangle = \langle y_p \rangle\}$, and $J_2 = \{i \in J: \langle y_i \rangle = -\langle y_p \rangle\}$.

If $J_2 = \varnothing$, then let $\underline{z} = \underline{y}_p$ and displace $\tilde{v}_k(i_0, \ldots, i_{k-1})$ using (3.5.1).

Suppose $J_2 \neq \emptyset$. First, let $\tilde{z}$ be such that $\underline{z} = \underline{y_p}$ and use (3.5.1) to displace $\tilde{v}_k(i_0, \ldots, i_{k-1})$. Next, let $\tilde{z}$ be such that $\underline{z} = -\underline{y_p}$ and use (3.5.1) to displace $\tilde{v}_k(i_0, \ldots, i_{k-1})$.

Case 3: $\dim \mathsf{L}\{y_i, i \in J\} > 1$ and $\mathsf{C}\{y_i, i \in J\}$ is pointed.

In this case, use the $\tilde{z}$ provided by the linear program of (2.3.33) to displace $\tilde{v}_k(i_0, \ldots, i_{k-1})$ via (3.5.1). Note that this LP also determines whether or not $\mathsf{C}\{y_i, i \in J\}$ is pointed.

Case 4: $\dim \mathsf{L}\{y_i, i \in J\} > 1$ and $\mathsf{C}\{y_i, i \in J\}$ is not pointed.

In this case, recursion is necessary. Call Algorithm (3.3.13) followed by Algorithm (3.5.7) for each boundary vector to provide interior vectors for the hills in the lower dimensional $\{y_i, i \in J\}$ problem. Displace $\tilde{v}_k(i_0, \ldots, i_{k-1})$ using (3.5.1) and each of the inverse images under ψ of the lower dimensional interior hill vectors.

-- *Algorithm (3.5.7) works* --

(3.5.8) Theorem: In order to find an interior vector for every hill, it is sufficient to use Algorithm (3.5.7) to displace all boundary vectors produced by the first phase of the tree algorithm. In order to determine if a given displaced boundary vector is in the interior of a hill, it is sufficient to find the frame of the dual of the cone it is in.

Proof: By (3.3.14), the first phase of the tree algorithm constructs a tree containing at least one vector in every hill. Since the second phase displaces all of the boundary vectors in its search for an interior vector for each hill, it suffices to show that if $\tilde{v}_k(i_0, \ldots, i_{k-1})$ is in a hill $C(\pi)^+$, then (3.5.7) will displace $\tilde{v}_k(i_0, \ldots, i_{k-1})$ into $\operatorname{int} C(\pi)^+$. (3.5.4) verifies this for Case 2 of (3.5.7). (3.5.3) verifies this for Case 3. (3.5.5) verifies this for Case 4 if the recursive process terminates in an acceptable manner.

First of all, the recursive process does terminate. Since $\{y_i, i \in J\} \subset \tilde{v}_k(i_0, \ldots, i_{k-1})^{\perp}$ and $\tilde{v}_k(i_0, \ldots, i_{k-1}) \neq 0$, $\dim \mathsf{L}\{y_i, i \in J\} < \dim \mathsf{L}\{y_i, i \in I\}$. Consequently, whenever (3.5.7) is called, it must start to work on a strictly lower dimensional problem. If ever the dimension of the lower dimensional problem reaches 2, then all of the boundary vectors will be in one-dimensional subspaces and will consequently be displaced by Case 2 of (3.5.7) and the recursion will end.

Also, note that the four cases in (3.5.7) are exhaustive so that there is never any doubt about what should be done with each $\tilde{v}_k(i_0, \ldots, i_{k-1})$ in the tree. □

-- *usually hills are not of primary interest* --

(3.5.9) Remark: In most cases, an interior vector in every hill is not really desired. For instance, when seeking to find all max-sum cones, only those hills which are max-sum cones are of interest. In this case, the first phase of the tree algorithm should save only those $\tilde{v}_k$ in the tree which have sufficiently high h values to qualify them as potentially being in a max-sum cone. Only these likely candidates are displaced by the second phase of the algorithm in order to discover which ones are in max-sum cones. This is the approach taken in Algorithm (3.2.12) and it saves an enormous amount of work over that which the tree algorithm designed to find all of the hills would take (cf., (3.4.2)).

Now, it is necessary to be careful here because if the algorithm is only saving those boundary vectors with large h-values, then the algorithm is only going to be finding all max-sum cones and not necessarily all of the hills *at any and all levels of recursion.* Therefore, for a given call for recursion in DISPLACE of (3.2.12), it must be shown that a max-sum cone boundary vector will be successfully displaced into the interior of all neighboring max-sum cones even if it is just the max-sum cones of the lower dimensional problem that are produced as a result of the recursive call. To do this, it suffices to show that any nonzero boundary face vector of every max-sum cone in the original problem generates a max-sum cone in the associated lower dimensional problem (cf., (3.5.5)). In symbols, it must be shown that if $C(\pi)^+$ is a max-sum cone and if, for some $\tilde{v}_k \neq 0$ and $I_0 \neq J \subset I$, $[\pi_i y_i, \tilde{v}_k] > 0$ for $i \in I$ w.o. J and $[\pi_i y_i, \tilde{v}_k] = 0$ for $i \in J$, then $\mathsf{C}\{\pi_i y_i, i \in J\}^+$ is a max-sum cone in $\tilde{R}$ where R is such that $R \oplus \mathsf{L}\{y_i, i \in J\} = X$ and the criterion function is

$$h_J(\tilde{r}) := \sum_J \sigma_i \, 1\{[y_i, \tilde{r}] > 0\}.$$

-- *max-sum cones generate lower-dimensional max-sum cones* --

This desired relationship between a max-sum cone $C(\pi)^+$ and the max-sum cones in the dual space of $\{y_i, i \in J\}$ will now be established when $F_J(C(\pi)^+) \neq \emptyset$.

(3.5.10) Theorem: Let $C(\pi)^+$ be a max-sum cone. Suppose $\tilde{v} \in \{\tilde{x} : [\pi_i y_i, \tilde{x}] > 0$ for $i \in I$ w.o. J, $[\pi_i y_i, \tilde{x}] = 0$ for $i \in J\}$ and that $\tilde{v} \neq 0$ so that $J \neq I$. Let $R = \mathsf{L}\{y_i, i \in J\}$ where it is assumed that $\dim R \geqslant 1$ so that $J \neq I_0$. Then $\mathsf{C}\{\pi_i y_i, i \in J\}^+$ is a max-sum cone in $\tilde{R}$.

Proof: In order to show $\mathsf{C}\{\pi_i y_i, i \in J\}$ is pointed, just recall that $\mathsf{C}\{\pi_i y_i, i \in J\}^+$ is a hill.

The claim here is that
$$\sum_{\substack{i \in J \text{ w.o. } I_0 \\ \pi_i = 1}} \sigma_i = \sup_{\tilde{r} \in \tilde{R}} \sum_{i \in J} \sigma_i \, 1\{[y_i, \tilde{r}] > 0\}.$$

Suppose to the contrary that there exists $\theta_i \in \{-1, 1\}$ for $i \in J$ and $\tilde{r}_0 \in \tilde{R}$ such that $[\theta_i y_i, \tilde{r}_0] > 0$ for $i \in J$ w.o. I_0 and
$$\sum_{\substack{i \in J \text{ w.o. } I_0 \\ \theta_i = 1}} \sigma_i > \sum_{\substack{i \in J \text{ w.o. } I_0 \\ \pi_i = 1}} \sigma_i .$$

Now $\tilde{r}_0 = \tilde{t}\,|_R$ for some $\tilde{t} \in \tilde{X}$ and there exists $\alpha > 0$ such that $[\pi_i y_i, \tilde{v} + \alpha\tilde{t}] > 0$ for $i \in I$ w.o. J and $[\theta_i y_i, \tilde{v} + \alpha\tilde{t}] > 0$ for $i \in J$ w.o. I_0. Hence, there exists $\alpha > 0$ such that

$$h(\tilde{v} + \alpha\tilde{t}) = \sum_{\substack{i \in I \text{ w.o. } J \\ \pi_i = 1}} \sigma_i + \sum_{\substack{i \in J \text{ w.o. } I_0 \\ \theta_i = 1}} \sigma_i > \sum_{\substack{i \in I \text{ w.o. } I_0 \\ \pi_i = 1}} \sigma_i$$ and this is a contradiction. □

-- and conversely --

Interestingly, the converse of this theorem holds whereas the converse of the analogous theorem for hills doesn't.

(3.5.11) Theorem: Let $C(\pi)^+$ be a max-sum cone and $\tilde{v} \in F_J(C(\pi)^+)$ where $I_0 \neq J \neq I$. Let $R = \mathsf{L}\{y_i, i \in J\}$ where $\dim R \geqslant 1$. Suppose $\mathsf{C}\{\theta_i y_i, i \in J\}^+$ is a max-sum cone in $\tilde{R}$. Then $\mathsf{C}\{\pi_i y_i, i \in I \text{ w.o. } J, \theta_i y_i, i \in J\}^+$ is a max-sum cone in $\tilde{X}$.

Proof: In order to show $\mathsf{C}\{\pi_i y_i, i \in I \text{ w.o. } J, \theta_i y_i, i \in J\}$ is pointed, first note that there exists $\tilde{r}_0 \in \tilde{R}$ and $\tilde{x}_0 \in \tilde{X}$ such that $\tilde{r}_0 = \tilde{x}_0|_R$ and $[\theta_i y_i, \tilde{r}_0] > 0$ for all $i \in J$ w.o. I_0. Then choose $\alpha > 0$ such that $[\pi_i y_i, \tilde{v} + \alpha\tilde{x}_0] > 0$ for $i \in I$ w.o. J and $[\theta_i y_i, \tilde{v} + \alpha\tilde{x}_0] > 0$ for $i \in J$ w.o. I_0.

Suppose $$\sum_{\substack{i \in I \text{ w.o. } J \\ \pi_i = 1}} \sigma_i + \sum_{\substack{i \in J \text{ w.o. } I_0 \\ \theta_i = 1}} \sigma_i < \sum_{\substack{i \in I \text{ w.o. } I_0 \\ \pi_i = 1}} \sigma_i .$$

Then $$\sum_{\substack{i \in J \text{ w.o. } I_0 \\ \theta_i = 1}} \sigma_i < \sum_{\substack{i \in J \text{ w.o. } I_0 \\ \pi_i = 1}} \sigma_i$$ which contradicts (3.5.10). □

So, every nonempty nonzero boundary face of a max-sum cone generates a max-sum cone in a suitably defined lower dimensional problem and in turn, every max-sum cone in this lower dimensional problem generates a max-sum cone back in the original problem.

-- obtaining lower-dimensional representations of vectors --

Algorithm (3.5.7) cannot be implemented on a computer unless there are methods for converting the representations of vectors in the original problem to those of the lower dimensional problem and vice-versa. Such methods are necessary because Theorem (3.3.14) assumes that $\mathsf{L}\{y_i, i \in I\} = X$, and so in order for the computer to run Algorithm (3.3.13) on a lower dimensional set $\{y_i, i \in J\}$ where $\dim \mathsf{L}\{y_i, i \in J\} = p < d$, it is certainly not possible to use any d-dimensional representations of these y_i but instead it is necessary to use their p-dimensional representations with respect to some basis for $\mathsf{L}\{y_i, i \in J\}$.

A straight-forward procedure yields the necessary representations for the lower dimensional problem. Suppose $\{y_i, i \in J\}$ for $J \subset I$ is given. Let $\{a_1, \ldots, a_d\}$ be a basis for X. Suppose $\dim \mathsf{L}\{y_i, i \in J\} = p$. The first thing to do is to find a basis $\{b_1, \ldots, b_p\}$ for $\mathsf{L}\{y_i, i \in J\}$. Let $\{\underline{y}_i, i \in J\}$ and $\{\underline{b}_i\}_1^p$ be the representations of these vectors with respect to $\{a_i\}_1^d$. Referring back to the notation of (3.4.8) and recalling the Gram-Schmidt algorithm (see Nering (1963)), observe that the following procedure determines an orthonormal $\{\underline{b}_i\}_1^p$:

(3.5.12) Algorithm: Without loss of generality, let $J = \{1, \ldots, m\}$ for some m and assume $y_1 \neq 0$.

Set $\underline{b}_1 = \underline{y}_1 / \| \underline{y}_1 \|$.
Set $j = 2$.
For $k = 2, \ldots, m$ do:
Set $\underline{b}_j = \underline{y}_k - P[\underline{y}_k \mid \mathsf{L}\{\underline{b}_1, \ldots, \underline{b}_{j-1}\}]$.
If $\underline{b}_j = \underline{0}$ then next k.
Set $\underline{b}_j = \underline{b}_j / \| \underline{b}_j \|$.
Set $j = j + 1$.
next k;

The computer program implementing the tree algorithm uses the numerically more stable Modified Gram-Schmidt algorithm. More precisely, the program uses a variant of the algorithm given on p. 217 of Stewart (1973) which may be obtained from (3.5.12) by replacing

"Set $\underline{b}_j = \underline{y}_k - P[\underline{y}_k \mid \mathrm{L}\{\underline{b}_1, \ldots, \underline{b}_{j-1}\}]$"

with " $\underline{b}_j = \underline{y}_k$

For $i = 1, \ldots, j-1$ do:

$\underline{b}_j = \underline{b}_j - P[\underline{b}_j \mid \mathrm{L}\{\underline{b}_i\}]$

next i; "

In order for the computer to work on the lower dimensional set $\{y_i, i \in J\}$, it is necessary for it to have the representations of the y_i, $i \in J$, with respect to $\{b_i\}_1^p$. In other words, γ_{ij} must be computed such that $y_i = \sum_1^p \gamma_{ij} b_j$. Since $y_i = \sum_1^p \gamma_{ij} b_j$ if and only if $\underline{y}_i = \sum_1^p \gamma_{ij} \underline{b}_j$, standard Hilbert space theory can be used to obtain $\gamma_{ij} = (\underline{y}_i, \underline{b}_j) = \underline{y}_i^T \underline{b}_j$.

So, the computer program takes these lower dimensional representations and finds elements of the p-dimensional dual space which lie in the interiors of the lower dimensional hills or max-sum cones. The problem of interest now is how to determine a d-dimensional representation of a linear functional given the p-dimensional representation of that linear functional restricted to a p-dimensional subspace.

(3.5.13) Theorem: Let $A = \{a_1, \ldots, a_d\}$ be a basis for X, $B = \{b_1, \ldots, b_p\}$ be a basis for a subspace $R \subset X$, and $\tilde{A}$ and $\tilde{B}$ be the corresponding dual bases. Suppose for $j = 1, \ldots, p$, $b_j = \sum_{i=1}^d \beta_{ij} a_i$ and so $\underline{b}_j = (\beta_{1j}, \ldots, \beta_{dj})$. Let $\underset{\sim}{B} = [\underline{b}_1 \cdots \underline{b}_p] = [\beta_{ij}]$ and suppose $\underset{\sim}{B}^T \underset{\sim}{B} = \underset{\sim}{I}_{p \times p}$, the $p \times p$ identity matrix. It is known that for each $\tilde{b}_j$ there exists $\tilde{u}_j \in S^\perp$ where $R \oplus S = X$ such that $\tilde{b}_j = \tilde{u}_j|_R$. Let

$$\tilde{v} = \sum_1^p \sigma_j \tilde{b}_j = \left(\sum_1^p \sigma_j \tilde{u}_j \right)\Big|_R .$$

Then

$$\psi^{-1}(\tilde{v}) = \sum_1^p \sigma_j \tilde{u}_j = \sum_{i=1}^d \left(\sum_{j=1}^p \sigma_j \beta_{ij} \right) \tilde{a}_i .$$

Proof: The first thing to do is to compute $\underset{\sim}{\tilde{u}_j}$ for $j = 1, \ldots, p$ in terms of known quantities. Now $\underset{\sim}{\tilde{u}_j} = \sum_1^d \Upsilon_{ij} \underset{\sim}{\tilde{a}_i}$ and since $\tilde{b}_j = \tilde{u}_j \big|_R$, it must be that for all j, k, $\underset{\sim}{\tilde{u}_j^T} \underset{\sim}{b_k} = \sum_{i=1}^d \Upsilon_{ij} \beta_{ik} = \delta_{jk}$ where δ_{jk} is the Kronecker δ. In other words, if $\underset{\sim}{U} := [\underset{\sim}{\tilde{u}_1} \cdots \underset{\sim}{\tilde{u}_p}]$, then $\underset{\sim}{U}^T \underset{\sim}{B} = \underset{\sim}{I}_{p \times p}$ is needed, which does occur if $\underset{\sim}{U} = \underset{\sim}{B}$.

Now take $\tilde{v} = \sum_1^p \sigma_j \tilde{b}_j$. Then

$$\sum_1^p \sigma_j \tilde{u}_j = \sum_{j=1}^p \sigma_j \left(\sum_{i=1}^d \beta_{ij} \tilde{a}_i \right) = \sum_{i=1}^d \left(\sum_{j=1}^p \sigma_j \beta_{ij} \right) \tilde{a}_i . \quad \Box$$

Summary For Section 3.5

The function of the second or displacement phase of the tree algorithm is to take the boundary vectors produced by the first phase and displace them into the interiors of neighboring cones in such a way as to produce an interior vector from every hill (or max-sum cone). Here is the basic procedure for displacing $\tilde{v}_k \in F_J(C(\pi)^+)$. (In what follows, when displacing $\tilde{v}_k$ in a certain direction, care should be taken so as not to displace $\tilde{v}_k$ so much that the displaced vector is not in any of the $C(\pi)^+$ containing $\tilde{v}_k$.)

If $\dim \mathrm{L}\{y_i, i \in J\} = 0$, then do nothing since $\tilde{v}_k = \tilde{v}_0 \in \mathrm{int}\, C(\pi)^+$.

If $\dim \mathrm{L}\{y_i, i \in J\} = 1$, then $\tilde{v}_k$ is in the relative interior of a $d - 1$ dimensional boundary face of $C(\pi)^+$. If this boundary face has a uniquely defined positive side, then displace $\tilde{v}_k$ in that direction. If not, then displace $\tilde{v}_k$ separately in both directions.

If $\dim \mathrm{L}\{y_i, i \in J\} > 1$ and $\mathrm{C}\{y_i, i \in J\}$ is pointed, then there is a positive halfspace containing $\{y_i, i \in J \text{ w.o. } I_0\}$ in its interior and $\tilde{v}_k$ should be displaced in the direction of the normal to that halfspace.

If $\dim \mathrm{L}\{y_i, i \in J\} > 1$ and $\mathrm{C}\{y_i, i \in J\}$ is not pointed, then call the entire tree algorithm recursively in order to obtain vectors in the interiors of the lower dimensional hills (or max-sum cones) corresponding to $\{y_i, i \in J\}$. Associate these solution vectors in the lower dimensional dual space with their inverse images under the isomorphism ψ in the original dual space. Then displace $\tilde{v}_k$ towards each of these images.

It is not necessary to displace all of the boundary vectors produced by the first phase of the tree algorithm when the problem is only to identify all max-sum cones. One of the reasons for this is that every max-sum cone in the original problem generates max-sum cones in its associated lower dimensional problems and conversely.

The algorithms used by the computer to set up the lower dimensional problem and then to re-express its solutions in terms of the original problem are also discussed.

Chapter 4:

Constrained And Unconstrained Optimization Of Functions Of Systems Of Linear Relations

This chapter introduces a class of problems concerned with extremizing functions of a system of linear relations with or without constraints. The common goal of each of these problems is that of seeking all those vectors which satisfy or don't satisfy elements of a system of linear relations in such a way as to maximize a given function. For example, the set of linear relations in the WOH problem is $\{y_i^T x > 0\}_1^n$ and the objective is to find all vectors $x \in \mathsf{R}^d$ which satisfy or don't satisfy these linear inequalities in such a way as to maximize $\sum_1^n \sigma_i \; 1\{y_i^T x > 0\}$ for given $\sigma_i > 0$.

A number of different types of problems of extremizing functions of systems of linear relations will be described in this chapter and it will be shown how all of them are equivalent to simpler-looking problems in what is called homogeneous canonical form. It will be pointed out that the tree algorithm of Chapter 3 can solve certain problems in homogeneous canonical form whereas it will be left to Chapter 5 to develop the apparently most general form of the tree algorithm which is capable of solving all problems in homogeneous canonical form as long as an associated function is nondecreasing.

In this chapter, problems of optimizing functions of systems of linear relations will be written in terms of vectors from R^d instead of in terms of vectors from some abstract vector space X. Certainly no generality is lost because for any problem of this sort expressed in terms of "$[y, \tilde{x}]$", it is sufficient to use the representations of the vectors and work with terms like "$\underline{y}^T \underline{x}$". What is gained by working in the context of R^d in this chapter is an ease of expression in writing down the operations a computer algorithm would have to go through in order to solve problems of this kind in practice. Just as before, however, and for the same reasons as before, all proofs concerning the tree algorithm proper in this chapter and the next will be set in the context of the abstract vector space X.

-- *sample problems of this type* --

For future reference, a few more examples of problems of optimizing functions of systems of linear relations will now be discussed. A problem of perennial interest in the literature is that of finding all solution vectors x to the system $\{a_i^T x > \mu_i : i \in J\} \cup \{a_i^T x \geqslant \mu_i : i \in I \text{ w.o. } J\}$ for fixed $I \neq \emptyset$, $\mu_i \in \mathrm{R}$, and $a_i \in \mathrm{R}^d$ under the condition that there are vectors x which satisfy all of the linear inequalities. This problem can be generalized to the case where no vector satisfies all of the inequalities by associating a positive reward with each linear inequality and then seeking those vectors which maximize the sum of the rewards of the inequalities they satisfy. In symbols, the problem is that of maximizing

$$\sum_J \sigma_i \, 1\{a_i^T x > \mu_i\} + \sum_{I \text{ w.o. } J} \sigma_i \, 1\{a_i^T x \geqslant \mu_i\},$$

where, with no increase in generality, the σ_i may be allowed to be negative. Observe, of course, that when $\sigma_i = 1$ for all i, then the problem is that of finding all vectors which satisfy as many of the linear inequalities as possible.

By setting all $\mu_i = 0$, the various hemisphere problems are obtained. These are, when all $\sigma_i = 1$, the Open Hemisphere (OH), Closed Hemisphere (CH), and the Mixed Hemisphere (MH) problems according to whether $J = I$, $J = \emptyset$, or $\emptyset \neq J \neq I$, respectively. The adjective "Weighted" is prepended to the name when the σ_i are allowed to be any real numbers. As hinted before, without loss of generality, it may be assumed that the σ_i are positive in the weighted hemisphere problems in that, for example, a WOH problem with all negative weights is equivalent to a WCH problem with all positive weights. The word "hemisphere" is used because if one introduces a norm $\| \cdot \|$ and divides, for each i, the ith inequality by $\| a_i \|$, then the resulting problem is one of finding all hemispheres of the unit sphere which contain points collecting the largest possible total reward.

One of the theorems of this chapter will show that in order to extremize $\sum_J \sigma_i \, 1\{a_i^T x > \mu_i\} + \sum_{I \text{ w.o. } J} \sigma_i \, 1\{a_i^T x \geqslant \mu_i\}$, it suffices to solve the WMH problem which the tree algorithm of Chapter 5 can do.

As a final example of what the procedures described in this chapter are capable of, it will be shown that for fixed $\sigma_i > 0$, $\mu_i \in \mathsf{R}$, and $a_i \in \mathsf{R}^d$, the tree algorithm of Chapter 3 is able to find all vectors x maximizing $\sum_{i=1}^{m} \sigma_i \, 1\{a_i^T x > \mu_i\}$

(i) subject to x being an element of a specified polyhedral set $\{z \in \mathsf{R}^d : \underset{\sim}{B} z > e\}$ where $\underset{\sim}{B}$ is an $n \times d$ matrix and $e \in \mathsf{R}^n$ or

(ii) subject to x being an element of a specified linear manifold $\{z \in \mathsf{R}^d : \underset{\sim}{C} z = w\}$ for $\underset{\sim}{C}$, a $p \times d$ matrix and $w \in \mathsf{R}^P$ or

(iii) subject to x maximizing a second function $\sum_{j=1}^{q} \tau_j \, 1\{b_j^T x > \nu_j\}$ where $\tau_j > 0$ or

(iv) subject to any number or none of the above.

Other problems of this kind drawn from certain applications will be described in Chapter 8.

-- *preliminary definitions* --

The theory begins with a few basic definitions.

(4.1.1) Definitions: For given $a \in \mathsf{R}^d$, $\mu \in \mathsf{R}$, and relational operator $R \in \{<, \leqslant, =, \neq, \geqslant, >\}$, $a^T x \; R \; \mu$ is a *linear relation* in $x \in \mathsf{R}^d$. A linear relation is said to be *homogeneous* if $\mu = 0$ and *inhomogeneous* if $\mu \neq 0$.

(4.1.2) Definitions: A *system of linear relations* in $x \in \mathsf{R}^d$ is defined to be $\{a_i^T x \; R_i \; \mu_i\}_1^m$ for given $a_i \in \mathsf{R}^d$, $\mu_i \in \mathsf{R}$, and relational operators $R_i \in \{<, \leqslant, =, \neq, \geqslant, >\}$. The object is usually to identify vectors $x \in \mathsf{R}^d$ which satisfy or don't satisfy the relations $a_i^T x \; R_i \; \mu_i$ in some desired way.

A system of linear relations $\{a_i^T x \; R_i \; \mu_i\}_1^m$ is said to be *homogeneous* if $\mu_i = 0$ for all i and *inhomogeneous* if for some i, $\mu_i \neq 0$.

A system of linear relations $\{a_i^T x \; R_i \; \mu_i\}_1^m$ is said to be *consistent* if there is some $y \in \mathsf{R}^d$ which satisfies all of the relations $a_i^T x \; R_i \; \mu_i$ and is said to be *inconsistent* otherwise.

(4.1.3) Definitions: A function $H : \mathsf{R}^d \to \mathsf{R}$ is a *function of the system of linear relations* $\{a_i^T x \; R_i \; \mu_i\}_1^m$ if and only if there is a function $g : \times_1^m \{0, 1\} \to \mathsf{R}$ such that for all $x \in \mathsf{R}^d$,

$$H(x) = g(1\{a_1^T x \; R_1 \; \mu_1\}, \ldots, 1\{a_m^T x \; R_m \; \mu_m\}).$$

The basic problem in this context is to develop ways to find vectors x which maximize (or minimize, if desired) specific functions of systems of linear relations. A few examples will illustrate the complexity of this problem.

-- *illustrating the complexity of this problem* --

(4.1.4) Examples: Consider the function $H : \mathsf{R}^2 \to \mathsf{R}$ where for $x = (\xi_1, \xi_2) \in \mathsf{R}^2$,

$$\begin{aligned} H(x) &= 1\{(1, 0)^T x \; R_1 \; 0\} + 1\{(0, -1)^T x \; R_2 \; 0\} \\ &\qquad + \alpha 1\{(-1, 1)^T x \; R_3 \; 0\} + \alpha 1\{(1, -1)^T x \; R_4 \; 0\} \\ &= 1\{\xi_1 \; R_1 \; 0\} + 1\{-\xi_2 \; R_2 \; 0\} + \alpha 1\{-\xi_1 + \xi_2 \; R_3 \; 0\} + \alpha 1\{\xi_1 - \xi_2 \; R_4 \; 0\} \end{aligned}$$

for specific relations R_1, R_2, R_3, R_4 and $\alpha > 0$. With regard to Definition (4.1.3), associate with H the homogeneous system of linear relations

$$\{(1,0)^T x \ R_1 \ 0, \ (0,-1)^T x \ R_2 \ 0, \ (-1,1)^T x \ R_3 \ 0, \ (1,-1)^T x \ R_4 \ 0\}$$

and the function $g : \bigtimes_1^4 \{0, 1\} \to \mathsf{R}$ where

$$g(\tau_1, \tau_2, \tau_3, \tau_4) = \tau_1 + \tau_2 + \alpha\tau_3 + \alpha\tau_4.$$

The problem of maximizing H will be considered for various choices of R_1, R_2, R_3, R_4 and α.

To begin with, let $R_1 = R_2 = R_3 = R_4 =$ ">" and set $\alpha = 1$. The resulting problem is a version of Problem (3.1.1) and Figure (4.1.5) shows how these linear inequalities partition up the solution space. Note that this system of linear inequalities is inconsistent since there is no vector x which satisfies all of them. As shown in Chapter 3, the maximum value of H must occur in the interior of a fully-dimensional cone and in this case, it can be seen that the sole max-sum cone is C which achieves a value of 3 in its interior. The other two hills in this example are A and E, each achieving a value of 2 in their interiors.

Now change R_3 and R_4 to be "$\geqslant$" and leave the other symbols the same. The values of H on the rays $\langle u_1 \rangle$, $\langle u_2 \rangle$ increase from 1 to 3 so that H is maximized not only in the interior of cone C but also on the rays $\langle u_1 \rangle$, $\langle u_2 \rangle$. If, in addition, α is changed from 1 to 2, then the maximum value of H becomes 5 and is assumed only on the rays $\langle u_1 \rangle$, $\langle u_2 \rangle$.

In the event that $R_1 = R_2 = R_3 = R_4 =$ "$\geqslant$" and $\alpha = 1$, the system becomes consistent and H assumes its maximum value of 4 on the vector 0 only. 0, of course, is not a very interesting solution. The remainder of this chapter is concerned only with finding nonzero solutions to the stated problems. The function value of any nonzero solution can always be compared to that corresponding to 0 to see which one is better. The nonzero solution vectors corresponding to $R_1 = R_2 = R_3 = R_4 =$ "$\geqslant$" in this example are contained in $\langle u_1 \rangle \cup \langle u_2 \rangle$ and are associated with the function value of $3 < 4$. If $R_1 =$ ">" and $R_2 = R_3 = R_4 =$ "$\geqslant$", then any vector in $\langle u_1 \rangle \cup \langle 0 \rangle \cup \langle u_2 \rangle$ has

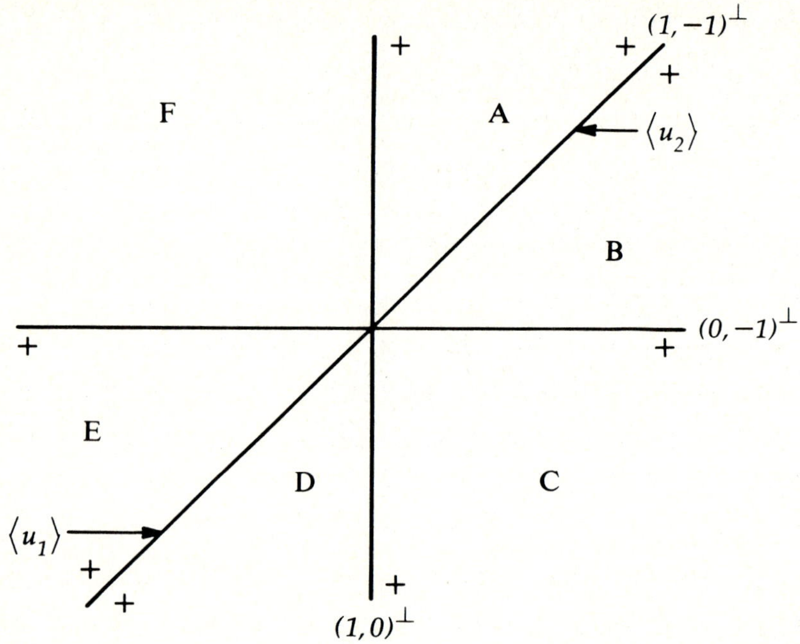

(4.1.5) Figure: Several cones in R^2.

function value 3, so that, in this instance, 0 is no better than the nonzero solutions.

So, it is clear that depending on the choice of α and ">" versus "$\geqslant$", the set of vectors maximizing H varies from being the interior of a fully-dimensional cone, to an interior along with a couple of nonzero rays, to the rays alone, and finally to the point 0 (if it has not otherwise been excluded from consideration).

Also, note that cone C which is the max-sum cone when $R_1 = R_2 = R_3 = R_4 =$ ">" and $\alpha = 2$, is nowhere near the optimal vectors when R_3 and R_4 are changed to "$\geqslant$" and so it is futile in general to hope that the nonzero faces of the max-sum cones in the strict inequality version of the problem will somehow contain the solutions to the version with strict and non-strict inequalities.

It is not even true that optimal vectors are restricted to lie only in rays or the interiors of cones. The vectors which maximize $1\{\xi_3 = 0\} + 1\{\xi_1 > 0\} + 1\{\xi_2 > 0\}$ over $x = (\xi_1, \xi_2, \xi_3)$ comprise the open positive quadrant of the ξ_1 - ξ_2 plane.

-- *on the way to a homogeneous canonical form* --

The reader may have noticed that the systems used in the preceding examples were all homogeneous. This turns out to be perfectly general as will be seen shortly in the theorem which proposes a homogeneous canonical form for the general optimization problem for systems of linear relations. The basic canonical form is presented first however and requires the following definitions:

(4.1.6) Definitions: Let $s = (\sigma_1, \ldots, \sigma_m)$ and $t = (\tau_1, \ldots, \tau_m)$ be elements of R^m. By definition, $s \leqslant t$ if and only if $\sigma_i \leqslant \tau_i$ for $i = 1, \ldots, m$ and at least one inequality is strict.

Let g be a real-valued function on $\overset{m}{\underset{1}{\times}} \{0, 1\}$. The jth variable of g is *nondecreasing* if and only if for all choices of $\sigma_1, \ldots, \sigma_{j-1}, \sigma_{j+1}, \ldots, \sigma_m \in \{0, 1\}$,

$$g(\sigma_1, \ldots, \sigma_{j-1}, 0, \sigma_{j+1}, \ldots, \sigma_m) \leqslant g(\sigma_1, \ldots, \sigma_{j-1}, 1, \sigma_{j+1}, \ldots, \sigma_m).$$

The jth variable of g is *nonincreasing* if and only if the jth variable of $-g$ is nondecreasing. The jth variable of g is *constant* if and only if it is nondecreasing and nonincreasing.

$g : \overset{m}{\underset{1}{\times}} \{0, 1\} \to \mathsf{R}$ is *nondecreasing* if and only if for all $s, t \in \overset{m}{\underset{1}{\times}} \{0, 1\}$ such that $s \leqslant t$, $g(s) \leqslant g(t)$. g is *strictly increasing* if and only if for all $s, t \in \overset{m}{\underset{1}{\times}} \{0, 1\}$ such that $s \leqslant t$, $g(s) < g(t)$.

It is easy to show that g is nondecreasing if and only if every variable of g is nondecreasing. The g function given for Examples (4.1.4) is nondecreasing. As an example of a g function with variables that are neither nondecreasing nor nonincreasing, consider g where $g(0, 0) = 2$, $g(0, 1) = 4$, $g(1, 0) = 7$, and $g(1, 1) = 3$. Note that the function $H(x) = g(1\{\xi_1 > 0\}, 1\{\xi_2 > 0\})$ is not maximized in the sole hill corresponding to the system $\{(1, 0)^T x > 0,\ (0, 1)^T x > 0\}$. Also note that the value of this g at arbitrary $t = (\tau_1, \tau_2)$ cannot be written as a linear combination of τ_1 and τ_2 so that the set of linear functions is not sufficiently general.

(4.1.7) Definitions: The problem of extremizing (i.e., maximizing or minimizing) a function H of a system of linear relations has a *canonical form* if and only if there is a system of linear inequalities $\{b_i^T x\ R_i\ \nu_i\}_1^n$ where $R_i \in \{>, \geqslant\}$ and a positive function g_2 with no nonincreasing variables such that any vector y which extremizes $H(y)$ can be obtained from some vector x which maximizes $g_2(1\{b_1^T x\ R_1\ \nu_1\}, \ldots, 1\{b_n^T x\ R_n\ \nu_n\})$ and vice-versa. The canonical form is *homogeneous* if and only if all $\nu_i = 0$ and is *inhomogeneous* otherwise.

Note that no b_i in the canonical form for a problem is 0.

In general, two optimization problems are said to be *equivalent* if and only if there are procedures whereby all the solutions of one can be obtained from all the solutions of the other and vice-versa. Consequently, in order to get all solutions to one of an equivalent pair of problems, it suffices to get all solutions to the other. By the Schroeder-Bernstein Theorem, in order to show that two problems are equivalent, it is sufficient to produce two one-to-one functions, one mapping the first solution set into the second solution set and the other mapping the second into the first. But the existence of such bijections is not necessary for two problems to be equivalent as will be seen.

-- the existence of a canonical form --

(4.1.8) Theorem: Every problem of extremizing a function H of a system of linear relations has a canonical form.

Proof: By definition, there exists a function g and a system of linear relations $\{a_i^T x \; R_i \; \mu_i\}_1^m$ such that for all $x \in \mathrm{R}^d$,

$$H(x) = g(1\{a_1^T x \; R_1 \; \mu_1\}, \ldots, 1\{a_m^T x \; R_m \; \mu_m\}).$$

The following procedure can be used to construct g_2 and its associated system of linear inequalities.

First, suppose there is some R_j which is "=". $1\{a_j^T x = \mu_j\}$ is replaced in the functional expression for $H(x)$ with its equivalent expression $1-1\{a_j^T x < \mu_j\} - 1\{a_j^T x > \mu_j\}$. This has the effect of replacing $a_j^T x = \mu_j$ in the system with $a_j^T x < \mu_j$ and $a_j^T x > \mu_j$ and redefining g appropriately so that $H(x)$ is now $g(1\{a_1^T x \; R_1 \; \mu_1\}, \ldots, 1\{a_{j-1}^T x \; R_{j-1} \; \mu_{j-1}\}, 1\{a_j^T x < \mu_j\}, 1\{a_j^T x > \mu_j\}, 1\{a_{j+1}^T x \; R_{j+1} \; \mu_{j+1}\}, \ldots, 1\{a_m^T x \; R_m \; \mu_m\})$. This procedure is followed for each R_j which is "=" and the analogous procedure is also followed for each R_j which is "$\neq$". At the end, g has been redefined sufficiently that H is now a function of a system of linear inequalities (i.e., the relations $=$ and $\neq$ do not appear.)

The standard trick of maximizing $-g$ in order to minimize g advocates multiplying the g obtained from the preceding step by -1 if necessary in order to convert the problem of extremizing H to an unequivocal maximization problem.

Next, any variable of g which is constant is deleted along with its corresponding linear relation in the system. Consequently, for new m, a_i, R_i, μ_i, and g, these modifications result in the construction of

$$H(x) = g(1\{a_1^T x \; R_1 \; \mu_1\}, \ldots, 1\{a_m^T x \; R_m \; \mu_m\}).$$

Suppose that the jth variable of g is nonincreasing. Define the function $e : \bigtimes_1^m \{0, 1\} \to \bigtimes_1^m \{0, 1\}$ by $e(t) := (\tau_1, \ldots, \tau_{j-1}, 1-\tau_j, \tau_{j+1}, \ldots, \tau_m)$. Define the relation S_j to be such that $\alpha\ R_j\ \beta$ is true if and only if $\alpha\ S_j\ \beta$ is false. Then the jth variable of $g \circ e$ is not nonincreasing and is nonconstant and nondecreasing. In addition, now

$$H(x) = (g \circ e)(1\{a_1^T x\ R_1\ \mu_1\}, \ldots, 1\{a_{j-1}^T x\ \ R_{j-1}\ \mu_{j-1}\},\ 1\{a_j^T x\ S_j\ \mu_j\},\\ 1\{a_{j+1}^T x\ \ R_{j+1}\ \mu_{j+1}\}, \ldots, 1\{a_m^T x\ R_m\ \mu_m\}).$$

In short, H has been expressed in terms of a new system of linear inequalities and a new g function whose j^{th} variable is not nonincreasing. Repeat this same procedure in order to eliminate all nonincreasing variables in favor of nondecreasing variables.

The next step is to change the current system of linear inequalities for defining H by multiplying any with the relations $<$ and $\leqslant$ by -1.

Since g can take on at most 2^m values, there is a positive constant function α such that $g + \alpha > 0$. The final step is to set $g_2 = g + \alpha$ and to let the inequalities in b_i for (4.1.7) be the current set of inequalities in a_i. □

-- *more general functions H* --

It is easy to see in the light of the proof of (4.1.8) that a problem in canonical form can also be derived for an H function which had constraints of the form $a_i^T x \in I(\alpha, \beta)$ or $a_i^T x \notin I(\alpha, \beta)$ where $I(\alpha, \beta)$ is one of (α, β), $[\alpha, \beta)$, $(\alpha, \beta]$, or $[\alpha, \beta]$ for $\alpha, \beta \in \mathrm{R}$. For example, $1\{a_i^T x \in (\alpha, \beta]\}$ could be replaced with $1\{a_i^T x > \alpha\} \cdot 1\{a_i^T x \leqslant \beta\}$. In fact, it is possible to work as well with finite unions of intervals since $1\{a_i^T x \in I(\alpha_1, \beta_1) \cup I(\alpha_2, \beta_2)\} = \max(1\{a_i^T x \in I(\alpha_1, \beta_1)\}, 1\{a_i^T x \in I(\alpha_2, \beta_2)\})$.

-- *homogeneous canonical form* --

It will now be shown that every problem of extremizing a function of a system of linear relations has a homogeneous canonical form. As will be seen, a useful corollary of this is that in order to maximize $\sum_J \sigma_i 1\{a_i^T x > \mu_i\} + \sum_{I \text{ w.o. } J} \sigma_i 1\{a_i^T x \geqslant \mu_i\}$ for arbitrary real σ_i, it is sufficient to be able to solve the Weighted Mixed Hemisphere problem for all positive σ_i. The reduction to homogeneous canonical form follows easily after it is shown for the following two problems that every instance of Problem I is equivalent to an instance of Problem II.

Problem I: Let $\{a_i\}_1^m \subset \mathsf{R}^d$ where $d \geqslant 2$. Define $f : \mathsf{R}^d \to \bigtimes_1^m \{0, 1\}$ via $f(x) := (1\{a_1^T x \ R_1 \ 0\}, \ldots, 1\{a_m^T x \ R_m \ 0\})$ where $R_i \in \{>, \geqslant\}$. Let $g : \bigtimes_1^m \{0, 1\} \to (0, \infty)$ be a positive function with no nonincreasing variables. Let $H_1 := g \circ f$. Let $e \in \mathsf{R}^d$ be nonzero. Find all those vectors $x \in \mathsf{R}^d$ such that $e^T x = 1$ which maximize H_1.

Problem II: Let $\{a_i\}_1^m \subset \mathsf{R}^d$ where $d \geqslant 2$. Define $f : \mathsf{R}^d \to \bigtimes_1^m \{0, 1\}$ via $f(x) := (1\{a_1^T x \ R_1 \ 0\}, \ldots, 1\{a_m^T x \ R_m \ 0\})$ where $R_i \in \{>, \geqslant\}$. Let $g : \bigtimes_1^m \{0, 1\} \to (0, \infty)$ be a positive function with no nonincreasing variables. Let θ be the largest of g's at most 2^m values. Let $e \in \mathsf{R}^d$ be nonzero. Define H_2 via $H_2(x) := g \circ f(x) + (\theta + 1)1\{e^T x > 0\}$. Find all those vectors $x \in \mathsf{R}^d$ which maximize H_2.

(4.1.9) Theorem: Every instance of Problem I is equivalent to an instance of Problem II. More specifically, given an instance of Problem I, define a corresponding instance of Problem II by using the same f and g. Then:

(a) Suppose x_0 is such that $e^T x_0 = 1$ and $g \circ f(x_0) \geqslant g \circ f(x)$ for all x such that $e^T x = 1$. Then for all $\alpha > 0$, $H_2(\alpha x_0) \geqslant H_2(x)$ for all $x \in \mathsf{R}^d$.

(b) Suppose x_0 is such that $H_2(x_0) \geqslant H_2(x)$ for all $x \in \mathsf{R}^d$. Then $g \circ f(x_0/e^T x_0) \geqslant g \circ f(x)$ for all x such that $e^T x = 1$.

Proof: (a): Fix $\alpha > 0$. Since for all $\beta > 0$ and all $x \in \mathsf{R}^d$, $f(\beta x) = f(x)$, it is clear that for all x such that $e^T x > 0$, $g \circ f(\alpha x_0) \geqslant g \circ f(x)$. Now take any $x \in \mathsf{R}^d$ and consider $H_2(x) = g \circ f(x) + (\theta + 1)\, 1\{e^T x > 0\}$. If $e^T x > 0$, then $H_2(\alpha x_0) - H_2(x) = g \circ f(x_0) - g \circ f(x) \geqslant 0$. If $e^T x \leqslant 0$, then $H_2(x) < \theta{+}1 \leqslant H_2(\alpha x_0)$.

(b): Since $H_2(x_0) \geqslant \theta + 1$, it must be that $e^T x_0 > 0$. Take x such that $e^T x = 1$. Then $0 \leqslant H_2(x_0) - H_2(x) = g \circ f(x_0/e^T x_0) - g \circ f(x)$.

To show that the two instances are equivalent, begin by letting x_1 be a solution to the instance of Problem II and suppose that it is not in $\langle x_0 \rangle$ for any solution x_0 of the instance of Problem I. But $\dfrac{x_1}{e^T x_1}$ is a solution of the instance of Problem I by (b) and so by (a), $x_1 \in \langle \dfrac{x_1}{e^T x_1} \rangle$ is a solution of the instance of Problem II, yielding a contradiction. The other direction follows in a similar way. □

(4.1.10) Theorem: Every problem of extremizing a function H of a system of linear relations has a homogeneous canonical form.

Proof: Let the given function H of a system of linear relations have the inhomogeneous canonical form defined by $g \circ f(x)$ where $g : \bigtimes_1^m \{0, 1\} \to (0, \infty)$ is a positive function with no nonincreasing variables and for each $x \in \mathsf{R}^d$, $f(x) := (1\{a_1^T x \; R_1 \; \mu_1\}, \ldots, 1\{a_m^T x \; R_m \; \mu_m\})$ where $R_i \in \{ > , \geqslant \}$.

Define $f_2 \colon \mathsf{R}^{d+1} \to \bigtimes_1^m \{0, 1\}$ via

$$f_2(z) := (1\{(a_1, -\mu_1)^T z \ R_1 \ 0\}, \ldots, 1\{(a_m, -\mu_m)^T z \ R_m \ 0\}).$$

Let $e_{d+1} := (0, \ldots, 0, 1) \in \mathrm{R}^{d+1}$. Let θ be the largest of g's at most 2^m values.

Clearly the problem of extremizing $H(x)$ is equivalent to the problem of maximizing $g \circ f_2(z)$ over all $z \in \mathrm{R}^{d+1}$ such that $e_{d+1}^T z = 1$. This latter problem is an instance of Problem I which is equivalent to an instance of Problem II by (4.1.9). □

-- *the WOH tree algorithm and homogeneous canonical form* --

This chapter will consider problems of extremizing a function H of a system of linear relations subject to the solutions being in a specified linear manifold or polyhedral set or being required to extremize another function of a system of linear relations or any number of the above. But before showing that these problems also have homogeneous canonical forms, two theorems will be proven which delineate the nature of the problems in homogeneous canonical form that the tree algorithm of Chapter 3 can solve.

The first step is to define a problem.

Problem III: Let $\{a_i\}_1^m \subset \mathrm{R}^d$. Suppose $\mathrm{L}\{a_i\}_1^m = \mathrm{R}^d$. Let the function $f : \mathrm{R}^d \to \times_1^m \{0, 1\}$ map x to $(1\{a_1^T x > 0\}, \ldots, 1\{a_m^T x > 0\})$. Let $g : \times_1^m \{0, 1\} \to \mathrm{R}$ be nondecreasing. Define $H := g \circ f$. Find all x which maximize H.

Note that the assumption $\mathrm{L}\{a_i\}_1^m = \mathrm{R}^d$ is no real restriction since if it doesn't hold, an equivalent lower dimensional problem can be obtained from the following procedure.

(4.1.11) Procedure: Suppose $1 \leqslant \dim \mathrm{L}\{a_i\}_1^m = k < d$. Find an orthonormal basis for R^d such that the first k components span $\mathrm{L}\{a_i\}_1^m$.

Compute the orthogonal change of basis matrix $\underset{\sim}{B} = \begin{bmatrix} \underset{\sim}{B}_1 \\ \underset{\sim}{B}_2 \end{bmatrix}$ from the standard basis to the newly selected orthonormal basis where the rows of the $k \times d$ matrix $\underset{\sim}{B}_1$ span $\mathrm{L}\{a_i\}_1^m$. Then note that for all i,

$$\begin{aligned} a_i^T x &= (\underset{\sim}{B} a_i)^T \underset{\sim}{B} x \\ &= \begin{bmatrix} \underset{\sim}{B}_1 a_i \\ 0 \end{bmatrix}^T \begin{bmatrix} \underset{\sim}{B}_1 x \\ \underset{\sim}{B}_2 x \end{bmatrix} \\ &= b_i^T (\underset{\sim}{B}_1 x) \end{aligned}$$

where $b_i := \underset{\sim}{B}_1 a_i$. For any $R_i \in \{>, \geqslant\}$, the problem of maximizing $g(1\{b_1^T y \; R_1 \; 0\}, \ldots, 1\{b_m^T y \; R_m \; 0\})$ over $y \in \mathrm{R}^k$ is a problem which is equivalent to the problem of maximizing over $x \in \mathrm{R}^d$, $g(1\{a_1^T x \; R_1 \; 0\}, \ldots, 1\{a_m^T x \; R_m \; 0\})$ in the sense that the set of solutions to one can be made to generate the set of solutions to the other. To be specific, if x_0 is a solution to the latter problem, then $y_0 = \underset{\sim}{B}_1 x_0$ solves the former. If y_0 is a solution to the former and z_0 is an arbitrary vector in R^{d-k}, then $\underset{\sim}{B}^T (y_0, z_0)$ solves the latter. All solutions to one problem can be generated using these procedures from all solutions to the other. The proofs of these statements are omitted. The reader may wish to compare this procedure with the comments on the situation $\mathrm{L}\{y_i, i \in I\} \neq X$ following Problem (3.1.1).

Theorem (4.1.13) will show that the following Procedure (4.1.12) solves Problem III.

(4.1.12) Procedure: In order to identify all solution vectors to Problem III, begin by using the tree algorithm of Chapter 3 to identify all hills whose interior vectors maximize H. If g is strictly increasing then this subset of hills contains all solution vectors.

In general though, when g is assumed only to be nondecreasing, it is necessary to do the following in order to find all solution vectors:

(i) for each of the maximizing hills, determine the corresponding boundary hyperspaces by determining the frames of their dual cones.

(ii) cross over into all neighboring cones whose interiors also achieve the maximum value of H. (Call any cone whose interior achieves the maximum value of H a *max cone.*)

(iii) Iterate this process for each of the newly found max cones. This will generate a finite number of finite sequences of max cones if one is careful to never cross the same boundary plane twice in any given sequence.

Carrying this process to completion will identify a set of cones which jointly contain all maximizing vectors.

The validity of Procedure (4.1.12) follows directly from the next theorem which relies heavily on notation from Chapter 3.

(4.1.13) Theorem: Consider the following problem:

Problem: Let $\{a_i\}_1^m \subset X$ where $\dim X \geqslant 1$. Let $I = \{0, 1, \ldots, m\}$ and $a_0 = 0$. Suppose $\mathsf{L}\{a_i, i \in I\} = X$. Let the function $f : \tilde{X} \to \mathsf{X}_1^m \{0, 1\}$ map $\tilde{x}$ to $(1\{[a_1, \tilde{x}] > 0\}, \ldots, 1\{[a_m, \tilde{x}] > 0\})$. Let the function $g : \mathsf{X}_1^m \{0, 1\} \to \mathsf{R}$ be nondecreasing. Let $H = g \circ f$. Find all $\tilde{x}$ which maximize H.

(Any definition made in Chapter 3 involving the y_i is considered to be made here with the y_i replaced by the a_i. So, for example, $C(\pi) = \mathsf{C}\{\pi_i a_i, i \in I\}$ in this theorem.)

With regard to the above problem, let $\tilde{x}_0$ be such that $H(\tilde{x}_0) \geqslant H(\tilde{x})$ for all $\tilde{x} \in \tilde{X}$. Then:

(a) If $\tilde{x}_0$ is not in the interior of a fully-dimensional cone, then there exists a pointed cone $C(\pi^o)$ such that $\tilde{x}_0 \in C(\pi^o)^+$ and such that $H(\text{int } C(\pi^o)^+) = H(\tilde{x}_0)$.

(b) If $C(\pi^o)^+$ is not a hill, then there exists a finite sequence of pointed cones $C(\pi^j)$ for $j = 1, \ldots, k$ such that:

(i) for each j, $H(\text{int } C(\pi^j)^+) = H(\tilde{x}_0)$.

(ii) $C(\pi^k)^+$ is a hill

(iii) for each j, $(\pi_1^{j-1}, \ldots, \pi_m^{j-1}) \leqslant (\pi_1^j, \ldots, \pi_m^j)$.

(c) If g is strictly increasing, then $\tilde{x}_0$ is in the interior of a hill.

Proof: (a) First, it is safe to assume that H is not constant and so, since H has at least two values, $H(0) < H(\tilde{x}_0)$ and $\tilde{x}_0 \neq 0$. Consequently, there is a nonempty set J_1 and corresponding π_i such that $[\pi_i a_i, \tilde{x}_0] > 0$ for $i \in J_1$. Suppose $J_2 := \{i \in I : [a_i, \tilde{x}_0] = 0\} \neq I_0$. By (2.3.35), there exists $\tilde{x}_2 \in \tilde{X}$ and for $i \in J_2$ w.o. I_0, $\theta_i \in \{-1, 1\}$, not all -1, such that $[\theta_i a_i, \tilde{x}_2] > 0$ for $i \in J_2$ w.o. I_0. Observe that if $\alpha > 0$ is chosen so that $[\pi_i a_i, \tilde{x}_0 + \alpha\tilde{x}_2] > 0$ for $i \in J_1$ and $[\theta_i a_i, \tilde{x}_0 + \alpha\tilde{x}_2] > 0$ for $i \in J_2$ w.o. I_0, then $f(\tilde{x}_0) \leqslant f(\tilde{x}_0 + \alpha\tilde{x}_2)$ and so $H(\tilde{x}_0) \leqslant H(\tilde{x}_0 + \alpha\tilde{x}_2) \leqslant H(\tilde{x}_0)$. Let $\pi_i^o = \pi_i$ for $i \in J_1$, $\pi_i^o = \theta_i$ for $i \in J_2$ w.o. I_0, and $\pi_i^o = 1$ for $i \in I_0$.

(b) Suppose $C(\pi^o)^+$ is not a hill. Let $\{\langle \pi_i^o a_i \rangle, i \in I^*\}$ be a frame for $C(\pi^o)$. There exists $j \in I^*$ such that $\langle -a_j \rangle \neq \langle 0 \rangle$ is an isolated ray of $C(\pi^o)$ and for all $i \in I$, $\langle -a_j \rangle \neq \langle a_i \rangle$. By (2.4.11), there exists $\tilde{x}_2$ such that $[\pi_i^o a_i, \tilde{x}_2] > 0$ for $i \in I$ w.o. $(I_0 \cup I_j(\pi^o))$ and $[a_i, \tilde{x}_2] > 0$ for $i \in I_j(\pi^o)$ w.o. I_0. Let $\pi_i^1 = \pi_i^o$ for $i \in I$ w.o. $(I_0 \cup I_j(\pi^o))$ and $\pi_i^1 = 1$ otherwise. Then $\pi^o \leqslant \pi^1$, $C(\pi^1)$ is pointed, $\tilde{x}_2 \in \text{int } C(\pi^1)^+$, $f(\tilde{x}_0) \leqslant f(\tilde{x}_2)$, and $H(\tilde{x}_0) \leqslant H(\tilde{x}_2) \leqslant H(\tilde{x}_0)$.

Now, if $C(\pi^1)^+$ is not a hill, then repeat this process and cross over into $C(\pi^2)^+$, a suitable neighboring cone of $C(\pi^1)^+$. Continue on in this fashion until a hill is reached. This must happen after a finite number of steps because there are only a finite number of cones and $\pi^{j-1} \leqslant \pi^j$ for each j in the sequence. □

-- *finding just the maximizing hills with the WOH algorithm* --

Naturally, when using Procedure (4.1.12), it would be nice to avoid enumerating all of the hills. To just find all of the maximizing hills, it is sufficient to displace only those boundary vectors with sufficiently high H values (cf., (3.4.2)) if it can only be shown that this process is recursively valid. Just as was done in Remark (3.5.9) with Theorem (3.5.10), it is necessary to show for a suitably defined lower dimensional problem associated with each nonzero boundary face vector of a maximizing hill that there is a lower dimensional maximizing hill generated in the expected way from the original maximizing hill.

In order to conveniently write this and the following argument in symbols, a slightly different notation must be introduced for indicating the structure of H. H is said to be a function of the system of linear inequalities $\{[a_i, \tilde{x}] > 0, i \in I\}$ if and only if for all $\tilde{x} \in \tilde{X}$,

$$H(\tilde{x}) = g(\{(i, 1\{[a_i, \tilde{x}] > 0\}) : i \in I\})$$

where g is a real-valued function of finite sequences of 0's and 1's with each element of g's domain being of the form $\{(i, \tau_i) : i \in I\}$ for $\tau_i \in \{0, 1\}$. (It is assumed here of course that the conditions of Problem III hold and so that g is nondecreasing.)

(4.1.14) Theorem: In using the tree algorithm of Chapter 3 to identify all maximizing hills in Problem III, it is sufficient to displace only those boundary vectors with sufficiently high H values where the H used in lower dimensional problems is the $H_{\pi,J}$ of the next paragraph. This will follow from the fact that boundary face vectors of maximizing hills generate maximizing hills in suitably defined lower dimensional problems.

In symbols, suppose $C(\pi)^+$ is a maximizing hill with respect to H. Suppose there exists $\tilde{v} \in F_J(C(\pi)^+)$ where $J \neq I_0$. Let $R := \mathsf{L}\{a_i, i \in J\}$ and S be such that $R \oplus S = X$. Then $\{\tilde{r} \in \tilde{R} : [\pi_i a_i, \tilde{r}] \geqslant 0, i \in J\}$ is a maximizing hill with respect to $H_{\pi,J}$ where $H_{\pi,J}(\tilde{r}) :=$

$$g(\{(i, 1\{[a_i, \tilde{v}] > 0\}) : i \in I \text{ w.o. } J\} \cup \{(i, 1\{[a_i, \tilde{r}] > 0\}) : i \in J\}).$$

Proof: First recall that $\{\tilde{r} \in \tilde{R} : [\pi_i a_i, \tilde{r}] \geqslant 0, i \in J\}$ is a hill by (3.5.5). So, suppose that $\tilde{r}_1$ is such that $[\pi_i a_i, \tilde{r}_1] > 0$ for $i \in J$ w.o. I_0 and that there exists $\tilde{r}_2 \in \tilde{R}$ such that $H_{\pi,J}(\tilde{r}_2) > H_{\pi,J}(\tilde{r}_1)$. Consequently, there exist $\beta_1, \beta_2 > 0$ such that for $k = 1, 2$,

$$\text{sgn}[a_i, \tilde{v}] = \text{sgn}\,[a_i, \tilde{v} + \beta_k \psi^{-1}(\tilde{r}_k)]$$

for $i \in I$ w.o. J and

$$H(\tilde{v} + \beta_2 \psi^{-1}(\tilde{r}_2)) > H(\tilde{v} + \beta_1 \psi^{-1}(\tilde{r}_1)),$$

which is a contradiction (cf., (3.5.10)). □

To summarize the results for Problem III, the tree algorithm can solve any problem in homogeneous canonical form as long as all of the variables of g are nondecreasing and all of the inequalities in the system are ">". Since, in transforming a problem to canonical form, all nonincreasing variables are removed from the g function, it is clear that it is precisely variables in the homogeneous canonical form which are neither nondecreasing nor nonincreasing which the tree algorithm is apparently incapable of handling. This is probably not a serious handicap since the author hasn't yet come across a situation in practice where a variable in the appropriate g function is neither nonincreasing nor nondecreasing although, of course, life being as rich as it is, there must be at least one.

-- *the WOH algorithm, both > and ⩾, and pointed position* --

The examples of (4.1.4) show how mixing ">" and "⩾" in the system of linear inequalities can lead to situations where the maximizing vectors are nowhere near the maximizing vectors for the corresponding problem obtained when all "⩾" are replaced by ">". There are some situations though when solving the latter problem (using the tree algorithm of Chapter 3) enables one to solve the former.

For notational convenience, the following problem is written in terms of the arbitrary vector space X. It can be reduced to the case $X = \mathsf{R}^d$ in the obvious way.

Problem IV: Let $\{a_i\}_1^m \subset X$. Suppose $\mathsf{L}\{a_i\}_1^m = X$ and all $a_i \neq 0$. Let the function $f_1 : \tilde{X} \to \bigtimes_1^m \{0, 1\}$ map $\tilde{x}$ to

$$(1\{[a_1, \tilde{x}]\ R_1\ 0\}, \ldots, 1\{[a_m, \tilde{x}]\ R_m\ 0\})$$

where for each i, $R_i \in \{ \geqslant , > \}$. Let $g : \bigtimes_1^m \{0, 1\} \to \mathsf{R}$ be nondecreasing. Define $H := g \circ f$. Find all $\tilde{x} \neq 0$ which maximize H.

(4.1.15) Theorem: In the context of Problem IV, define $f_2 : \tilde{X} \to \bigtimes_1^m \{0, 1\}$ via $f_2(\tilde{x}) := (1\{[a_1, \tilde{x}] > 0\}, \ldots, 1\{[a_m, \tilde{x}] > 0\})$. Suppose $\{a_i\}_1^m$ is in pointed position (cf., (2.3.34)). Then:

(a) Every solution to Problem IV is in a face of a cone whose interior maximizes $g \circ f_2$.

(b) $\sup_{\tilde{x} \neq 0} g \circ f_1(\tilde{x}) = \sup_{\tilde{x} \neq 0} g \circ f_2(\tilde{x})$.

(c) Problem IV can be solved by using the tree algorithm of Chapter 3 to produce the cones whose interiors maximize $g \circ f_2$ (cf., Problem III) and then if desired, enumerate the faces of these cones.

Proof: Let $\tilde{x}_0 \neq 0$ be such that $g \circ f_1(\tilde{x}_0) = \sup_{\tilde{x} \neq 0} g \circ f_1(\tilde{x})$. Let $I = \{0, \ldots, m\}$, $a_0 = 0$, and $J = \{i \in I : [a_i, \tilde{x}_0] = 0\} \neq \varnothing$. Then $\tilde{x}_0 \in F_J(C(\pi)^+)$ for some $C(\pi)^+$. Since $\tilde{x}_0 \neq 0$, $C\{a_i, i \in J\}$ is pointed. Using Theorem (3.5.1), $\tilde{x}_0$ can be displaced to $\tilde{x}_1 \in \text{int}\, C\{\pi_i a_i, i \in I \text{ w.o. } J, a_i, i \in J\}^+$. Since for all $i \in J$, $1\{[a_i, \tilde{x}_0]\, R_i\, 0\} \leqslant 1\{[a_i, \tilde{x}_1]\, R_i\, 0\}$, $g \circ f_1(\tilde{x}_0) \leqslant g \circ f_1(\tilde{x}_1) \leqslant g \circ f_1(\tilde{x}_0)$. Next, observe that

$$\begin{aligned} g \circ f_2(\tilde{x}_1) &= g \circ f_1(\tilde{x}_1) = g \circ f_1(\tilde{x}_0) \\ &= \sup_{\tilde{x} \neq 0} g \circ f_1(\tilde{x}) \geqslant \sup_{\tilde{x} \neq 0} g \circ f_2(\tilde{x}) \geqslant g \circ f_2(\tilde{x}_1). \quad \square \end{aligned}$$

One of the easiest ways for $\{a_i\}_1^m$ to fail to be in pointed position is for $\langle a_i \rangle = \langle -a_j \rangle \neq \langle 0 \rangle$ for some $i \neq j$. This situation arises when an equation appears in the original system of linear relations: any relation $a_i^T x = \beta_i$ generates the relations $(a_i, -\beta_i)^T z \geqslant 0$ and $(-a_i, \beta_i)^T z \leqslant 0$ when the problem is being transformed to homogeneous canonical form.

The general tree algorithm developed in Chapter 5 will solve Problem IV in its generality and, for that matter, will actually solve any problem in homogeneous canonical form with a nondecreasing g function.

-- *introducing constraints* --

The remainder of this chapter is concerned with showing how various constrained problems of extremizing a function H of a system of linear relations can be reduced to homogeneous canonical form. These constrained problems shall be stated without loss of generality in a manner consistent with inhomogeneous canonical form. Once the homogeneous canonical forms of these problems have been obtained, then the tree algorithm of Chapter 3 can be used to solve them if they are versions of Problem III or if they satisfy the hypotheses of Theorem (4.1.15). The tree algorithm of Chapter 5 can be used to solve these problems if the appropriate g function is nondecreasing.

-- *when solutions are constrained to lie in a linear manifold* --

The first constrained optimization problem to be considered is that of extremizing a function of a system of linear relations subject to requiring the solution vectors to lie in a specified linear manifold.

Problem V: Let $\{a_i\}_1^m$, $\{s_i\}_1^q \subset \mathsf{R}^d$ where $d \geqslant 2$ and $1 \leqslant \dim \mathsf{L}\{s_i\}_1^q \leqslant d-1$. Let $\{\mu_i\}_1^m$, $\{\omega_i\}_1^q \subset \mathsf{R}$. Define $f : \mathsf{R}^d \to \displaystyle\times_1^m \{0, 1\}$ via $f(x) := (1\{a_1^T x \; R_1 \; \mu_1\}, \ldots, 1\{a_m^T x \; R_m \; \mu_m\})$ where $R_i \in \{>, \geqslant\}$. Let $g : \displaystyle\times_1^m \{0, 1\} \to (0, \infty)$ be a positive function with no nonincreasing variables. Let $H = g \circ f$. Find all those nonzero vectors x which maximize H over $\{x \in \mathsf{R}^d : s_1^T x = \omega_1, \ldots, s_q^T x = \omega_q\}$ which is assumed to be nonempty.

Theorem (4.1.17) will show that, for any instance of Problem V, the following Procedure (4.1.16) will produce an equivalent instance of Problem I. Since every instance of Problem I is equivalent to an instance of Problem II, Problem V has a homogeneous canonical form.

(4.1.16) Procedure: A given instance of Problem V is equivalent to the problem of maximizing

$$g(1\{(a_1, -\mu_1)^T z \; R_1 \; 0\}, \ldots, 1\{(a_m, -\mu_m)^T z \; R_m \; 0\})$$

over

$$\{z \in \mathsf{R}^{d+1} : (s_1, -\omega_1)^T z = 0, \ldots, (s_q, -\omega_q)^T z = 0, e_{d+1}^T z = 1\}$$

where $e_{d+1} := (0, \ldots, 0, 1) \in \mathsf{R}^{d+1}$. Obtain an orthonormal basis for $\{(s_1, -\omega_1), \ldots, (s_q, -\omega_q)\}^\perp$ which is the orthogonal complement of $\mathsf{L}\{(s_1, -\omega_1), \ldots, (s_q, -\omega_q)\}$ (cf. Definition (3.4.8)). Extend this basis to an orthonormal basis for R^{d+1}. Let $\underset{\sim}{B} = \begin{bmatrix} \underset{\sim}{B}_1 \\ \underset{\sim}{B}_2 \end{bmatrix}$ be the orthogonal change of basis matrix from the standard basis for R^{d+1} to this newly constructed orthonormal basis where the rows of the $k \times (d+1)$ matrix $\underset{\sim}{B}_1$ form a basis for $\{(s_1, -\omega_1), \ldots, (s_q, -\omega_q)\}^{\perp}$. For $i = 1, \ldots, m$, define $b_i := \underset{\sim}{B}_1(a_i, -\mu_i)$. Define $f_2\colon \mathsf{R}^k \to \overset{m}{\underset{1}{\times}} \{0, 1\}$ via $f_2(y) := (1\{b_1^T y \; R_1 \; 0\}, \ldots, 1\{b_m^T y \; R_m \; 0\})$. Define Problem A to be the problem of maximizing $g \circ f_2(y)$ over $y \in \mathsf{R}^k$ such that $(\underset{\sim}{B}_1 e_{d+1})^T y = 1$.

Then, if x_0 solves a given instance of Problem V, $\underset{\sim}{B}_1(x_0, 1)$ solves the associated Problem A and if y_0 solves Problem A, then $\underset{\sim}{B}_1^T y_0 =: (x_0, 1)$ and x_0 solves the given instance of Problem V.

(4.1.17) Theorem: Problem V has a homogeneous canonical form. To be specific, Procedure (4.1.16) produces an instance of Problem I which is equivalent to the given instance of Problem V.

Proof: The proof rests on the central fact that for all $z \in \{(s_1, -\omega_1), \ldots, (s_q, -\omega_q)\}^{\perp}$,

$$
\begin{aligned}
(a_i, -\mu_i)^T z &= (a_i, -\mu_i)^T \underset{\sim}{B}^T \underset{\sim}{B} z \\
&= \begin{bmatrix} \underset{\sim}{B}_1 (a_i, -\mu_i) \\ \underset{\sim}{B}_2 (a_i, -\mu_i) \end{bmatrix}^T \begin{bmatrix} \underset{\sim}{B}_1 z \\ 0 \end{bmatrix} \\
&= (\underset{\sim}{B}_1 (a_i, -\mu_i))^T \underset{\sim}{B}_1 z
\end{aligned}
$$

Suppose x_0 is a solution to a given instance of Problem V so that $(s_i, -\omega_i)^T (x_0, 1) = 0$ for $i = 1, \ldots, q$ and, for all $z \in \mathsf{R}^{d+1}$ such that $(s_i, -\omega_i)^T z = 0$ for $i = 1, \ldots, q$ and $e_{d+1}^T z = 1$,

$$
\begin{aligned}
&g(1\{(a_1, -\mu_1)^T (x_0, 1) \; R_1 \; 0\}, \ldots, 1\{(a_m, -\mu_m)^T (x_0, 1) \; R_m \; 0\}) \\
&\qquad \geqslant g(1\{(a_1, -\mu_1)^T z \; R_1 \; 0\}, \ldots, 1\{(a_m, -\mu_m)^T z \; R_m \; 0\}).
\end{aligned}
$$

Then for all $z \in \mathsf{R}^{d+1}$ such that $(s_i, -\omega_i)^T z = 0$ for $i = 1, \ldots, q$ and

$(B_1 e_{d+1})^T(B_1 z) = 1$, $g \circ f_2(B_1(x_0, 1)) \geqslant g \circ f_2(B_1 z)$. It is easy to see that $\{B_1 z : z \in \mathrm{R}^{d+1}$ and $(s_i, -\omega_i)^T z = 0$ for $i = 1, \ldots, q\} = \mathrm{R}^k$. Hence for all $y \in \mathrm{R}^k$ such that $(B_1 e_{d+1})^T y = 1$, $g \circ f_2(B_1(x_0, 1)) \geqslant g \circ f_2(y)$.

To see that the map which maps each solution x_0 of the given instance of Problem V to the solution $B_1(x_0, 1)$ of the associated problem A is one-to-one, first observe that for all $z \in \mathrm{R}^{d+1}$, $z = B^T B z = B_1^T B_1 z + B_2^T B_2 z$. Next take two solutions x_1 and x_2 of the given instance of Problem V. It is known that $(x_1, 1), (x_2, 1) \in \{(s_1, -\omega_1), \ldots, (s_q, -\omega_q)\}^{\perp}$;
hence, $B_2(x_1, 1) = B_2(x_2, 1) = 0$. So,

$$
\begin{aligned}
& B_1(x_1, 1) = B_1(x_2, 1) \\
\Rightarrow \quad & B_1^T B_1(x_1, 1) = B_1^T B_1(x_2, 1) \\
\Rightarrow \quad & B^T B(x_1, 1) = B^T B(x_2, 1) \\
\Rightarrow \quad & x_1 = x_2
\end{aligned}
$$

For the other direction, it must first be shown that Problem A has a solution. This will be the case if the constraint set is nonempty or, in other words, if $B_1 e_{d+1} \neq 0$. Observe that $B_1 e_{d+1} = 0$ if and only if $e_{d+1} \in \mathrm{L}\{(s_i, -\omega_i)\}_1^q$ which is true if and only if there is a nonzero vector a such that

$$
[0, \ldots, 0, 1] = a^T \begin{bmatrix} s_1^T & -\omega_1 \\ & \vdots \\ s_q^T & -\omega_q \end{bmatrix}
$$

This latter condition is true (see Theorem 2.7.2, Nering (1963)) if and only if the system of linear equations $\{s_i^T x = \omega_i\}_1^q$ does not have a solution. Since it is assumed in Problem V that it does have a solution, $B_1 e_{d+1} \neq 0$.

So assume $y_0 \in \mathrm{R}^k$ is such that $(B_1 e_{d+1})^T y_0 = 1$ and $g \circ f_2(y_0) \geqslant g \circ f_2(y)$ for all $y \in \mathrm{R}^k$ such that $(B_1 e_{d+1})^T y = 1$. Since

$b_i^T y = (a_i, -\mu_i)^T (\underset{\sim}{B}_1^T y)$, for all $y \in \mathsf{R}^k$ such that $e_{d+1}^T (\underset{\sim}{B}_1^T y) = 1$,

$$g(1\{(a_1, -\mu_1)^T (\underset{\sim}{B}_1^T y_0)\ R_1\ 0\}, \ldots, 1\{(a_m, -\mu_m)^T (\underset{\sim}{B}_1^T y_0)\ R_m\ 0\}) \geqslant g(1\{(a_1, -\mu_1)^T (\underset{\sim}{B}_1^T y)\ R_1\ 0\}, \ldots, 1\{(a_m, -\mu_m)^T (\underset{\sim}{B}_1^T y)\ R_m\ 0\}).$$

It is easy to see that $\{\underset{\sim}{B}_1^T y : y \in \mathsf{R}^k\} = \{(s_1, -\omega_1), \ldots, (s_q, -\omega_q)\}^\perp$. Consequently, x_0 defined by $(x_0, 1) := \underset{\sim}{B}_1^T y_0$ solves the given instance of Problem V.

To see that the map which maps each solution of Problem A to a solution of the given instance of Problem V is one-to-one, let y_1 and y_2 solve Problem A and suppose $\underset{\sim}{B}_1^T y_1 = \underset{\sim}{B}_1^T y_2$. Then $y_1 = \underset{\sim}{B}_1 \underset{\sim}{B}_1^T y_1 = y_2$.

Problem A is an instance of Problem I since $\underset{\sim}{B}_1 e_{d+1} \neq 0$. □

-- *the most general constrained problem treated here* --

The last problem that is discussed in this chapter is a constrained maximization problem for a function of a system of linear inequalities where the solution vectors are required to lie in specific linear manifolds or specific polyhedral sets or are required to maximize auxiliary problems or any number of the above. Theorem (4.1.19) shows that such problems have homogeneous canonical forms.

Problem VI: Let $\{a_i\}_1^m$, $\{b_i\}_1^n$, $\{c_i\}_1^p$, $\{s_i\}_1^q \subset \mathsf{R}^d$ where $d \geqslant 2$ and $\{\mu_i\}_1^n$, $\{\nu_i\}_1^m$, $\{\tau_i\}_1^p$, $\{\omega_i\}_1^q \subset \mathsf{R}$. For $i = 1, 2, 3$, let $R_{ij} \in \{>, \geqslant\}$ for appropriate j. Define $f_1\colon \mathsf{R}^d \to \bigtimes_1^m \{0, 1\}$ via

$$f_1(x) := (1\{a_1^T x\ R_{11}\ \mu_1\}, \ldots, 1\{a_m^T x\ R_{1m}\ \mu_m\}),$$

$f_2\colon \mathsf{R}^d \to \bigtimes_1^n \{0, 1\}$ via

$$f_2(x) := (1\{b_1^T x\ R_{21}\ \nu_1\}, \ldots, 1\{b_n^T x\ R_{2n}\ \nu_n\}),$$

and $f_3\colon \mathsf{R}^d \to \bigtimes_1^p \{0, 1\}$ via

$$f_3(x) := (1\{c_1^T x \; R_{31} \; \tau_1\}, \ldots, 1\{c_p^T x \; R_{3p} \; \tau_p\}).$$

Define $\underset{\sim}{S} := [s_1 \cdots s_q]$ and $w := (\omega_1, \ldots, \omega_q)$.

Let $g_1 \colon \overset{m}{\underset{1}{\times}} \{0, 1\} \to (0, \infty)$, $g_2 \colon \overset{n}{\underset{1}{\times}} \{0, 1\} \to (0, \infty)$, and

$g_3 \colon \overset{p}{\underset{1}{\times}} \{0, 1\} \to (0, \infty)$ be positive functions with no nonincreasing variables. For $i = 1, 2, 3$, let $H_i = g_i \circ f_i$. Let $\lambda \in \mathsf{R}$.

Find all $x_0 \neq 0$ such that

$$\underset{\sim}{S}^T x_0 = w$$

and

$$H_3(x_0) \geqslant \lambda$$

and

$$H_2(x_0) = \max\{H_2(x) : \underset{\sim}{S}^T x = w, \; H_3(x) \geqslant \lambda\}$$

and

$$\begin{aligned} H_1(x_0) = \max\{H_1(x) : \underset{\sim}{S}^T x = w, \; H_3(x) \geqslant \lambda, \\ \text{and } H_2(x) = \max\{H_2(y) : \underset{\sim}{S}^T y = w, \; H_3(y) \geqslant \lambda\}\} \end{aligned}$$

where reference to any of $\underset{\sim}{S}$, H_2, and H_3 may be omitted.

-- *comments on Problem VI* --

A few comments might be helpful in understanding the nature of Problem VI. As can be seen by setting $\underset{\sim}{S} = 0$ and $w = 0$ or H_2 and H_3 to appropriate constant functions or any number of the above, Problem VI contains as special cases the problems arising when all references to any of $\underset{\sim}{S}$, H_2, or H_3 are dropped.

Note if $\operatorname{rank} \underset{\sim}{S} = d$, then it is necessary to examine at most one vector.

Since H_3 has a finite number of values, the condition $H_3(x) \geqslant \lambda$ could as well be $H_3(x) > \lambda$.

Suppose it is desired to maximize H_1 subject to satisfying at least k of the inequalities $\{c_i^T x \ R_{3i} \ \tau_i\}_1^p$ where $R_{3i} \in \{>, \geqslant\}$. This is a special case of Problem VI with H_2 and $\underset{\sim}{S}$ omitted, $\lambda = k$, and $H_3(x) = \sum_1^p 1\{c_i^T x \ R_{3i} \ \tau_i\}$.

If all references to H_3 and $\underset{\sim}{S}$ are omitted, then the resulting problem is an analogue of the problem of finding a minimum-norm solution to a least-squares problem. To be more specific, x_0 is the minimum-norm least-squares solution to the system $\underset{\sim}{A}x = b$ if and only if for all x, $\|\underset{\sim}{A}x_0 - b\| \leqslant \|\underset{\sim}{A}x - b\|$ and for all y such that $\|\underset{\sim}{A}y - b\| = \|\underset{\sim}{A}x_0 - b\|$, $\|x_0\| \leqslant \|y\|$. When no reference is made to H_3 or $\underset{\sim}{S}$ in Problem VI then the problem is to find all vectors x_0 which maximize H_2 as well as maximizing H_1 among all vectors which maximize H_2.

Suppose it is desired to maximize H_1 subject to the solution vector lying in $\bigcap_1^n \{b_i^T x \ R_{2i} \ \nu_i\}$ where $R_{2i} \in \{>, \geqslant\}$. This can be accomplished by maximizing H_1 subject to satisfying as many of the inequalities $\{b_i^T x \ R_{2i} \ \nu_i\}_1^n$ as possible which is a special case of Problem VI with $H_2(x) = \sum_1^p 1\{b_i^T x \ R_{2i} \ \nu_i\}$ and all references to H_3 and $\underset{\sim}{S}$ omitted.

-- a procedure for solving Problem VI --

When an instance of Problem VI has a solution and $1 \leqslant \text{rank } \underset{\sim}{S} \leqslant d-1$, then Procedure (4.1.18) reduces it to an instance of Problem V.

(4.1.18) Procedure: Given an instance of Problem VI, let θ_1 be the largest of g_1's at most 2^m values and let θ_2 be the largest of g_2's at most 2^n values. If g_2 is not constant, let Δ_2 be the smallest absolute difference between two distinct values of g_2; otherwise, let $\Delta_2 = \theta_2$. Define Problem B which is an instance of Problem V when $\underset{\sim}{S}^T x = w$ has a solution and $1 \leqslant \text{rank } \underset{\sim}{S} \leqslant d-1$.

Problem B: Define H_4 via

$$H_4(x) := H_1(x) + \frac{(\theta_1+1)}{\Delta_2} H_2(x) + (\theta_1+1)\left(\frac{\theta_2}{\Delta_2} + 1\right) 1\{H_3(x) \geqslant \lambda\}.$$

Find all nonzero x such that $\underset{\sim}{S}^T x = w$ which maximize H_4.

Then:

(a) Suppose there is no solution to the instance of Problem VI. If $\underset{\sim}{S}^T x = w$ is inconsistent, then there is no solution to Problem B. If $\underset{\sim}{S}^T x = w$ is consistent and there is no nonzero x such that $H_3(x) \geqslant \lambda$, then $\max_{x \neq 0} H_4(x) < (\theta_1+1)\left(\frac{\theta_2}{\Delta_2} + 1\right)$. Finally, if x_0 is a solution to the given instance of Problem VI, then x_0 is a solution to Problem B.

(b) If Problem B does not have a solution, then $\underset{\sim}{S}^T x = w$ is inconsistent and the given instance of Problem VI does not have a solution. Let x_0 be a solution to Problem B. If $H_4(x_0) < (\theta_1+1)\left(\frac{\theta_2}{\Delta_2} + 1\right)$, then there is no solution to the instance of Problem VI, while, if otherwise, x_0 solves the instance of Problem VI.

(4.1.19) Theorem: When Problem VI has a solution, Problem VI has a homogeneous canonical form. More specifically, when an instance of Problem VI has a solution and $1 \leqslant \text{rank}\, \underset{\sim}{S} \leqslant d-1$, then Procedure (4.1.18) constructs an instance of Problem V which is equivalent to the instance of Problem VI. If an instance of Problem VI has a solution and $\underset{\sim}{S} = 0$ and $w = 0$ then the given instance of Problem VI is equivalent to Problem B as constructed by (4.1.18) and this Problem B in turn is either already in homogeneous canonical form or is equivalent to an instance of Problem I.

Proof: This proof refers to (a) and (b) of Procedure (4.1.18).

(a): Let x_0 be a solution to the given instance of Problem VI. It is necessary to show that for all $x \in \mathrm{R}^d$, $H_4(x_0) \geqslant H_4(x)$. In the case when $H_3(x) \geqslant \lambda$ and $H_2(x) < H_2(x_0)$, then

$$H_4(x_0) - H_4(x) = \frac{(\theta_1+1)}{\Delta_2}(H_2(x_0) - H_2(x)) + H_1(x_0) - H_1(x) > 0.$$

(b): Let x_0 be a solution to Problem B. Suppose $H_4(x_0) \geqslant (\theta_1+1)\left(\frac{\theta_2}{\Delta_2} + 1\right)$. Then $H_3(x_0) \geqslant \lambda$. Take any nonzero x such that $\underset{\sim}{S}^T x = w$ and $H_3(x) \geqslant \lambda$. Then

$$0 \leqslant H_4(x_0) - H_4(x) = \frac{(\theta_1+1)}{\Delta_2}(H_2(x_0) - H_2(x)) + H_1(x_0) - H_1(x).$$

If $H_2(x) > H_2(x_0)$, then $\theta_1 + 1 \leqslant H_1(x_0) - H_1(x)$ which is impossible (remember $g_i > 0$). So $H_2(x) \leqslant H_2(x_0)$ and if $H_2(x) = H_2(x_0)$, then $H_1(x) \leqslant H_1(x_0)$. □

This chapter concludes with noting that if an instance of Problem VI has a solution, then the tree algorithm of Chapter 3 will find all of its solutions if all the g_i are nondecreasing functions and if either R_{ij} = ">" for all i, j or $\{(a_i, -\mu_i), (b_j, -\nu_j), (c_k, -\tau_k) : \text{all } i, j, k\}$ is in pointed position when transformed as in (4.1.16). A sufficient condition for the tree algorithm of Chapter 5 to find all solutions is that all of the g_i must be nondecreasing functions.

Summary for Chapter 4

In this chapter, a general framework is introduced for expressing problems which seek to determine how to satisfy a given system of linear inequalities and equalities in some desired way. The problem is posed of finding all vectors $x \in \mathbf{R}^d$ which extremize (i.e., maximize or minimize) a given function H of a system of linear relations $\{a_i^T x \ R_i \ \mu_i\}_1^m$ where
$R_i \in \{ <, \leqslant, =, \neq, \geqslant, > \}$ and

$$H(x) := g(1\{a_1^T x \ R_1 \ \mu_1\}, \ldots, 1\{a_m^T x \ R_m \ \mu_m\})$$

for some $g : \bigtimes_1^m \{0, 1\} \to \mathbf{R}$. (At the end of this summary, an extension of this problem will be considered where the solution vectors x are required to lie in specified linear manifolds or polyhedral sets or to maximize other functions of systems of linear relations or any number of the above.)

Examples of unconstrained optimization problems include the problem of maximizing over $x \in \mathbf{R}^d$, $\sum_J \sigma_i 1\{a_i^T x > \mu_i\} + \sum_{I \text{ w.o. } J} \sigma_i 1\{a_i^T x \geqslant \mu_i\}$ for finite index sets $J \subset I$. Note in the latter problem that if all $\sigma_i = 1$, then the object is to find those vectors x which satisfy as many of the linear inequalities as possible. The conditions under which tree algorithms can solve these and other problems will be mentioned shortly. Further examples of such problems will be given in Chapter 8.

An introductory set of examples in this chapter shows how the location and nature of the solutions to problems of extremizing functions of systems of linear relations are extremely sensitive to the choice of the function g (e.g., the choice of the weights σ_i in the example above), to linear degeneracies in the a_i, and to such choices among relational conditions as that between "$>$" and "$\geqslant$".

All problems of extremizing functions of systems of linear relations are equivalent to certain problems expressed in a canonical form. To define this, it is necessary to define nondecreasing and nonincreasing variables. The jth variable of $g : \bigtimes_1^m \{0, 1\} \to \mathsf{R}$ is nondecreasing if for all choices of $\sigma_1, \ldots, \sigma_{j-1}, \sigma_{j+1}, \ldots, \sigma_m \in \{0, 1\}$,

$$g(\sigma_1, \ldots, \sigma_{j-1}, 0, \sigma_{j+1}, \ldots, \sigma_m) \leqslant g(\sigma_1, \ldots, \sigma_{j-1}, 1, \sigma_{j+1}, \ldots, \sigma_m).$$

The jth variable of g is nonincreasing if the jth variable of $-g$ is nondecreasing. g is a nondecreasing function if all of its variables are nondecreasing.

The problem of extremizing a specific function H of a system of linear relations has a canonical form if there is a system of linear inequalities $\{b_i^T x \; R_i \; \nu_i\}_1^n$ where $R_i \in \{ >, \geqslant \}$ and a positive function g_2 with no nonincreasing variables such that any vector y which extremizes $H(y)$ can be obtained from some vector x which maximizes

$$g_2(1\{b_1^T x \; R_1 \; \nu_1\}, \ldots, 1\{b_n^T x \; R_n \; \nu_n\})$$

and vice versa. The first theorem of this chapter shows that every problem of extremizing a function of a system of linear relations has a canonical form and the second theorem shows that this form can be taken to be homogeneous, i.e., all $\nu_i = 0$.

Once a problem has been reduced to homogeneous canonical form, the WOH tree algorithm of Chapter 3 can solve it if all the variables of the appropriate g_2 function are nondecreasing and either all of the inequalities are ">" or $\{b_i\}_1^n$ is in pointed position. $\{b_i\}_1^n$ is in pointed position if for any $J \subset \{1, \ldots, n\}$ such that $\{b_i, i \in J\}^\perp \neq \langle 0 \rangle$, $\mathsf{C}\{b_i, i \in J\}$ is pointed. Recall that every set in general position is in pointed position. It should be noted that the nondecreasing requirement does not appear to be a restriction in practice. On the other hand, the general tree algorithm developed in Chapter 5 solves all such problems in homogeneous canonical form if the associated g_2 function is nondecreasing.

The last two theorems in this chapter show that the following problem of maximizing a function of a system of linear relations subject to any of a variety of constraints can be reduced to homogeneous canonical form. If it then satisfies the appropriate conditions, then the tree algorithm can solve it.

Problem: Let H_1, H_2, H_3 be arbitrary functions of systems of linear relations in $x \in \mathsf{R}^d$. Let $M = \{x \in \mathsf{R}^d : \underset{\sim}{S}^T x = w\}$ be a nonempty linear manifold. Let $\lambda \in \mathsf{R}$.

Find all $x_0 \neq 0$ such that

$$x_0 \in M$$

and

$$H_3(x_0) \geqslant \lambda$$

and

$$H_2(x_0) = \max\{H_2(x) : x \in M, H_3(x) \geqslant \lambda\}$$

and

$$H_1(x_0) = \max\{H_1(x) : x \in M, H_3(x) \geqslant \lambda, \text{ and } H_2(x) = \max\{H_2(y) : y \in M, H_3(y) \geqslant \lambda\}\}$$

where reference to any of M, H_2, and H_3 may be omitted.

Chapter 5:

Tree Algorithms For Extremizing Functions Of Systems Of Linear Relations Subject To Constraints

This chapter shows how the tree algorithm described in Chapter 3 can be extended so as to maximize any constrained or unconstrained function of a system of linear relations so as long as the g function associated with its homogeneous canonical form is nondecreasing. Since it appears that virtually all (the author has seen no exceptions) practical problems of this sort are associated with nondecreasing g functions, the extended tree algorithm is seen to be quite general for solving applied problems of this kind.

From a geometric standpoint, this general tree algorithm is distinguished from the tree algorithm of Chapter 3 by its ability to find and identify lower dimensional equivalence classes of solutions. For example, the general tree algorithm will identify the positive quadrant of the ξ_1 - ξ_2 plane as the solution set for the problem of finding all (ξ_1, ξ_2, ξ_3) which maximize $1\{\xi_3 = 0\} + 1\{\xi_1 > 0\} + 1\{\xi_2 > 0\}$. The WOH tree algorithm will not solve this problem. As another example, the general tree algorithm is capable of identifying all of those vectors which satisfy as many of a system of linear equations as possible whereas the WOH tree algorithm cannot.

By way of review, H is a function of the system of linear relations $\{a_i^T x \;\; R_i \;\; \mu_i\}_1^m$ where $R_i \in \{<, \leqslant, =, \neq, \geqslant, >\}$ and where x is a vector in R^d if and only if there is a function $g: \times_1^m \{0,1\} \rightarrow \mathsf{R}$ such that for all $x \in \mathsf{R}^d$, $H(x) = g(1\{a_1^T x \;\; R_1 \;\; \mu_1\}, \ldots, 1\{a_m^T x \;\; R_m \;\; \mu_m\})$. The problem is to maximize (or minimize) H over $x \in \mathsf{R}^d$ subject to

(i) requiring the maximizing vectors to lie in a designated linear manifold or polyhedral set or both or

(ii) maximizing another function H_2 of a system of linear relations or

(iii) maintaining the value of yet another function H_3 of a system of linear relations greater than some preset constant or

(iv) any or none of the above.

It was shown in Chapter 4 that any constrained or unconstrained problem of this sort is equivalent to an unconstrained problem in homogeneous canonical form which occurs when (a) all $\mu_i = 0$, (b) all $R_i \in \{>, \geqslant\}$, and (c) g is a positive function with no nonincreasing variables. It will be shown in this chapter that if g is in addition a nondecreasing function, then the general version of the tree algorithm will perform the required optimization.

The first section of this chapter discovers the geometry of the set of solution vectors to problems in homogeneous canonical form with nondecreasing g functions and as a consequence, discovers the appropriate analogs of max-sum cones and hills. It also includes a programming language type summary description of what is basically the complete general tree algorithm and then leaves to subsequent sections the development of the individual pieces of this algorithm. The second section develops the relative boundary vector collection part of the general tree algorithm. The third section shows that all improvements included in Section 3.4 for the WOH tree algorithm carry over to the general tree algorithm. The fourth section concludes this chapter with a discussion of the displacement phase of the general tree algorithm.

Section 5.1:

The Geometry Of The Solution Space

In order to show that a tree algorithm exists for solving any problem in homogeneous canonical form with nondecreasing g function, this chapter describes a tree algorithm which solves the following problem:

(5.1.1) Problem: Let X be a d-dimensional vector space over R with $d \geqslant 1$ and let $I = I^{\#} \cup I^{\#\#}$ be a finite index set of nonnegative integers containing 0 where $I^{\#}$ and $I^{\#\#}$ are disjoint and either may be empty. Let $\{y_i, i \in I\} \subset X$ be such that $y_0 := 0$ and $\mathsf{L}\{y_i, i \in I\} = X$.

Let g be a real-valued function of finite sequences of 0's and 1's where each element of g's domain is of the form $\{(i, \tau_i): i \in I\}$ for $\tau_i \in \{0, 1\}$. Assume g is nondecreasing in that for $s, t \in \bigtimes_I \{0, 1\}$ with $s \leqslant t$ (i.e., $\sigma_i \leqslant \tau_i$ for $i \in I$ with at least one strict inequality),

$$g(\{(i, \sigma_i): i \in I\}) \leqslant g(\{(i, \tau_i): i \in I\}).$$

Define $H: \tilde{X} \to \mathsf{R}$ via

$$H(\tilde{x}) := g(\{(i, 1\{[y_i, \tilde{x}] > 0\}): i \in I^{\#}\} \cup \{(i, 1\{[y_i, \tilde{x}] \geqslant 0\}): i \in I^{\#\#}\}).$$

Find all $\tilde{x}$ which maximize H.

Note that $I^{\#}$ is the set of all indices i such that R_i = " $>$ " and $I^{\#\#}$ is the set of all indices i such that R_i = " $\geqslant$ ".

As was indicated after (3.1.1), the assumption $\mathsf{L}\{y_i, i \in I\} = X$ is not restrictive.

For an indication of the idiosyncrasies of this problem, recall Examples (4.1.4).

This section will characterize the geometrical nature of the set of solutions to Problem (5.1.1).

-- *the $C(\pi,M)$ class of cones* --

The first step before giving a generalized definition of hill and max-cone is to prove a theorem which will help to characterize a certain class of cones.

(5.1.2) Theorem: Let $\{a_i, i \in I\} \subset X$ be a set of vectors. Let $I_0 := \{i \in I: a_i = a_0 := 0\}$ and let M be an index set such that $I_0 \subset M \subset I$.

Define $D_M := \mathsf{C}\{a_i, i \in M\}$. The following two statements are equivalent (recall $\mathsf{C}\varnothing = \langle 0 \rangle$):

(a) $\text{Lin } \mathsf{C}\{a_i, i \in I\} = D_M$

(b) D_M is a subspace and $\mathsf{C}\{u_i, i \in I \text{ w.o. } M\}$ is pointed where $u_i := P[a_i | U, D_M]$ for any subspace U such that $U \oplus D_M = X$.

Proof: The extreme cases $M = I_0$ and $M = I$ are easily taken care of. For the remaining case, consider:

$(a) \Rightarrow (b)$: If $\mathsf{C}\{u_i, i \in I \text{ w.o. } M\}$ is not pointed, then there exists $v \neq 0$ such that

$$\langle v \rangle \cup \langle -v \rangle \subset \mathsf{C}\{u_i, i \in I \text{ w.o. } M\}.$$

Let $z_1, z_2 \in \mathsf{C}\{a_i, i \in I\}$ be such that $P[z_1 | U, D_M] = v$ and $P[z_2 | U, D_M] = -v$. Since $P[z_1 + z_2 | U, D_M] = 0$,

$$z_1 + z_2 \in \text{Lin } \mathsf{C}\{a_i, i \in I\}.$$

By (2.3.46), $z_1 \in D_M$ but $z_1 \notin D_M$ since $v \neq 0$.

$(b) \Rightarrow (a)$: Certainly $LHS \supset RHS$. So, suppose there exists $z \neq 0$ such that $z, -z \in (\text{Lin } \mathsf{C}\{a_i, i \in I\}) \text{ w.o. } D_M$. Select U such that $U \oplus D_M = X$ and let $t = P[z | U, D_M] \neq 0$. Then, since

$$z = \sum_{I \text{ w.o. } M} \lambda_i a_i + \sum_M \lambda_i a_i \text{ for suitable } \lambda_i \geqslant 0, \; t \in \mathrm{C}\{u_i, i \in I \text{ w.o. } M\}$$

and so also is $-t$ by a similar argument. This contradicts the assumed pointedness of $\mathrm{C}\{u_i, i \in I \text{ w.o. } M\}$. □

Now, by taking some care in defining the index set M used in defining D_M, the $C(\pi, M)$ class of cones is defined.

(5.1.3) Definition: Let $I_0 \subset M \subset I$ be an index set where $k \in M$ if and only if $y_k \in C_M := \mathrm{C}\{y_i, i \in M\}$. Suppose C_M is a subspace.

If $M \neq I$, then let $\pi_i \in \{-1, 1\}$ for $i \in I$ w.o. M and suppose $\mathrm{C}\{\pi_i u_i, i \in I \text{ w.o. } M\}$ is pointed where $u_i := P[y_i | U, C_M]$ for any subspace U such that $U \oplus C_M = X$.

If $M = I$, then $C(\pi, M) := \mathrm{C}\{y_i, i \in M\}$.

If $M \neq I$, then $C(\pi, M) := \mathrm{C}\{\pi_i y_i, i \in I \text{ w.o. } M, y_i, i \in M\}$.

Note that $C(\pi, I_0) = C(\pi)$ if and only if $C(\pi)$ is pointed. ($C(\pi)$ is defined in (3.2.1).)

-- *useful facts about* $C(\pi, M)$ --

A number of useful facts about $C(\pi, M)$ and its dual follow.

(5.1.4) Theorem: Given $C(\pi, M) \subset X$. Then:

(a) $\mathrm{Lin}\, C(\pi, M) = C_M$

(b) $C(\pi, M)^+$

$$= \{\tilde{x} \in \tilde{X}: [\pi_i y_i, \tilde{x}] \geqslant 0, i \in I \text{ w.o. } M, [y_i, \tilde{x}] = 0, i \in M\}$$
$$\subset C_M^{\perp}$$

(c) $C(\pi, M)^+$ is pointed.

(d) $\mathrm{rel\ int}\, C(\pi, M)^+ = \{\tilde{x} \in C_M^{\perp}: [\pi_i y_i, \tilde{x}] > 0, i \in I \text{ w.o. } M\}$

$$= F_M(C(\pi, M)^+)$$

(e) For $k = 0, \ldots, d$, $\dim C(\pi,M)^+ = k$ if and only if $\dim C_M = d-k$

(f) $M = I$ if and only if $C(\pi,M)^+ = \langle 0 \rangle$.

(g) If $F_J(C(\pi,M)^+) \neq \emptyset$, then $J \supset M$.

Proof: (b): Certainly $LHS \supset RHS$. For the other inclusion, observe that for any $\tilde{x} \in \tilde{X}$ such that $[y_i, \tilde{x}] \geqslant 0$ for all $i \in M$, it is true that for each $k \in M$, $[-y_k, \tilde{x}] \geqslant 0$ since $-y_k$ is in the subspace $\mathrm{C}\{y_i, i \in M\}$.

(c): Use (2.4.7), (2.1.13), and the assumption that $\mathrm{L}\{y_i, i \in I\} = X$.

(d): See (2.4.9).

(e): See (2.4.7).

(f): If $M = I$, then $\mathrm{C}\{y_i, i \in M\} = \mathrm{C}\{y_i, i \in I\}$ is a subspace of dimension d since $\mathrm{L}\{y_i, i \in I\} = X$ by assumption and so $C_M = X$ implies $C(\pi,M)^+ = \langle 0 \rangle$. If $C(\pi,M)^+ = \langle 0 \rangle$ then $\mathrm{Lin}\, C(\pi,M) = \mathrm{C}\{y_i, i \in M\} = X$. Suppose there exists $i \in I$ w.o. M. Then $y_i \notin C_M = X$ which is impossible.

(g): Suppose M is not a subset of J. Then there exists $k \in M$ such that $k \notin J$. Since each face of $C(\pi,M)^+$ is in the dimensionality space of the cone, there exists $\tilde{x} \in C_M^{\perp}$ such that $[y_k, \tilde{x}] \neq 0$ which is impossible. □

-- *max-cones and hills in terms of $C(\pi,M)$* --

It is now possible to define max-cone and hill in this more general setting. That neither the max-cone nor hill concepts are vacuous is a consequence of Theorem (5.1.10).

(5.1.5) Definition:

$$\{\tilde{x} \in \tilde{X} : [\pi_i y_i, \tilde{x}] \geqslant 0, i \in I \text{ w.o. } K, [y_i, \tilde{x}] = 0, i \in K\}$$

is a *max-cone* if and only if

(i) $\mathsf{C}\{y_i\,,\ i\in K\}$ is a subspace.

(ii) There exists $\tilde{x}_0 \in \tilde{X}$ such that $[\pi_i y_i\,,\ \tilde{x}_0] > 0$ for $i \in I$ w.o. K and $[y_i\,,\ \tilde{x}_0] = 0$ for $i \in K$.

(iii) $H(\tilde{x}_0) = \sup_{\tilde{x}\in\tilde{X}} H(\tilde{x})$

(5.1.6) Theorem: Every max-cone of Definition (5.1.5) is $C(\pi,M)^+$ for some M (namely, K). The maximum H value is assumed in rel int $C(\pi,M)^+$.

(5.1.7) Definition: $C(\pi,M)^+$ is a *hill* if and only if for any subspace U such that $U \oplus C_M = X$ and $u_i := P[y_i\,|\,U,\ C_M]$, there exists $K^+ \subset I$ w.o. M such that $\{\langle 0\rangle, \langle u_i\rangle,\ i\in K^+\}$ is a frame for $\mathsf{C}\{\pi_i u_i\,,\ i\in I \text{ w.o. } M\}$. (Note K^+ could be empty.)

-- showing that a cone is a hill --

In order to show that $\mathsf{C}\{\pi_i y_i\,,\ i\in M,\ y_i\,,\ i\in M\}^+$ is a hill, it is sufficient to produce only one subspace U which satisfies the required two conditions. For suppose T is any other subspace such that $X = T\oplus C_M$. It is true that for all i, $y_i - u_i$, $y_i - t_i \in C_M$ for some $u_i \in U$ and $t_i \in T$. Hence for all i, $u_i = t_i + b_i$ for some $b_i \in C_M$. Then if there exist $\lambda_i \geqslant 0$, not all 0, such that $0 = \sum_{I \text{ w.o. } M} \lambda_i\pi_i t_i = \sum_{I \text{ w.o. } M} \lambda_i\pi_i u_i + \sum_{I \text{ w.o. } M} \lambda_i\pi_i b_i$, then $\mathsf{C}\{\pi_i u_i\,,\ i\in I \text{ w.o. } M\}$ is not pointed. Similarly, suppose that $\mathsf{C}\{\pi_i u_i\,,\ i\in I \text{ w.o. } M\} = \mathsf{C}\{u_i\,,\ i\in K^+(\pi)\}$. Then for any set of $\lambda_i \geqslant 0$ for $i\in I$ w.o. M, there exists $\alpha_i \geqslant 0$ for $i\in K^+(\pi)$ such that

$$\sum_{I \text{ w.o. } M} \lambda_i\pi_i t_i - \sum_{K^+(\pi)} \alpha_i t_i \in C_M.$$

Notice that $C(\pi,M)^+$ is a hill if and only if $M = \{i\in I\colon y_i \in C_M\}$, C_M is a subspace, and for any U such that $X = U\oplus C_M$, $P[C(\pi,M)\,|\,U,\ C_M]^+$ is a hill according to Definition (3.2.5).

Note also that if $C(\pi, M)^+$ is a hill, then there exists K^+ such that $C(\pi, M)^+ = \{\tilde{x} \in C_M^\perp : [y_i, \tilde{x}] \geqslant 0, i \in K^+\}$.

-- *an example of a lower dimensional hill* --

For an example of a lower dimensional hill, consider the problem of maximizing over $x \in \mathsf{R}^3$,

$$
\begin{aligned}
1\{(0, 0, 1)^T x = 0\} &+ 1\{(1, 0, 0)^T x > 0\} + 1\{(0, 1, 0)^T x > 0\} \\
&= 1\{(0, 0, 1)^T x \geqslant 0\} + 1\{(0, 0, -1)^T x \geqslant 0\} \\
&\qquad + 1\{(1, 0, 0)^T x > 0\} + 1\{(0, 1, 0)^T x > 0\} - 1.
\end{aligned}
$$

This last expression is in homogeneous canonical form. Let $y_0 = (0, 0, 0)$, $y_1 = (0, 0, 1)$, $y_2 = (0, 0, -1)$, $y_3 = (1, 0, 0)$, and $y_4 = (0, 1, 0)$. Let $M = \{0, 1, 2\}$ and $\pi_3 = \pi_4 = 1$. The claim is that $C(\pi, M)^+ = \mathsf{C}\{y_i, i = 0, \ldots, 4\}^+$ is a hill. Observe $C_M = \mathsf{L}\{(0, 0, 1)\}$ and $M = \{i : y_i \in C_M\}$. Next, let $U = \mathsf{L}\{(1, 0, 0), (0, 1, 0)\}$. Then $u_0 = u_1 = u_2 = (0, 0, 0)$, $u_3 = (1, 0, 0)$, and $u_4 = (0, 1, 0)$. Finally $\mathsf{C}\{u_3, u_4\}$ is pointed and has frame $\{\langle 0\rangle, \langle u_3\rangle, \langle u_4\rangle\}$.

-- *when max-cones are hills* --

(5.1.8) Definition: A max-cone which is also a hill is called a *max-hill.*

The next theorem characterizes the nature of maximizing vectors and shows that max-cones are necessarily hills when g is strictly increasing. But first it is necessary to discuss in more detail the manner in which g can be nondecreasing.

(5.1.9) Definition: Let g be a real-valued function with domain $\{\{(i, \sigma_i) : i \in I\} : \sigma_i \in \{0, 1\}\}$ where I is some index set.

The jth variable of g is *nondecreasing* if and only if for all choices of $\sigma_i \in \{0, 1\}$ with $i \in I$ w.o. j,

$$g\left[\{(i,\sigma_i): i \in I \text{ w.o. } j\} \cup \{(j,0)\}\right] \leqslant g\left[\{(i,\sigma_i): i \in I \text{ w.o. } j\} \cup \{(j,1)\}\right]$$

The jth variable of g is *strictly increasing* if the "$\leqslant$" in the preceding sentence is replaced with "$<$". Analogous substitutions yield definitions for *nonincreasing*, *strictly decreasing*, and *constant* variables.

The function g is *nondecreasing* if and only if for all $s,t \in \bigtimes_I \{0, 1\}$ such that $s \leqslant t$ (i.e., $\sigma_i \leqslant \tau_i$ for $i \in I$ with at least one strict inequality),

$$g\left[\{(i,\sigma_i): i \in I\}\right] \leqslant g\left[\{(i,\tau_i): i \in I\}\right]$$

g is *strictly increasing* if and only if the last "$\leqslant$" in the previous sentence is replaced with "$<$".

It is easy to show that g is nondecreasing (strictly increasing) if and only if every variable of g is nondecreasing (strictly increasing).

(5.1.10) Theorem: Let $\tilde{x}_0$ be such that $H(\tilde{x}_0) = \sup_{\tilde{x} \in \tilde{X}} H(\tilde{x})$.

Then $\tilde{x}_0$ is in a face of a max-cone which is either a hill or leads through a finite sequence of adjacent max-cones to a max-cone which is a hill.

In symbols, there exists $I_0 \subset J \subset I$ such that $[y_i, \tilde{x}_0] = 0$ for $i \in J$ and for $i \in I$ w.o. J, $[\pi_i^0 y_i, \tilde{x}_0] > 0$ for suitable $\pi_i^0 \in \{-1, 1\}$ (if $J \neq I$). There also exists $M \subset J$ such that $C(\pi^0, M)^+$ is a max-cone with $\pi_i^0 := 1$ for $i \in J$ w.o. M (if $J \neq M$) and $\tilde{x}_0 \in F_J(C(\pi^0, M)^+)$.

It is also true that

$$A := \{i \in J \text{ w.o. } M: \text{ the } i^{th} \text{ variable of g is strictly increasing }\} \subset I^{\#\#}.$$

Hence if g is strictly increasing (or as a weaker condition, if $A = J$ w.o. M), then J w.o. $M \subset I^{\#\#}$. Recall that $I^{\#\#}$ is the set of all indices in I such that $R_i =$ "$\geqslant$".

In addition, if g is strictly increasing and all of the inequalities are strict, then $M = I_0$ and there are no max-cones of dimension less than d. In more detail, if $M \neq I_0$, then for all $j \in M$ w.o. I_0, either $j \in I^{\#\#}$ or the jth

variable of g is not strictly increasing.

Also, if $\{y_i, i \in I \text{ w.o. } I_0\}$ is in general position (or weaker still, pointed position (cf., (2.3.34))), then $M = I_0$ or $M = I$.

If $M = I$, then $C(\pi^0, M)^+ = \langle 0 \rangle$ and is a hill.

Now suppose $M \neq I$ so that $C(\pi^0, M)^+ \neq \langle 0 \rangle$. If g is strictly increasing, then $C(\pi^0, M)^+$ is a hill.

Suppose $C(\pi^0, M)^+$ is not a hill. Then there exists a finite sequence of cones $C(\pi^j, M)$ for $j = 1, \ldots, k$ such that:

(i) $C(\pi^j, M)^+$ is a max-cone for $j = 1, \ldots, k$.

(ii) $C(\pi^k, M)^+$ is a hill.

(iii) For $j = 0, \ldots, k-1$, and for each $i \in I$ w.o. M, $\pi_i^j \leqslant \pi_i^{j+1}$ with at least one strict i^{th} inequality for each j.

Proof: By computing $[y_i, \tilde{x}_0]$ for $i \in I$, J can be determined such that $I_0 \subset J \subset I$ and $[y_i, \tilde{x}_0] = 0$ for $i \in J$ and, if $J \neq I$, for $i \in I$ w.o. J, $[\pi_i^0 y_i, \tilde{x}_0] > 0$ for suitable $\pi_i^0 \in \{-1, 1\}$.

If $\mathsf{C}\{y_i, i \in J\}$ is a subspace, then let $M = J$ and obtain the max-cone $C(\pi^0, M)^+$ with $\tilde{x}_0 \in F_M(C(\pi^0, M)^+) = \text{rel int } C(\pi^0, M)^+$.

Suppose $\mathsf{C}\{y_i, i \in J\}$ is not a subspace so that

$$\mathsf{C}\{y_i, i \in J\} \neq \text{Lin } \mathsf{C}\{y_i, i \in J\} =: \mathsf{C}\{y_i, i \in M\} =: C_M$$

where $M \subset J$ is chosen such that $i \in M$ if and only if $y_i \in \text{Lin } \mathsf{C}\{y_i, i \in J\}$ (cf., (2.3.47)). Note that $M \neq I$ in this case. Let U be such that $U \oplus C_M = X$ and for $i \in J$, $u_i := P[y_i \,|\, U, C_M]$. Note J w.o. $M \neq \varnothing$ and by Theorem (5.1.2), $\mathsf{C}\{u_i, i \in J \text{ w.o. } M\}$ is pointed. Hence, there exists $\tilde{r} \in \tilde{U}$ such that $[u_i, \tilde{r}] > 0$ for all $i \in J$ w.o. M and so there exists $\tilde{t} \in C_M^\perp$ such that $\tilde{r} = \psi(\tilde{t}) = \tilde{t}\big|_R$ and $[y_i, \tilde{t}] > 0$ for all $i \in J$ w.o. M (cf., (2.1.27)).

Next choose $\alpha > 0$ such that

$$[\pi_i^0 y_i, \tilde{x}_0 + \alpha \tilde{t}] > 0 \text{ for } i \in I \text{ w.o. } J \text{ (possibly null)},$$

$$[y_i, \tilde{x}_0 + \alpha \tilde{t}] > 0 \text{ for } i \in J \text{ w.o. } M \neq \emptyset,$$

$$\text{and } [y_i, \tilde{x}_0 + \alpha \tilde{t}] = 0 \text{ for } i \in M \neq \emptyset.$$

Let $\pi_i^0 := 1$ for $i \in J$ w.o. M. Since g is nondecreasing, $H(\tilde{x}_0) \leqslant H(\tilde{x}_0 + \alpha \tilde{t}) \leqslant H(\tilde{x}_0)$ and so $\{\tilde{x} \in \tilde{X}: [\pi_i^0 y_i, \tilde{x}] \geqslant 0$ for $i \in I$ w.o. M, $[y_i, \tilde{x}] = 0$ for $i \in M\}$ is a max-cone $C(\pi^0, M)^+$ and $\tilde{x}_0 \in F_J(C(\pi^0, M)^+)$.

Suppose there was $i \in (J$ w.o. $M) \cap I^{\#}$ such that the ith variable of g is strictly increasing. Then $H(\tilde{x}_0) < H(\tilde{x}_0 + \alpha \tilde{t}) \leqslant H(\tilde{x}_0)$ which is impossible.

So at this point, a max-cone $C(\pi^0, M)^+$ has been identified and a vector $\tilde{x}_1 \in$ rel int $C(\pi^0, M)^+$ has been obtained.

Suppose M w.o. $I_0 \neq \emptyset$ and for some $j \in M$ w.o. I_0, the jth variable of g is strictly increasing and $j \in I^{\#}$. Let $\tilde{x}_2$ be such that $[y_j, \tilde{x}_2] > 0$. Observe that there exists $\beta > 0$ such that $[\pi_i^0 y_i, \tilde{x}_1 + \beta \tilde{x}_2] > 0$ for all $i \in I$ w.o. M and $[y_j, \tilde{x}_1 + \beta \tilde{x}_2] > 0$ and so

$$H(\tilde{x}_1) < H(\tilde{x}_1 + \beta \tilde{x}_2) \leqslant H(\tilde{x}_1)$$

which is impossible.

If $\{y_i, i \in I$ w.o. $I_0\}$ is in pointed position and $M \neq I$, then $C_M^{\perp} \neq \langle 0 \rangle$ and so by hypothesis, C_M is pointed which is impossible unless $C_M = \langle 0 \rangle$.

Suppose that $M \neq I$ so that $C(\pi^0, M)^+ \neq \langle 0 \rangle$ and suppose that $C(\pi^0, M)^+$ is not a hill. Let U be a subspace such that $U \oplus C_M = X$ and $u_i := P[y_i | U, C_M]$ for $i \in I$ w.o. M. Let u_k be such that $\langle -u_k \rangle \neq \langle 0 \rangle$ is an isolated ray of $\mathrm{C}\{\pi_i^0 u_i, i \in I$ w.o. $M\}$ and be such that for all $i \in I$ w.o. M, $\langle u_i \rangle \neq \langle -u_k \rangle$. Then by (2.4.11), there exists

$$\tilde{r}_2 \in \{\tilde{r} \in \tilde{U}: [\pi_i^0 u_i, \tilde{r}] > 0 \text{ for } i \in I \text{ w.o. } (M \cup I_k(\pi^0)),$$
$$[\pi_i^0 u_i, \tilde{r}] < 0 \text{ for } i \in I_k(\pi^0)\}$$

where $I_k(\pi^0) := \{i \in I \text{ w.o. } M: \langle \pi_i^0 u_i \rangle = \langle \pi_k^0 u_k \rangle = \langle -u_k \rangle\} \neq \emptyset$. Note that for all $i \in I_k(\pi^0)$, $\pi_i^0 = -1$ since if some $\pi_i^0 = 1$ for $i \in I_k(\pi^0)$, then it would be the case that $\langle u_i \rangle = \langle -u_k \rangle$ which contradicts the choice of u_k.

Consequently for $\tilde{t}_2 = \psi^{-1}(\tilde{r}_2) \in C_M^{\perp}$, $[\pi_i^0 y_i, \tilde{t}_2] > 0$ for $i \in I$ w.o. $(M \cup I_k(\pi^0))$ and $[y_i, \tilde{t}_2] > 0$ for $i \in I_k(\pi^0)$. For $i \in I$ w.o. $(M \cup I_k(\pi^0))$, let $\pi_i^1 = \pi_i^0$ and for $i \in I_k(\pi^0)$, let $\pi_i^1 = 1 = -\pi_i^0$. Then $C(\pi^1, M)^+$ is a max-cone since $H(\tilde{x}_1) \leqslant H(\tilde{t}_2) \leqslant H(\tilde{x}_1)$. Note also that for each $i \in I$ w.o. M, $\pi_i^0 \leqslant \pi_i^1$ with at least one strict inequality.

Note also that if g was strictly increasing, then $H(\tilde{x}_1) < H(\tilde{t}_2)$ which would have implied that $C(\pi^0, M)^+$ must have been a hill in the first place.

Now if $C(\pi^1, M)^+$ is not a hill, then repeat this process by crossing over into $C(\pi^2, M)^+$, a suitable neighboring max-cone of $C(\pi^1, M)^+$. Continue on in this fashion until a hill is reached. This must happen after a finite number of steps because there are only a finite number of cones and because for each j in the sequence for all $i \in I$ w.o. M, $\pi_i^{j-1} \leqslant \pi_i^j$ with at least one strict inequality. □

-- *how to identify all maximizing vectors* --

Theorem (5.1.10) justifies the following procedure for identifying *all* vectors which maximize $H(\tilde{x})$:

(5.1.11) Procedure:

(i) Identify all hills

(ii) By evaluating H on the relative interior of each hill, identify all max-hills.

(iii) For each nonzero max-hill $C(\pi, M)^+$, determine the $(d-k-1)$-dimensional boundary faces of $C(\pi, M)^+$ where $k = \dim C_M$ by determining the frame of $\mathrm{C}\{u_i := P[y_i | U, C_M]: i \in I \text{ w.o. } M\}$. Call a $(d-k-1)$-dimensional boundary face an *unequivocably positive boundary face* if for all rays $\langle \pi_i u_i \rangle$ generating this boundary face, $\pi_i = 1$. Cross each unequivocably positive boundary face staying in $C_M^{\perp}$ and determine if the cone on the other side is a max-cone.

(iv) Construct a finite tree of cones for each max-hill in the following way. The root node contains the max-hill. The first level of the tree consists of the neighboring max-cones (if any) associated with unequivocably positive boundary faces determined in step (iii). In order to determine the next level, take each cone in the first level and say its children are all those max-cones that are on the opposite side of its unequivocably positive boundary faces that were *not* generated by $\langle u_i \rangle$ generating unequivocably positive boundary faces on the path down to the max-cone in question. Iterate this cone-hopping until the tree can grow no further.

(v) Every vector maximizing $H(\tilde{x})$ is in a face of some cone in the forest resulting from step (iv).

So, there is an algorithm for identifying all solution vectors if it is but possible to produce all hills or better yet all max-hills. The tree algorithms developed in the next two sections will do this. Just as the WOH tree algorithms have two phases, so also do the more general tree algorithms have two phases: i.e., a relative boundary vector collection phase and a displacement phase. On the whole, the WOH tree algorithms and the more general tree algorithms are remarkably similar although of course there are significant differences.

-- a statement of a general tree algorithm --

For the sake of the reader's convenience in assimilating the material of later sections, it is now time to present a general tree algorithm which will find all vectors maximizing a function of a system of linear relations in homogeneous canonical form with nondecreasing g function. Other variants of this algorithm will be discussed later. Since the assertion that this tree algorithm solves problems in homogeneous canonical form with nondecreasing g functions is only discussed and validated in the following three sections, the reader should not expect to fully understand the algorithm at this point. The reader might want to refer back to this algorithm statement after reading each of the following sections in order to see how each of the subpieces of the algorithm fit back into the whole.

(5.1.12) defines the variables that will appear in the algorithm.

(5.1.12) Definition: Recall that Problem (5.1.1) seeks to produce all of those vectors $\tilde{x} \in \tilde{X}$ which maximize

$$H(\tilde{x}) := g(\{(i, 1\{[y_i, \tilde{x}]\ R_i\ 0\}) : i \in I\})$$

for given $\{y_i, i \in I\} \subset X$, $R_i \in \{>, \geqslant\}$, and nondecreasing $g : \times_I \{0, 1\} \to \mathbb{R}$.

For any nonempty subset J of I, let $R_J := \mathrm{L}\{y_i, i \in J\}$. Let $J_0 := \{i \in J : y_i = 0\}$. Let $\{\pi_i : i \in I \text{ w.o. } J\} \subset \{-1, 1\}$. Define $H_{\pi,J} : \tilde{R}_J \to (0, \infty)$ via for all $\tilde{r} \in \tilde{R}_J$, $H_{\pi,J}(\tilde{r}) :=$

$$g(\{(i, 1\{\pi_i > 0\}) : i \in I \text{ w.o. } J\} \cup \{(i, 1\{[y_i, \tilde{r}]\ R_i\ 0\}) : i \in J\})$$

By understandable convention, $H = H_{\pi,I}$ and by the assumption in (5.1.1), $\tilde{X} = \tilde{R}_I$.

Also, for $\varnothing \neq J \subset I$, for all $\tilde{r} \in \tilde{R}_J$, let $N_J(\tilde{r}) := \{i \in J : [y_i, \tilde{r}] < 0\}$ and $Z_J(\tilde{r}) := \{i \in J \text{ w.o. } J_0 : [y_i, \tilde{r}] = 0\}$. y_{i_k} will be written as $y(i_k)$. Also, at times, "$\tilde{v}_k$" will be used to represent some $\tilde{v}_k(i_0, \ldots, i_{k-1})$ either generically or individually as the context will indicate. Similarly, "$\tilde{w}_k$" will be used to represent $\tilde{w}_k(i_0, \ldots, i_{k-1})$ where "$\tilde{w}_k(i_0, \ldots, i_{k-1})$" itself ambiguously represents one of $\tilde{v}_k(i_0, \ldots, i_{k-1})$ and $-\tilde{v}_k(i_0, \ldots, i_{k-1})$ whichever is desired at the moment.

From Chapter 2, recall that $\#A$ is the cardinality of the set A, $\underline{x}$ is the representation in R^p of the vector $x \in R_J$ according to some fixed arbitrary basis of size $p = \dim R_J$, and $\psi_{J,K}$ is the vector space isomorphism mapping $S^\perp$ onto $S^\perp|_{R_K}$ where $S \oplus R_K = R_J$ for $K \subset J$ (cf., (2.1.27)).

Similar to the EXPLORE of (3.2.12), the EXPLORE of (5.1.13) is the procedure which constructs and searches the relative boundary vector tree and periodically calls upon its subroutine UPDATE_B to update a set B_J which contains the most promising relative boundary vectors found so far. Once certain conditions regarding B_J have been satisfied, EXPLORE calls its subroutine DISPLACE to initiate the second phase of this tree algorithm where the relative boundary vectors in B_J are displaced and the resulting solution vectors are saved in the set A_J.

To help DISPLACE do its job, the subroutine COMP_DISP computes the α necessary to satisfactorily displace a given relative boundary vector $\tilde{w}_k(i_0, \ldots, i_{k-1})$ in the direction of a given $\tilde{z}$. The subroutine UPDATE_A updates A_J with candidate solution vectors as they are found.

The following algorithm is written in a hopefully self-explanatory hybrid of Fortran, BASIC, PL/I, and English.

(5.1.13) Algorithm:

Obtain I, $\{y_i, i \in I\} \subset X$ and $H = H_{\pi,I}$ where $\mathsf{L}\{y_i, i \in I\} = X$. If desired, modify the preceding to eliminate any $y_i = 0$ and to eliminate all ties among the $\langle y_i \rangle$. By way of convention, any set indexed by the null set is null itself.

Obtain some nonzero $\tilde{v}_0 \in R_I = \tilde{X}$.

Call EXPLORE $(I, \{y_i, i \in I\}, H_{\pi,I}, \tilde{v}_0, A_I)$.

EXPLORE: Procedure $(J, \{y_i, i \in J\}, H_{\pi,J}, \tilde{v}_0, A_J)$;

Step 1: Set $B_J = \{0\}$.

Step 2: If $\#N_J(-\tilde{v}_0) < \#N_J(\tilde{v}_0)$ then set $\tilde{v}_0 = -\tilde{v}_0$.

If $N_J(\tilde{v}_0) = \emptyset$ then do:

Set $B_J = \{\tilde{v}_0\}$.

Call DISPLACE $(H_{\pi,J}, B_J, A_J)$.

return from EXPLORE.

end;

Call UPDATE_B $(\tilde{v}_0, \emptyset, B_J)$;

Call UPDATE_B $(-\tilde{v}_0, \emptyset, B_J)$;

Step 3: For $k = 1, \ldots, d-1$, do:

For each $i_0 \in N_J(\tilde{v}_0)$, $i_1 \in N_J(\tilde{v}_1(i_0))$, ...,

$i_{k-1} \in N_J(\tilde{v}_{k-1}(i_0, \ldots, i_{k-2}))$, do:

Obtain $\tilde{x} \in y(i_0)^\perp \cap \cdots \cap y(i_{k-1})^\perp$ where $\tilde{x} \neq 0$.

If $\#N_J(-\tilde{x}) < \#N_J(\tilde{x})$ then set $\tilde{x} = -\tilde{x}$.

Set $\tilde{v}_k(i_0, \ldots, i_{k-1}) = \tilde{x}$.

If $N_J(\tilde{v}_k(i_0, \ldots, i_{k-1})) = \emptyset$ then do:

Set $B_J = \{\tilde{v}_k(i_0, \ldots, i_{k-1})\}$.

Call DISPLACE $(H_{\pi,J}, B_J, A_J)$.

return from EXPLORE.

end;

If $\#N_J(\tilde{v}_k) + \#Z_J(\tilde{v}_k) - k < \#N_J(\tilde{v}_0)$ then do:

Set $B_J = \{\tilde{v}_k(i_0, \ldots, i_{k-1})\}$.

Let $H''(\tilde{r}) = \sum_J 1\{[y_i, \tilde{r}] > 0\}$.

Call DISPLACE $(H''_{\pi,J}, B_J, A_J)$.

Set $\tilde{v}_0$ equal to any element of A_J.

Go to "Step 1" of EXPLORE.

end;

Call UPDATE_B $(\tilde{v}_k(i_0, \ldots, i_{k-1}), \{i_0, \ldots, i_{k-1}\}, B_J)$.

Call UPDATE_B $(-\tilde{v}_k(i_0, \ldots, i_{k-1}), \{i_0, \ldots, i_{k-1}\}, B_J)$.

next i_{k-1}; . . . ; next i_0;

next k;

Step 4: Call DISPLACE $(H_{\pi,J}, B_J, A_J)$.

return from EXPLORE;

UPDATE_B: Procedure $(\tilde{x}, \{i_0, \ldots, i_{k-1}\}, B_J)$;

If $g(\{(i, 1\{\pi_i > 0\}) : i \in I \text{ w.o. } J\}$

$\cup\{(i, 1\{[y_i, \tilde{x}]\ R_i\ \ 0\}) : i \in J \text{ w.o. } Z_J(\tilde{x})\}$

$\cup\{(i, 1) : i \in Z_J(\tilde{x})\})$

$\geqslant \sup\limits_{\tilde{w}_j(i_0, \ldots, i_{j-1}) \in B_J} \max\{H_{\pi,J}(\tilde{w}_j), g(\{(i, 1\{\pi_i > 0\}) : i \in I \text{ w.o. } J\}$

$\cup\{(i, 1\{[y_i, \tilde{w}_j]\ R_i\ \ 0\}) : i \in J \text{ w.o. } \{i_0, \ldots, i_{j-1}\}$

and either $y_i \notin \tilde{w}_j^{\perp}$ or R_i = ">" or $i \in J_0\}$

$\cup\{(i, 0) : i \in J \text{ w.o. } \{i_0, \ldots, i_{j-1}\}$ and $y_i \in \tilde{w}_j^{\perp}$

and R_i = "⩾" and $i \notin J_0\}$

$\cup\{(i, 1) : i \in \{i_0, \ldots, i_{j-1}\}\})\}$ then do:

For each $\tilde{w}_j \in B_J$ do:

If $\max\{H_{\pi,J}(\tilde{x}), g(\{(i, 1\{\pi_i > 0\}) : i \in I \text{ w.o. } J\}$

$\cup\{(i, 1\{[y_i, \tilde{x}]\ R_i\ \ 0\}) : i \in J \text{ w.o. } \{i_0, \ldots, i_{k-1}\}$ and

either $y_i \notin \tilde{x}^{\perp}$ or R_i = ">" or $i \in J_0\}$

$\cup\{(i, 0) : i \in J \text{ w.o. } \{i_0, \ldots, i_{k-1}\}$ and $y_i \in \tilde{x}^{\perp}$

and R_i = "⩾" and $i \notin J_0\}$

$\cup\{(i, 1) : i \in \{i_0, \ldots, i_{k-1}\}\})\}$

$> g(\{(i, 1\{\pi_i > 0\}) : i \in I \text{ w.o. } J\}$

$\cup\{(i, 1\{[y_i, \tilde{w}_j]\ R_i\ \ 0\}) : i \in J \text{ w.o. } Z_J(\tilde{w}_j)\}$

$\cup\{(i, 1) : i \in Z_J(\tilde{w}_j)\})$

then set $B_J = B_J$ w.o. $\{\tilde{w}_j\}$.

next $\tilde{w}_j$;

Set $B_J = B_J \cup \{\tilde{x}\}$.

end;

end UPDATE_B;

DISPLACE: Procedure $(H_{\pi,J}, B_J, A_J)$;

Step 1: Set $A_J = \varnothing$.

Step 2: For each $\tilde{w}_k(i_0, \ldots, i_{k-1}) \in B_J$, do:

Call UPDATE_A $(\tilde{w}_k(i_0, \ldots, i_{k-1}), A_J)$.

Set $K = \{i \in J: [y_i, \tilde{w}_k(i_0, \ldots, i_{k-1})] = 0\}$.

If dim $\mathsf{L}\{y_i, i \in K\} = 1$ then do:

Take $p \in K$ w.o. J_0.

Set $K_1 = \{i \in K: \langle y_i \rangle = \langle y_p \rangle\}$.

Set $K_2 = \{i \in K: \langle y_i \rangle = -\langle y_p \rangle\}$.

Set $\underline{\tilde{z}} = \underline{y}_p$.

Call COMP_DISP $(\tilde{w}_k(i_0, \ldots, i_{k-1}), K_1, \tilde{z}, \alpha)$.

Call UPDATE_A $((1-\alpha)\tilde{w}_k(i_0, \ldots, i_{k-1}) + \alpha\tilde{z}, A_J)$.

If $K_2 \neq \varnothing$ then do:

Set $\underline{\tilde{z}} = -\underline{y}_p$.

Call COMP_DISP $(\tilde{w}_k(i_0, \ldots, i_{k-1}), K_2, \tilde{z}, \alpha)$.

Call UPDATE_A $((1-\alpha)\tilde{w}_k(i_0, \ldots, i_{k-1}) + \alpha\tilde{z}, A_J)$.

end;

end;

If dim $\mathsf{L}\{y_i, i \in K\} > 1$ then do:

Solve the linear program:

maximize γ subject to $\gamma \leqslant 1$ and $\gamma \leqslant \underline{y}_i^T \underline{x}$ for

$i \in K$ w.o. K_0 and $\underline{x} \in \mathsf{R}^p$ where $p =$ dim J.

If $\gamma = 1$ then do:

Set $\underline{\tilde{z}} = \underline{x}$.

Call COMP_DISP $(\tilde{w}_k(i_0, \ldots, i_{k-1}), K, \tilde{z}, \alpha)$.

Call UPDATE_A $((1-\alpha)\tilde{w}_k(i_0, \ldots, i_{k-1}) + \alpha\tilde{z}, A_J)$.

end;

If $\gamma = 0$ then do:

Select some $\tilde{v}_0(K) \in \tilde{R}_K$.

Obtain for each $i \in J$ w.o. K, $\pi_i \in \{-1, 1\}$ such

that $[\pi_i y_i, \tilde{w}_k(i_0, \ldots, i_{k-1})] > 0$.

```
            Call EXPLORE (K, {y_i, i ∈ K}, H_{π,K}, ṽ_0(K), A_K).
            For each r̃_K ∈ A_K do:
                Call COMP_DISP (w̃_k(i_0, ..., i_{k-1}), K, ψ^{-1}_{J,K}(r̃_K), α).
                Call UPDATE_A
                        ((1 − α)w̃_k(i_0, ..., i_{k-1}) + αψ^{-1}_{J,K}(r̃_K), A_J).
            end;
        end;
    end;
next w̃_k(i_0, ..., i_{k-1});
```

COMP_DISP: Procedure $(\tilde{x}, K, \tilde{z}, \alpha)$;

Obtain for each $i \in J$ w.o. K, $\pi_i \in \{-1, 1\}$ such that $[\pi_i y_i, \tilde{x}] > 0$.

Set $L = \{i \in J \text{ w.o. } K : [\pi_i y_i, \tilde{z}] < 0\}$.

If $L = \emptyset$ then set $\alpha = \frac{1}{2}$.

If $L \neq \emptyset$ then set $\alpha = \frac{1}{2} \min_{i \in L} \frac{[y_i, \tilde{x}]}{[y_i, \tilde{x}] - [y_i, \tilde{z}]}$.

end COMP_DISP;

UPDATE_A: Procedure $(\tilde{x}, A_J)$;

If $A_J = \emptyset$ then set $A_J = \{\tilde{x}\}$.

If $H_{\pi,J}(\tilde{x}) \geqslant \sup\{H_{\pi,J}(\tilde{w}) : \tilde{w} \in A_J\}$ then do:

If $H_{\pi,J}(\tilde{x}) > \sup\{H_{\pi,J}(\tilde{w}) : \tilde{w} \in A_J\}$ then set $A_J = \emptyset$.

Set $A_J = A_J \cup \{\tilde{x}\}$.

end;

end UPDATE_A;

end DISPLACE;

end EXPLORE;

Algorithm (5.1.13) does not incorporate the major improvements of trimming, depth-first searching, and the projection method of determining $\tilde{v}_k(i_0, \ldots, i_{k-1})$.

Summary For Section 5.1

This section initiates the development of the general tree algorithm by characterizing the geometry of the solution space for maximizing functions H of systems of linear relations in homogeneous canonical form with nondecreasing g functions. In contrast to the WOH tree algorithm, the general tree algorithm is capable of identifying lower dimensional equivalence classes of solutions when they exist. As a simple example, the general tree algorithm will discover that the positive quadrant of the $\xi_1 - \xi_2$ plane is the sole solution equivalence class to the problem of maximizing over $(\xi_1, \xi_2, \xi_3) \in \mathrm{R}^3$ the function $1\{\xi_3 = 0\} + 1\{\xi_1 > 0\} + 1\{\xi_2 > 0\}$.

Two things are surprising about the general tree algorithm. The first is that it is almost identical to the WOH tree algorithm. The second is that a great deal more mathematics is necessary to show that it is valid.

The theory rests on the use of cones of the form

$$C(\pi, M) := \mathrm{C}\{\pi_i y_i, i \in I \text{ w.o. } M, y_i, i \in M\}$$

where $\pi_i \in \{-1, 1\}$, $C_M := \mathrm{C}\{y_i, i \in M\}$ is a subspace, $M = \{i : y_i \in C_M\}$, and for any subspace U such that $U \oplus C_M = X$, $P[C(\pi,M) \mid U, C_M]$ is pointed.

A cone $C(\pi,M)^+$ is called a max-cone if any relative interior vector of $C(\pi,M)^+$ achieves the maximum value of H. $C(\pi,M)^+$ is called a hill if for any U such that $X = U \oplus C_M$, $P[C(\pi,M) \mid U, C_M]^+$ is a hill according to the earlier definition (3.2.5). A max-cone which is also a hill is called a max-hill.

The solution space geometry is such that any vector which maximizes H is in a face of a max-cone which is either a hill or leads through a finite sequence of adjacent max-cones to a max-cone which is a hill.

This section concludes with a programming language type description of what is basically the complete general tree algorithm. Hopefully, it will serve as a useful reference when the reader tries to get a global picture of how the individual algorithm pieces described subsequently fit together into a unified whole.

Section 5.2:

The Construction Of A Tree Of Relative Boundary Vectors

In this section, the boundary vector collection algorithm of Chapter 3 will be extended to the more general situation so that it will construct a tree of vectors containing at least one vector in any given hill. Since the overwhelming majority but not necessarily all of the vectors in this tree will be in the relative boundaries of cones $\mathrm{C}\{\pi_i y_i,\ i \in I\}^+$ for $\pi_i \in \{-1,\ 1\}$, any vector in this tree will be called for simplicity's sake a relative boundary vector even though it could conceivably be a relative interior vector for some cone.

-- *signals from hills* --

Following the basic approach of Chapter 3, the first step is to show that whenever a vector is not in a hill, then the hill will signal that condition.

(5.2.1) Theorem: Let $C(\pi, M)^+$ be a nonzero hill which implies that $M \neq I$ and $C_M \neq X$. Suppose $\tilde{x}_0 \notin C(\pi, M)^+$. Then:

(i) If $\tilde{x}_0 \notin C_M^\perp$, then there exists $j \in M$ w.o. $I_0 \neq \emptyset$ such that $[y_j,\ \tilde{x}_0] < 0$.

(ii) If $\tilde{x}_0 \in C_M^\perp$, then there exists $k \in I$ w.o. $M \neq \emptyset$ such that $\langle 0 \rangle \neq \langle u_k \rangle$ is in the frame of $\mathrm{C}\{\pi_i u_i,\ i \in I \text{ w.o. } M\}$ and such that $[y_k,\ \tilde{x}_0] < 0$.

Proof: (i): Since $\tilde{x}_0 \notin C_M^\perp$, $M \neq I_0$ and $C_M \neq \langle 0 \rangle$. Suppose for all $i \in M$ w.o. I_0, $[y_i,\ \tilde{x}_0] \geqslant 0$. Now there must be some $j \in M$ w.o. I_0 such that $[y_j,\ \tilde{x}_0] > 0$ or else $\tilde{x}_0 \in C_M^\perp$. By (2.3.38) then, C_M is not a subspace which is a contradiction.

(ii): Given $\tilde{x}_0 \in C_M^{\perp}$. There exists $K^+ \subset I$ w.o. M such that $\mathrm{C}\{\pi_i u_i, i \in I \text{ w.o. } M\} = \mathrm{C}\{u_i, i \in K^+\}$ and where each $\langle u_i \rangle$ is an isolated ray of the cone. If for all $j \in K^+$, $[y_j, \tilde{x}_0] = [u_j, \tilde{x}_0] \geqslant 0$ then $[\pi_i y_i, \tilde{x}_0] \geqslant 0$ for all $i \in I$ w.o. M. This yields the contradiction that $\tilde{x}_0 \in C(\pi, M)^+$. □

-- when the answer is simple --

The next theorem provides a useful sufficient condition to halt construction of the relative boundary vector tree.

(5.2.2) Theorem: Let $\tilde{x}_0$ be such that $[y_i, \tilde{x}_0] \geqslant 0$ for all $i \in I$. Then $\tilde{x}_0$ is in every nonzero hill.

Proof: Let $\tilde{x}_0 \notin C(\pi, M)^+$, a nonzero hill. Then there exists j such that $[y_j, \tilde{x}_0] < 0$, a contradiction. □

-- lower dimensional problems --

In order to prove the validity of the upcoming relative boundary vector collection algorithm, it is necessary to inductively relate the hills of one problem to the hills of associated lower dimensional problems. The following definitions and theorems parallel corresponding ones in Chapter 3.

(5.2.3) Definition: Let $K \subset I$. Let $S := \mathrm{L}\{y_i, i \in K\}$. Suppose $1 \leqslant \dim S \leqslant d-1$. Let R be any subspace such that $R \oplus S = X$. For all $i \in I$, let $z_i := P[y_i \mid R, S]$.

(5.2.4) Theorem: The set $\{z_i, i \in I\} \subset R$ is a set of vectors which satisfies all of the assumptions listed for $\{y_i, i \in I\} \subset X$ in problem statement (5.1.1), namely:

(i) $\dim R \geqslant 1$

(ii) $z_0 = 0$

(iii) $\mathsf{L}\{z_i, i \in I\} = R$

(5.2.5) Theorem:

$$P[C(\pi,M) \mid R, S] = \mathsf{C}\{\pi_i z_i, i \in I \text{ w.o. } M, z_i, i \in M\}.$$

Also, $P[C(\pi,M) \mid R, S]^+ = (C(\pi,M)^+ \cap S^\perp)|_R$.

The next definition provides notation for "$C(\pi,M)$" in the $\{z_i, i \in I\}$ setting.

(5.2.6) Definition: Let $\{z_i, i \in I\}$ be determined as in (5.2.3). Let $I_0 \subset M \subset I$ be an index set where $k \in M$ if and only if $z_k \in \dot{C}_M := \mathsf{C}\{z_i, i \in M\}$. Suppose $\dot{C}_M$ is a subspace.

If $M \neq I$, then let $\pi_i \in \{-1, 1\}$ for $i \in I$ w.o. M and suppose $\mathsf{C}\{\pi_i t_i, i \in I \text{ w.o. } M\}$ is pointed where $t_i := P[z_i \mid T, \dot{C}_M]$ for any subspace T such that $T \oplus \dot{C}_M = R$.

If $M = I$, then $\dot{C}(\pi,M) := \mathsf{C}\{z_i, i \in M\}$.

If $M \neq I$, then $\dot{C}(\pi,M) := \mathsf{C}\{\pi_i z_i, i \in I \text{ w.o. } M, z_i, i \in M\}$.

Notation is also needed for certain subsets of I relative to the $\{z_i, i \in I\}$ context.

(5.2.7) Definition: Let $\{z_i, i \in I\}$ be determined as in (5.2.3) and let $M \neq I$ be as in (5.2.6). For $i \in I$ w.o. M, let $\pi_i \in \{-1, 1\}$. For $j \in I$ w.o. M, $\dot{I}_j(\pi) := \{i \in I \text{ w.o. } M : \langle \pi_i z_i \rangle = \langle \pi_j z_j \rangle\}$. (Recall from before that $I_j(\pi) := \{i \in I \text{ w.o. } M : \langle \pi_i y_i \rangle = \langle \pi_j y_j \rangle\}$.)

-- one way to propagate hills to lower dimensions --

The next theorem shows one way a hill in the original problem can generate a hill in a suitable lower dimensional problem.

(5.2.8) Theorem: Let $\dim X \geqslant 2$. Suppose $C(\pi,M)^+$ is a hill such that $I_0 \neq M \neq I$ (or, in other words, $C(\pi,M)^+ \subset \tilde{X}$ is a nonzero lower dimensional hill.) Let $S \subset C_M$ be a subspace of dimension $\geqslant 1$. Let R be any subspace such that $R \oplus S = X$ and let $z_i := P[y_i \mid R, S]$ for $i \in I$. Then:

(i) $(P[C(\pi, M) \mid R, S])^+$

$$= \{\tilde{r} \in \tilde{R} : [\pi_i z_i, \tilde{r}] \geqslant 0,\ i \in I \text{ w.o. } M,\ [z_i, \tilde{r}] = 0,\ i \in M\}$$

is a hill $\dot{C}(\pi, M)^+$ for the lower dimensional set $\{z_i, i \in I\}$.

(ii) $\dot{C}_M \oplus S = C_M$ and so $\dim \dot{C}_M = \dim C_M - \dim S$.

(iii) $\dot{C}(\pi, M)^+ = C(\pi, M)^+|_R$ and so $\dim \dot{C}(\pi, M)^+ = \dim C(\pi, M)^+$ by virtue of the isomorphism ψ of (2.1.27).

Proof:

(i): In order to use the notation $\dot{C}(\pi, M)^+$ for $(P[C(\pi, M) \mid R, S])^+$, it is necessary to first verify that the index set M has the requisite property. Observe for $j \in I$:

$$z_j \in \dot{C}_M$$

$$\Leftrightarrow \quad \text{for some } \lambda_i \geqslant 0,\ z_j = \sum_{i \in M} \lambda_i z_i$$

$$\Leftrightarrow \quad \text{for some } \lambda_i \geqslant 0 \text{ and } s \in S,\ y_j = \sum_{i \in M} \lambda_i y_i + s$$

$$\Leftrightarrow \quad j \in M$$

Next, note $\dot{C}_M$ is a subspace since it is the projection using $P[\cdot \mid R, S]$ of the subspace C_M.

To continue with the proof of (i), it is necessary to prove (ii) which states that $\dot{C}_M \oplus S = C_M$. To see that $\mathrm{L}\{z_i, i \in M\} + S = \mathrm{L}\{y_i, i \in M\}$, recall that for all $i \in M$, there exists $s_i \in S$ such that $y_i = z_i + s_i$. Finally, if for some $s \in S$, $s \in \dot{C}_M \subset R$, then $s = 0$.

Now to satisfy the rest of Definition (5.2.6) and furthermore, to show that $\dot{C}(\pi, M)^+$ is a hill for the $\{z_i, i \in I\}$ problem, let T be any subspace such that $T \oplus \dot{C}_M = R$. (Note that since

$$d - \dim S = \dim R = \dim T + \dim C_M - \dim S,$$

$\dim T = d - \dim C_M \geqslant 1$.) Let $t_i := P[z_i \,|\, T, \dot{C}_M]$ for $i \in I$ w.o. M. It is necessary to show:

(a) $C\{\pi_i t_i, \, i \in I \text{ w.o. } M\}$ is pointed.

(b) There exists $\dot{K}^+(\pi) \subset I$ w.o. M such that $\{\langle 0 \rangle, \langle t_i \rangle, \, i \in \dot{K}^+(\pi)\}$ is the frame for $C\{\pi_i t_i, \, i \in I \text{ w.o. } M\}$.

Observe that $X = R \oplus S = T \oplus \dot{C}_M \oplus S = T \oplus C_M$. Letting $U = T$ and $u_i := P[y_i \,|\, U, C_M]$ for $i \in I$ w.o. M, observe that since $C(\pi, M)^+$ is a hill,

(a) $C\{\pi_i u_i, \, i \in I \text{ w.o. } M\}$ is pointed.

(b) There exists $K^+(\pi) \subset I$ w.o. M such that $\{\langle 0 \rangle, \langle u_i \rangle, \, i \in K^+(\pi)\}$ is the frame for $C\{\pi_i u_i, \, i \in I \text{ w.o. } M\}$.

It now suffices to show that $u_i = t_i$ for $i \in I$ w.o. M. Fix $i \in I$ w.o. M. There exists $s_i, w_i \in S$ and $v_i \in \dot{C}_M$ such that $u_i + v_i + s_i = y_i = z_i + w_i$. Consequently, $z_i - u_i - v_i \in S \cap R = \langle 0 \rangle$ and so $t_i = P[u_i + v_i \,|\, U, \dot{C}_M] = u_i$.

(iii): Observe

$$\begin{aligned} &\{\tilde{r} \in \tilde{R} : [\pi_i z_i, \tilde{r}] \geqslant 0 \text{ for } i \in I \text{ w.o. } M, [z_i, \tilde{r}] = 0 \text{ for } i \in M\} \\ &= \{\tilde{x}|_R : \tilde{x} \in S^\perp, [\pi_i y_i, \tilde{x}] \geqslant 0 \text{ for } i \in I \text{ w.o. } M, [y_i, \tilde{x}] = 0 \text{ for } i \in M\}. \end{aligned}$$

□

In particular, the previous theorem shows that when a nonzero lower dimensional hill $C(\pi, M)^+$ in the $\{y_i, \, i \in I\}$ problem is intersected with the hyperspace $y_q^\perp$ for $0 \neq y_q \in \operatorname{Lin} C(\pi, M)$, the resulting set of linear functionals is isomorphic to a certain hill in a lower dimensional problem. This second hill is of the same dimension as the first and one dimension closer to being a fully dimensional cone for its problem.

-- *another way to propagate hills to lower dimensions* --

The next theorem shows that for each nonzero hill $C(\pi,M)^+$ in the $\{y_i, i \in I\}$ problem and for each y_q associated with an isolated ray $\langle u_q \rangle$ for $u_q = P[y_q | U, C_M]$, $C(\pi,M)^+ \cap y_q^\perp$ is isomorphic to a certain hill in a lower dimensional problem. This second hill is of one dimension less than the first and no closer to being a fully dimensional cone for its problem than the first one was (if the first one wasn't). For future purposes, Theorem (5.2.9) is stated in a more general setting.

(5.2.9) Theorem: Suppose $\dim X \geqslant 2$. Let $C(\pi,M)^+ \neq \langle 0 \rangle$ be a hill. Let S be any subspace such that $1 \leqslant \dim S \leqslant d-1$ and $S \cap C_M = \langle 0 \rangle$. Let R be any subspace such that $R \oplus S = X$ and let $z_i := P[y_i | R, S]$ for each $i \in I$. Let $L := \{i \in I : y_i \in S \oplus C_M\} \supset M$ and $\dot{C}_L := \mathrm{C}\{z_i, i \in L\}$. Then:

(i)

$$P[C(\pi,M) | R, S]^+ = \{\tilde{r} \in \tilde{R} : [\pi_i z_i, \tilde{r}] \geqslant 0, i \in I \text{ w.o. } M, \\ [z_i, \tilde{r}] = 0, i \in M\}$$

$$= (C(\pi,M)^+ \cap S^\perp)|_R .$$

(ii) If $\mathrm{Lin}\, P[C(\pi,M) | R, S] = P[\mathrm{Lin}\, C(\pi,M) | R, S]$ then $\dot{C}(\pi,L)$ is well-defined and $\dot{C}(\pi,L) = P[C(\pi,M) | R, S]$.

(iii) If $\mathrm{Lin}\, P[C(\pi,M) | R, S] = P[\mathrm{Lin}\, C(\pi,M) | R, S]$, then $P[C(\pi,M) | R, S]^+ = \dot{C}(\pi,L)^+$ and is a hill of dimension $\dim C(\pi,M)^+ - \dim S$ for the lower dimensional set $\{z_i, i \in I\}$. Also, $\dot{C}(\pi,L)^+ = \langle 0 \rangle$ if and only if $\dim C_M = d-1$.

(iv) In particular, if $S = \mathrm{L}\{y_q\}$ for any $q \in I$ w.o. M such that $\langle u_q \rangle$ is an isolated ray of $\mathrm{C}\{\pi_i u_i = \pi_i P[y_i | U, C_M] : i \in I \text{ w.o. } M\}$, then $P[C(\pi,M) | R, S]^+ = \dot{C}(\pi,L)^+$ and is a hill of dimension $\dim C(\pi,M)^+ - 1$ for $\{z_i, i \in I\}$.

(v) $\dot{C}_L = P[C_M \mid R, S]$.

(vi) $\dim \dot{C}_L = \dim C_M$.

(vii) $\dot{C}_L \subset C_M \oplus S$ and $C_M \subset \dot{C}_L \oplus S$.

Proof: The first thing to show is that the index set L is such that $j \in L$ if and only if $z_j \in \dot{C}_L$. Observe for any $j \in I$:

$$z_j \in \dot{C}_L$$

$$\Longleftrightarrow \text{ there exist } \lambda_i \geqslant 0 \text{ such that } z_j = \sum_L \lambda_i z_i$$

$$\Longleftrightarrow \text{ there exist } \lambda_i \geqslant 0 \text{ and } s \in S \text{ such that } y_j = \sum_L \lambda_i y_i + s$$

$$\Longleftrightarrow y_j \in C_M \oplus S$$

$$\Longleftrightarrow j \in L$$

(v): Note $\dot{C}_L$ is a subspace since C_M is a subspace and

$$\dot{C}_L = P[\mathrm{C}\{y_i, i \in L\} \mid R, S] = P[\mathrm{C}\{y_i, i \in M\} \mid R, S]$$

where the last "$\subset$" inclusion holds since for any choice of $\lambda_i \geqslant 0$, $i \in L$, there exist $s \in S$ and for each $i \in L$, $\alpha_{ij} \geqslant 0$ for all $j \in M$ such that

$$\sum_{i \in L} \lambda_i y_i = \sum_{j \in M} \left(\sum_{i \in L} \lambda_i \alpha_{ij} \right) y_j + s$$

and so $\mathrm{C}\{y_i, i \in L\} \subset C_M \oplus S$.

(vi): Next, it is shown that $\dim \dot{C}_L = \dim C_M$. If $C_M = \langle 0 \rangle$, then the result follows.

When $C_M \neq \langle 0 \rangle$, there exists $K \neq \varnothing$ such that $\{y_i, i \in K\}$ is a basis for $C_M = \mathrm{L}\{y_i, i \in K\}$. Clearly $\dot{C}_L = \mathrm{L}\{z_i, i \in K\}$. Suppose there exist β_i such that $\sum_{i \in K} \beta_i z_i = 0$. Then $\sum_{i \in K} \beta_i y_i \in S \cap C_M = \langle 0 \rangle$ and so all $\beta_i = 0$.

(vii): Consider now showing half of (vii), namely $\dot{C}_L \subset C_M \oplus S$.

Using the basis $\{z_i, i \in K\}$ from (vi), observe that for all $\alpha_i \in \mathrm{R}$, $\sum_{i \in K} \alpha_i z_i = \sum_{i \in K} \alpha_i y_i + s$ for some $s \in S$.

(ii): It is given that $\mathrm{Lin}\ P[C(\pi, M) \mid R, S] = P[C_M \mid R, S]$. First observe:

$$P[C(\pi,M) \mid R, S] = \mathrm{C}\{\pi_i z_i, i \in I \text{ w.o. } M, z_i, i \in M\}$$

$$= \mathrm{C}\{\pi_i z_i, i \in I \text{ w.o. } L, \pi_i z_i, i \in L \text{ w.o. } M, z_i, i \in M\}$$

$$= \mathrm{C}\{\pi_i z_i, i \in I \text{ w.o. } L, z_i, i \in M\}$$

$$= \mathrm{C}\{\pi_i z_i, i \in I \text{ w.o. } L, z_i, i \in L\}$$

where both of the last equalities use the fact that $\dot{C}_L = \mathrm{C}\{z_i, i \in M\}$ is a subspace.

In order to complete the requirements of Definition (5.2.6) for using "$\dot{C}(\pi,L)$" to represent $P[C(\pi,M) \mid R, S]$, observe that if $L \neq I$, T is any subspace such that $T \oplus \dot{C}_L = R$, and $t_i := P[z_i \mid T, \dot{C}_L]$ for $i \in I$ w.o. L, then by Theorem (5.1.2), $\mathrm{C}\{\pi_i t_i, i \in I \text{ w.o. } L\}$ is pointed since

$$\mathrm{Lin}\ \mathrm{C}\{\pi_i z_i, i \in I \text{ w.o. } M, z_i, i \in M\} = \dot{C}_L.$$

(iii): As in (ii), it is given that $\mathrm{Lin}\ P[C(\pi,M) \mid R, S] = \dot{C}_L$. To show that $\dot{C}(\pi,L)^+$ is a hill, it must be shown that for any subspace T such that $T \oplus \dot{C}_L = R$ and $t_i := P[z_i \mid T, \dot{C}_L]$, there exists $\dot{K}^+(\pi) \subset I$ w.o. L such that $\{\langle 0 \rangle, \langle t_i \rangle, i \in \dot{K}^+(\pi)\}$ is a frame for $\mathrm{C}\{\pi_i t_i, i \in I \text{ w.o. } L\}$.

Suppose $\dim T \geqslant 1$ and consequently I w.o. $L \neq \varnothing$. Observe that $X = T \oplus \dot{C}_L \oplus S$. So take $U = T \oplus S$.

Claim: $U \oplus C_M = X$.

Since $U \oplus \dot{C}_L = X$, $\dim U = d - \dim C_M$. It suffices now to show $U \cap C_M = \langle 0 \rangle$. Suppose $a = t + s$ where $a \in C_M$, $t \in T$, and $s \in S$.

Then $t = P[t \mid R, S] = P[a \mid R, S] \in \dot{C}_L$. This implies $t = 0$ and so $a \in S \cap C_M = \langle 0 \rangle$.

Now, let $u_i := P[y_i \mid U, C_M]$ for $i \in I$ w.o. M.

Claim: For all $i \in I$ w.o. L, there exists $s_i \in S$ such that $t_i = u_i + s_i$.

Fix $i \in I$ w.o. L. For this i, there exists $w_i \in S$, $b_i \in \dot{C}_L$, and $c_i \in C_M$ such that $y_i = z_i + w_i$, $z_i = t_i + b_i$, and $y_i = u_i + c_i$. Consequently, $u_i + c_i = t_i + b_i + w_i$. Now, $b = e_i + v_i$ where $e_i \in C_M$ and $v_i \in S$. It follows that $t_i - u_i + (v_i + w_i) = c_i - e_i \in U \cap C_M = \langle 0 \rangle$.

Now, since $C(\pi, M)^+ \neq \langle 0 \rangle$ is a hill, there exists $\varnothing \neq K^+ \subset I$ w.o. M such that $\mathsf{C}\{\pi_i u_i, i \in I \text{ w.o. } M\} = \mathsf{C}\{u_i, i \in K^+\}$. Consequently,

$$\mathsf{C}\{\pi_i P[u_i \mid T, S], i \in I \text{ w.o. } M\} = \mathsf{C}\{P[u_i \mid T, S], i \in K^+\}.$$

Now I w.o. $M = (I$ w.o. $L) + (L$ w.o. $M)$. From before, it is known that $P[u_i \mid T, S] = t_i$ for $i \in I$ w.o. L. For $i \in L$ w.o. M, $P[u_i \mid T, S] = 0$ since for any such i, $y_i = e_i + s_i$ for some $e_i \in C_M$ and $s_i \in S$ and so $u_i = s_i$. So, in short, $\mathsf{C}\{\pi_i t_i, i \in I \text{ w.o. } L\} = \mathsf{C}\{t_i, i \in K^+ \text{ w.o. } L\}$.

Observe:

$$\begin{aligned}
\dim \dot{C}(\pi, L)^+ &= (d - \dim S) - \dim \operatorname{Lin} \dot{C}(\pi, L) \\
&= (d - \dim S) - \dim \dot{C}_L \\
&= (d - \dim C_M) - \dim S \\
&= \dim C(\pi, M)^+ - \dim S
\end{aligned}$$

(iv): In view of (iii), it suffices to show that $\operatorname{Lin} \mathsf{C}\{\pi_i z_i, i \in I \text{ w.o. } L, z_i, i \in L\} = \dot{C}_L$ when $S = \mathsf{L}\{y_q\}$. By Theorem (5.1.2), it suffices to show for any subspace T such that $T \oplus \dot{C}_L = R$ and $t_i := P[z_i \mid T, \dot{C}_L]$ that $\mathsf{C}\{t_i, i \in I \text{ w.o. } L\}$ is pointed.

Let $U = T \oplus \mathrm{L}\{y_q\}$ and $u_i := P[y_i | U, C_M]$ for $i \in I$ w.o. M. By the proof of (iii), $U \oplus C_M = X$ and for all $i \in I$ w.o. L, there exists $\alpha_i \in \mathbf{R}$ such that $t_i = u_i + \alpha_i y_q$.

Now suppose there exists $\lambda_i \geqslant 0$, not all 0, for $i \in I$ w.o. L such that $0 = \sum_{I \text{ w.o. } L} \lambda_i \pi_i t_i$. Noting that $y_q = u_q$ and $q \in L$ w.o. M, it follows that for $\beta = \sum_{I \text{ w.o. } L} \lambda_i \pi_i \alpha_i$, $0 = \sum_{I \text{ w.o. } L} \lambda_i \pi_i u_i + \beta u_q$. Observe that if $\beta \geqslant 0$, then $\mathrm{C}\{\pi_i u_i, i \in I \text{ w.o. } M\}$ is not pointed. If $\beta < 0$, then $\langle u_q \rangle$ is not isolated. □

-- *the general hill boundary vector collection algorithm* --

The following algorithm constructs a tree of vectors containing at least one vector belonging to each hill and is thus the first phase of the (extended) tree algorithm.

(5.2.10) Definition: $N(\tilde{x}) := \{i \in I : [y_i, \tilde{x}] < 0\}$. $y(i_k) := y_{i_k}$.

(5.2.11) Algorithm:

For $k = 0, \ldots, d-1$, do: Set $V_k = \varnothing$. next k;

Step 0: If $\mathrm{C}\{y_i, i \in I\}$ is a subspace, set $V_0 = \{0\}$.

Step 1: Obtain $\tilde{v}_0 \neq 0$.
Set $V_0 = V_0 \cup \{\tilde{v}_0, -\tilde{v}_0\}$.
If $N(\tilde{v}_0) = \varnothing$ or $N(-\tilde{v}_0) = \varnothing$ then exit.

Step 2: For $k = 1, \ldots, d-1$, do:
For each $i_0 \in N(\tilde{v}_0)$, $i_1 \in N(\tilde{v}_1(i_0)), \ldots,$
$i_{k-1} \in N(\tilde{v}_{k-1}(i_0, \ldots, i_{k-2}))$, do:
Obtain $\tilde{x} \in y(i_0)^\perp \cap \cdots \cap y(i_{k-1})^\perp$ where $\tilde{x} \neq 0$.
Set $\tilde{v}_k(i_0, \ldots, i_{k-1}) = \tilde{x}$.
Set $V_k = V_k \cup \{\tilde{v}_k(i_0, \ldots, i_{k-1}), -\tilde{v}_k(i_0, \ldots, i_{k-1})\}$.
If $N(\tilde{v}_k(i_0, \ldots, i_{k-1})) = \varnothing$
or $N(-\tilde{v}_k(i_0, \ldots, i_{k-1})) = \varnothing$ then exit.

next i_{k-1}; . . . ; next i_0;

next k;

(5.2.12) Definition: In the sequel, "$\tilde{w}_k(i_0, \ldots, i_{k-1})$" will be used to ambiguously represent one of $\tilde{v}_k(i_0, \ldots, i_{k-1})$ and $-\tilde{v}_k(i_0, \ldots, i_{k-1})$ whichever is desired at the moment. The tree constructed by (5.2.11) is $\bigcup_0^{d-1} V_k$ w.o. $\{0\}$.

-- *(5.2.10)'s hill boundary vector collection works* --

(5.2.13) Theorem: Let X be a d-dimensional vector space over R with $d \geq 2$. Let $\{y_i, i \in I\} \subset X$ be as in (5.1.1) with $y_0 := 0$ and $\mathsf{L}\{y_i, i \in I\} = X$. Suppose the V_k are created as in (5.2.11). Then:

(a) For $k = 0, \ldots, d-2$, for $i_0 \in N(\tilde{v}_0), \ldots,$ $i_k \in N(\tilde{v}_k(i_0, \ldots, i_{k-1}))$, $\{y(i_0), \ldots, y(i_k)\}$ is linearly independent.

(b) Algorithm (5.2.11) is well-defined.

(c) If $C(\pi, M)^+$ is a hill, then there exists $\tilde{v} \in \bigcup_0^{d-1} V_k$ such that $\tilde{v} \in C(\pi, M)^+$.

Proof: (a) and (b) are proved in (3.3.22).

If $\langle 0 \rangle$ is a hill $C(\pi, M)^+$, then $M = I$ and C_I is a subspace. Step 0 saves 0 in this case.

Next, it is shown that if $C(\pi, M)^+$ is a nonzero hill, then there is a vector in (5.2.11)'s tree which is in this hill.

-- *when* $\dim X = 2$ --

The proof proceeds by induction on the dimension of X.

Suppose $\dim X = 2$ and $C(\pi,M)^+$ is a nonzero hill. If there exists $\tilde{w}_k(i_0, \ldots, i_{k-1}) \in \bigcup_0^{d-1} V_j$ w.o. 0 ($\tilde{w}_k$ for short) such that $N(\tilde{w}_k) = \varnothing$, then by Theorem (5.2.2), $\tilde{w}_k$ is in every nonzero hill including $C(\pi,M)^+$.

So suppose for all $\tilde{w}_k \in \bigcup_0^{d-1} V_j$ w.o. 0 that $N(\tilde{w}_k) \neq \varnothing$. Suppose additionally that no $\tilde{w}_k \in C(\pi,M)^+$. Observe that since $C(\pi,M)^+ \neq \langle 0 \rangle$, $C_M \neq X$ (and $M \neq I$). Hence $\dim C_M = 0$ or 1.

Case 1: $\tilde{v}_0 \notin C_M^\perp$.

Now if $\dim C_M = 0$, then $\tilde{v}_0 \in C_M^\perp$; hence $\dim C_M = 1$. So there exist p, $q \in M$ w.o. I_0 such that $C_M = \langle y_p \rangle \cup \langle 0 \rangle \cup \langle y_q \rangle$. Using Theorem (5.2.1), without loss of generality, it is permissible to assume $p \in N(\tilde{v}_0)$. Step 2 of Algorithm (5.2.11) obtains

$$\tilde{v}_1(p), -\tilde{v}_1(p) \in y_p^\perp = \langle \tilde{v}_1(p) \rangle \cup \langle 0 \rangle \cup \langle -\tilde{v}_1(p) \rangle.$$

But since $C(\pi,M)^+$ is a pointed one-dimensional cone contained in $y_p^\perp$, either $C(\pi,M)^+ = \langle \tilde{v}_1(p) \rangle \cup \langle 0 \rangle$ or $C(\pi,M)^+ = \langle -\tilde{v}_1(p) \rangle \cup \langle 0 \rangle$ and a nonzero vector in the tree has been found in $C(\pi,M)^+ \neq \langle 0 \rangle$, which is a contradiction.

Case 2: $\tilde{v}_0 \in C_M^\perp$

Subcase 2.1: $\dim C_M = 0$

By Theorem (5.2.1), there exists $j \in N(\tilde{v}_0)$ such that $\langle y_j \rangle \neq \langle 0 \rangle$ is in the frame of $C\{\pi_i y_i, i \in I \text{ w.o. } M\}$ and so by (2.4.11), $F_j(C(\pi,M)^+) \neq \varnothing$ or $\langle 0 \rangle$. Now $F_j(C(\pi,M)^+)$ is a (one-dimensional) open ray in $y_j^\perp$ and since $\tilde{v}_1(j)$ and $-\tilde{v}_1(j)$ are saved by Algorithm (5.2.11) and generate both of the one-dimensional open rays in $y_j^\perp$, Algorithm (5.2.11) saves a nonzero vector in $C(\pi,M)^+$ which is a contradiction.

Subcase 2.2: $\dim C_M = 1$

Since $C(\pi,M)^+$ is a nonzero pointed cone in $C_M^\perp = \langle \tilde{v}_0 \rangle \cup \langle 0 \rangle \cup \langle -\tilde{v}_0 \rangle$, either $\tilde{v}_0 \in C(\pi,M)^+$ or $-\tilde{v}_0 \in C(\pi,M)^+$ which is a contradiction.

-- *when* dim $X \geqslant 3$ --

Suppose that dim $X \geqslant 3$ and that Theorem (5.2.13) holds for all vector spaces of strictly smaller dimension. It will be shown that if $C(\pi, M)^+$ is a nonzero hill, then there is a vector in Algorithm (5.2.11)'s tree which is in this hill. As before, if there exists $\tilde{w}_k(i_0, \ldots, i_{k-1}) \in \bigcup_0^{d-1} V_j$ w.o. 0 such that $N(\tilde{w}_k) = \emptyset$, then by Theorem (5.2.2), $\tilde{w}_k$ is in every nonzero hill including $C(\pi, M)^+$.

So suppose for all $\tilde{w}_k \in \bigcup_0^{d-1} V_j$ w.o. 0 that $N(\tilde{w}_k) \neq \emptyset$. Suppose additionally that no $\tilde{w}_k \in C(\pi, M)^+$. Observe since $C(\pi, M)^+ \neq \langle 0 \rangle$, dim $C_M \leqslant d-1$.

Case 1: $\tilde{v}_0 \in C_M^\perp$ and dim $C_M = d-1$.

Since $\langle 0 \rangle \neq C(\pi, M)^+ \subset C_M^\perp$, dim $C(\pi, M)^+ = 1$. But $C_M^\perp = \langle \tilde{v}_0 \rangle \cup \langle 0 \rangle \cup \langle -\tilde{v}_0 \rangle$, so that either $\tilde{v}_0 \in C(\pi, M)^+$ or $-\tilde{v}_0 \in C(\pi, M)^+$ which is a contradiction.

For the other two cases, it will be necessary to construct a tree for a lower dimensional problem out of the tree which Algorithm (5.2.11) has provided. The following procedure shows how to do this:

Procedure: Take $i_0 \in N(\tilde{v}_1(i_0)), \ldots, i_{k-1} \in N(\tilde{v}_{k-1}(i_0, \ldots, i_{k-2}))$ for some $1 \leqslant k \leqslant d-2$. Let $S = \mathsf{L}\{y(i_0), \ldots, y(i_{k-1})\}$. Let R be any subspace such that $R \oplus S = X$. Let $z_i := P[y_i \mid R, S]$ for all $i \in I$. For $\tilde{r} \in \tilde{R}$, $\dot{N}(\tilde{r}) := \{i \in I : [z_i, \tilde{r}] < 0\}$. Since for $i \in I$ and $\tilde{v} \in S^\perp$, $[z_i, \tilde{v}] = [y_i, \tilde{v}]$, it is clear that $\dot{N}(\tilde{v}|_R) = N(\tilde{v})$ for all $\tilde{v} \in S^\perp$.

Here is the procedure:

For $k = 0, \ldots, d - \dim S - 1$, do: Set $\dot{V}_k = \varnothing$; next k;

Step 1: Set $\dot{\tilde{v}}_0 = \tilde{v}_k(i_0, \ldots, i_{k-1})\big|_R$.
Set $\dot{V}_0 = \dot{V}_0 \cup \{\dot{\tilde{v}}_0, -\dot{\tilde{v}}_0\}$.

Step 2: For $q = 1, \ldots, d - \dim S - 1$, do:
For each $j_0 \in \dot{N}(\dot{\tilde{v}}_0)$, $j_1 \in \dot{N}(\dot{\tilde{v}}_1(j_0)), \ldots,$
$j_{q-1} \in \dot{N}(\dot{\tilde{v}}_{q-1}(j_0, \ldots, j_{q-2}))$, do:
Set $\dot{\tilde{v}}_q(j_0, \ldots, j_{q-1}) = \tilde{v}_{k+q}(i_0, \ldots, i_{k-1}, j_0, \ldots, j_{q-1})\big|_R$.
Set $\dot{V}_q = \dot{V}_q \cup \{\dot{\tilde{v}}_q(j_0, \ldots, j_{q-1}), -\dot{\tilde{v}}_q(j_0, \ldots, j_{q-1})\}$.
next j_{q-1}; ... ; next j_0;
next q;

Claim: The above procedure is well-defined in the sense that it is able to find all of the necessary vectors in the original tree in order to construct the new tree. The new tree satisfies the requirements of Algorithm (5.2.11) for the $\{z_i, i \in I\}$ problem.

Proof: As far as Step 1 goes, $\tilde{v}_k(i_0, \ldots, i_{k-1})$ is certainly in the original tree and since $\tilde{v}_k(i_0, \ldots, i_{k-1}) \in S^\perp$, $\tilde{v}_k(i_0, \ldots, i_{k-1})\big|_R$ is a valid choice for $\dot{\tilde{v}}_0$. Note that $\dot{N}(\dot{\tilde{v}}_0)$ and $\dot{N}(-\dot{\tilde{v}}_0) \neq \varnothing$ and so Algorithm (5.2.11) for the $\{z_i, i \in I\}$ problem would continue on to Step 2 at this point.

To handle Step 2, proceed by finite induction on q. Observe since $d \geqslant 3$, Step 2 is always executed for $q = 1$. Suppose $q = 1$. Take any

$$j_0 \in \dot{N}(\dot{\tilde{v}}_0) = \dot{N}(\tilde{v}_k(i_0, \ldots, i_{k-1})\big|_R) = N(\tilde{v}_k(i_0, \ldots, i_{k-1})) \neq \varnothing.$$

Then $0 \neq \tilde{v}_{k+1}(i_0, \ldots, i_{k-1}, j_0) \in S^\perp \cap y(j_0)^\perp$ is an element of the original tree. Observe

$$[z(j_0), \tilde{v}_{k+1}(i_0, \ldots, i_{k-1}, j_0)\big|_R] = [y(j_0), \tilde{v}_{k+1}(i_0, \ldots, i_{k-1}, j_0)] = 0.$$

In short, $\tilde{v}_{k+1}(i_0, \ldots, i_{k-1}, j_0)\big|_R$ exists and is a valid choice for $\dot{\tilde{v}}_1(j_0)$. Note $\dot{N}(\dot{\tilde{v}}_1(j_0))$ and $\dot{N}(-\dot{\tilde{v}}_1(j_0)) \neq \varnothing$ so that Algorithm (5.2.11) for the $\{z_i, i \in I\}$

problem would not exit at this point.

Assume that the procedure works as claimed for $q = 1, \ldots, p$. To show that it works for $q = p + 1 \leqslant d - \dim S - 1$, begin by taking

$$j_0 \in \dot{N}(\dot{v}_0) = N(\tilde{v}_k(i_0, \ldots, i_{k-1})) \neq \varnothing,$$
$$j_1 \in \dot{N}(\dot{v}_1(j_0)) = N(\tilde{v}_{k+1}(i_0, \ldots, i_{k-1}, j_0)) \neq \varnothing, \ldots,$$
$$j_p \in \dot{N}(\dot{v}_p(j_0, \ldots, j_{p-1})) = N(\tilde{v}_{k+p}(i_0, \ldots, i_{k-1}, j_0, \ldots, j_{p-1})) \neq \varnothing.$$

Then $\tilde{v}_{k+p+1}(i_0, \ldots, i_{k-1}, j_0, \ldots, j_p) \neq 0$ is in the original tree. Note that $\tilde{v}_{k+p+1}(i_0, \ldots, i_{k-1}, j_0, \ldots, j_p)\big|_R \in \tilde{R}$ and for $m = 0, \ldots, p$,

$$\begin{aligned}[z(j_m), \tilde{v}_{k+p+1}(i_0, \ldots, i_{k-1}, j_0, \ldots, j_p)\big|_R] \\ &= [y(j_m), \tilde{v}_{k+p+1}(i_0, \ldots, i_{k-1}, j_0, \ldots, j_p)] \\ &= 0\end{aligned}$$

so that $\tilde{v}_{k+p+1}(i_0, \ldots, i_{k-1}, j_0, \ldots, j_p)\big|_R$ is a valid choice for $\dot{v}_{p+1}(j_0, \ldots, j_p)$. Note also that $\dot{N}(\dot{v}_{p+1}(j_0, \ldots, j_p))$ and $\dot{N}(-\dot{v}_{p+1}(j_0, \ldots, j_p)) \neq \varnothing$ so that Algorithm (5.2.11) for the $\{z_i, i \in I\}$ problem would not exit at this point.

Here are the last two cases:

Case 2: $\tilde{v}_0 \notin C_M^{\perp}$.

By Theorem (5.2.1), there exists $p \in M$ w.o. $I_0 \neq \varnothing$ such that $p \in N(\tilde{v}_0)$. Use the preceding procedure to construct a tree satisfying all requirements of Algorithm (5.2.11) for the $\{z_i, i \in I\}$ problem when $S = \mathrm{L}\{y_p\}$. By Theorem (5.2.8), $\dot{C}(\pi, M)^+ = C(\pi, M)^+\big|_R$ is a hill for the lower dimensional $\{z_i, i \in I\}$ problem. It is a nonzero hill since $\dim \dot{C}(\pi, M)^+ = \dim C(\pi, M)^+ \neq 0$. Consequently, the induction hypothesis asserts that there exists $\dot{w}_k \in \bigcup_0^{d-2} \dot{V}_k$ w.o. 0 such that $\dot{w}_k \in \dot{C}(\pi, M)^+$. As a result, there exists $\tilde{u} \in C(\pi, M)^+$ such that $\psi(\tilde{w}_{k+1}(p, i_0, \ldots, i_{k-1})) = \psi(\tilde{u})$ which implies that $\tilde{w}_{k+1}(p, i_0, \ldots, i_{k-1}) \in C(\pi, M)^+$ and that the original tree contained a vector in $C(\pi, M)^+$ after all. This is a contradiction.

Case 3: $\tilde{v}_0 \in C_M^{\perp}$ and $\dim C_M < d - 1$.

By Theorem (5.2.1), there exists $p \in I$ w.o. $M \neq \emptyset$ such that $\langle 0 \rangle \neq \langle u_p \rangle$ is in the frame of $\mathsf{C}\{\pi_i u_i, i \in I \text{ w.o. } M\}$ and such that $p \in N(\tilde{v}_0)$. Use the preceding procedure to construct a tree satisfying all requirements of Algorithm (5.2.11) for the $\{z_i, i \in I\}$ problem when $S = \mathsf{L}\{y_p\}$. By Theorem (5.2.9), $\dot{C}(\pi, L)^+ = (C(\pi, M)^+ \cap y_p^{\perp})\big|_R$ is a hill in the lower dimensional $\{z_i, i \in I\}$ problem. $\dot{C}(\pi, L)^+$ is nonzero since $\dim C_M < d - 1$. Consequently, the induction hypothesis asserts that there exists $\dot{\tilde{w}}_k \in \bigcup_0^{d-2} \dot{V}_k$ w.o. 0 such that $\dot{\tilde{w}}_k \in \dot{C}(\pi, L)^+$. As a result, there exists $\tilde{u} \in C(\pi, M)^+ \cap y_p^{\perp}$ such that $\psi(\tilde{w}_{k+1}(p, i_0, \ldots, i_{k-1})) = \psi(\tilde{u})$ which implies that $\tilde{w}_{k+1}(p, i_0, \ldots, i_{k-1}) \in C(\pi, M)^+$ and that the original tree contained a vector in $C(\pi, M)^+$ after all. This is a contradiction. $\square$

Summary For Section 5.2

Similar to the WOH tree algorithm, the general tree algorithm constructs a tree of relative boundary vectors in such a way that for each nonzero hill $C(\pi,M)^+$, the tree contains at least one vector in that hill. It does this by starting with an arbitrary initial vector $\tilde{v}_0 \neq 0$ and identifying all y_i such that $[y_i, \tilde{v}_0] < 0$. For each such y_i, a nonzero $\tilde{v}_1 \in y_i^{\perp}$ is computed. The process now becomes recursive with the children of each node $\tilde{v}_k$ in the tree being determined in two steps. The first step identifies all y_i such that $[y_i, \tilde{v}_k] < 0$; the second obtains a child $\tilde{v}_{k+1}$ for each such y_i by computing some $\tilde{v}_{k+1} \neq 0$ constrained not only to lie in $y_i^{\perp}$ but also constrained to lie in all of the hyperspaces constraining its parent. This process continues until nodes are obtained which cannot have children according to the preceding rules. The tree is completed by expanding each node to contain the negative of each $\tilde{v}_k$ in it. This tree then plus $\{0\}$ if $\mathsf{C}\{y_i, i \in I\}$ is a subspace contains at least one nontrivial vector for each hill. This collection of vectors differs from a valid WOH tree algorithm tree only in the presence of $-\tilde{v}_k$ for each $\tilde{v}_k$ not in the bottom level of the tree and in the possible presence of 0. It is to be stressed that the relative boundary vector collection algorithm recommended for use in practice is considerably more sophisticated than this one.

One of the principal results validating the relative boundary vector collection algorithm is that if some $\tilde{v}$ is not an element of a nonzero hill $C(\pi,M)^+$, then:

(i) if $\tilde{v}$ is not in the dimensionality space of $C(\pi,M)^+$, then there is a $y_i \in \mathsf{Lin}\, C(\pi,M) = C_M$ such that $[y_i, \tilde{v}] < 0$, whereas

(ii) if $\tilde{v}$ is in the dimensionality space of $C(\pi,M)^+$, then there is a y_i with $\langle P[y_i \,|\, U, C_M] \rangle$ in the frame of $P[C(\pi,M) \,|\, U, C_M]$ for any U such that $U \oplus C_M = X$ such that $[y_i, \tilde{v}] < 0$.

These are the "signals" which hills send to $\tilde{v}$ which are not in them.

One of the other two principal results justifying the relative boundary vector collection algorithm states that for any nonzero hill $C(\pi, M)^+$, if S is a subspace such that $S \subset C_M$ and R is any subspace such that $X = R \oplus S$, then $P[C(\pi, M) \mid R, S]^+$ is a nonzero hill for the associated lower dimensional problem in R. Its companion result states that if $C(\pi, M)^+$ is a nonzero hill and $S \neq X$ is any nonzero subspace such that $S \cap C_M = \langle 0 \rangle$ then for any R such that $X = R \oplus S$, $P[C(\pi, M) \mid R, S]^+$ is a hill in the lower dimensional problem if $\text{Lin}\, P[C(\pi, M) \mid R, S] = P[\text{Lin}\, C(\pi, M) \mid R, S]$.

The proof validating the relative boundary vector collection algorithm proceeds in much the same way as its predecessor in Section 3.3. If $\dim X = 2$ then the result follows by direct brute force. If $\dim X > 2$, then for any nonzero hill $C(\pi, M)^+$ which has none of its vectors in the first level of the tree, one identifies a subtree of the original tree which is isomorphic to a tree corresponding to a lower dimensional problem. By an induction hypothesis, this second tree contains a nonzero vector in a lower dimensional hill which is itself isomorphic to a nonzero subset of $C(\pi, M)^+$. Thus, there is a subtree of the tree in the original problem which contains a nonzero vector of $C(\pi, M)^+$.

Section 5.3:

Improvements To The Relative Boundary Vector Collection Phase Of The General Tree Algorithm

Many of the suggestions given in Section 3.4 for improving the first phase of the WOH tree algorithm carry over without need of change or much discussion to improvements for the general tree algorithm. These include eliminating all conical ties and all $y_i = 0$, exploring the tree in a depth-first search manner, making sure that the algorithm stops if $N(\pm\tilde{v}_{d-1}) = \varnothing$, making sure $\#N(\tilde{v}_k) < \#N(-\tilde{v}_k)$, using the projection method of determining $\tilde{v}_k$, restarting whenever possible, and using best-first depth-first searching to produce good vectors quickly.

This section will discuss the logic behind UPDATE_B of (5.1.13) which saves promising $\tilde{v}_k$ for displacement. It will also present a theorem which proves that precisely the same method of trimming can be used for the general tree algorithm as was used for the WOH tree algorithm.

-- *the logic behind UPDATE_B in the general tree algorithm* --

(The following discussion of UPDATE_B will become more understandable after reading Section 5.4.)

There are essentially two operative expressions in UPDATE_B:

I: $g(\{(i, 1\{\pi_i > 0\}) : i \in I \text{ w.o. } J\}$
$\cup\{(i, 1\{[y_i, \tilde{x}] \; R_i \; 0\}) : i \in J \text{ w.o. } Z_J(\tilde{x})\}$
$\cup\{(i, 1) : i \in Z_J(\tilde{x})\})$

and

$$\text{II:}\quad g(\{(i,\ 1\{\pi_i > 0\}) : i \in I \text{ w.o. } J\}$$
$$\cup\{(i,\ 1\{[y_i,\ \tilde{x}]\ R_i\ 0\}) : i \in J \text{ w.o. } \{i_0, \ldots, i_{k-1}\}$$
$$\text{and either } y_i \notin \tilde{x}^{\perp} \text{ or } R_i = \text{"}>\text{" or } i \in J_0\}$$
$$\cup\{(i,\ 0) : i \in J \text{ w.o. } \{i_0, \ldots, i_{k-1}\} \text{ and } y_i \in \tilde{x}^{\perp} \text{ and}$$
$$R_i = \text{"}\geqslant\text{" and } i \notin J_0\}$$
$$\cup\{(i,\ 1) : i \in \{i_0, \ldots, i_{k-1}\}\})$$

If it were possible to displace $\tilde{x}$ in such a way as for the displaced vector to be on the positive side of all of its constraining hyperplanes, then Expression I would be the H-value of the displaced vector. Expression I therefore is the largest H-value that a vector (legally) displaced from $\tilde{x}$ could possible have. The first set of ordered pairs in Expression I is that which the current problem inherits from any higher dimensional problems calling it during the recursion part of the displacement phase. The other two sets are self-explanatory.

Expression II gives a lower bound on the set of H-values corresponding to the set of vectors (legally) displaced from $\tilde{x}$. This is based on the knowledge that $\tilde{x}$ can be displaced to some $\tilde{z}$ such that $[y_i, \tilde{z}] > 0$ for $i \in \{i_0, \ldots, i_{k-1}\}$ and that for any $i \in J$ w.o. $\{i_0, \ldots, i_{k-1}\}$ such that $y_i \notin \tilde{x}^{\perp}$ or $R_i = \text{"}>\text{"}$ or $i \in J_0$, $1\{[y_i, \tilde{x}]\ R_i\ 0\} \leqslant 1\{[y_i, \tilde{z}]\ R_i\ 0\}$. However, if $y_i \in \tilde{x}^{\perp}$ and $R_i = \text{"}\geqslant\text{"}$ and $i \notin J_0$, then it is necessary to settle for $0 \leqslant 1\{[y_i, \tilde{z}] \geqslant 0\}$.

The overall strategy here is that UPDATE_B does *not* save $\tilde{x}$ if and only if the best possible H-value that a vector (legally) displaced from $\tilde{x}$ could have is known to be less than some H-value which is known to be obtainable from some vector already saved in B_J either by leaving it alone or by displacing it. And, secondly, any vector in B_J is to be deleted if the largest H-value possible for it is known to less than the best H-value guaranteed to be obtainable from the newcomer $\tilde{x}$ in the event that $\tilde{x}$ is added to B_J.

-- *the WOH trimming procedure carries over to the general setting* --

It will now be shown that the trimming procedure described in Section 3.4 can be used without change by the general tree algorithm. A careful look at the theorems justifying WOH tree algorithm trimming shows that if it can be but proven that Theorem (3.4.15) holds for hills of the form $C(\pi, M)^+$, then the general tree algorithm can successfully use trimming algorithm (3.4.12) as long as no y_i are in $\tilde{v}_k(i_0, \ldots, i_{k-1})^\perp$ when they needn't be.

(5.3.1) Theorem: Let $d = \dim X \geqslant 3$. Suppose the tree constructed by (5.2.11) has been constructed in such a way that for all $k = 1, \ldots, d-2$, for all $i_0 \in N(\tilde{v}_0)$, $i_1 \in N(\tilde{v}_1(i_0)), \ldots, i_{k-1} \in N(\tilde{v}_{k-1}(i_0, \ldots, i_{k-2}))$,

$$\{i \in I : y_i \in \tilde{v}_k(i_0, \ldots, i_{k-1})^\perp \text{ and } y_i \notin \mathsf{L}\{y(i_0), \ldots, y(i_{k-1})\}\} = \varnothing.$$

Let $C(\pi, M)^+ \neq \langle 0 \rangle$ be a hill. Suppose for some fixed $k = 1, \ldots, d-2$ and $i_0, \ldots, i_{k-1} \in I$ that $\tilde{v}_k(i_0, \ldots, i_{k-1})$ is in the untrimmed tree and

$$y(i_0)^\perp \cap \cdots \cap y(i_{k-1})^\perp \cap C(\pi, M)^+ \neq \langle 0 \rangle.$$

Then $\mathsf{Tr}(\tilde{v}_k(i_0, \ldots, i_{k-1}))$ contains a vector of $C(\pi, M)^+$.

Proof: Just as in the proof of Theorem (3.4.15), it will be shown that if the lineality space of $\dot{D}$ where $\dot{D} := P[C(\pi, M) \mid R; \mathsf{L}\{y(i_0), \ldots, y(i_{k-1})\}]$ is the same as $P[C_M \mid R, \mathsf{L}\{y(i_0), \ldots, y(i_{k-1})\}]$ or if $k = d - 2$, then the result follows directly. Otherwise, if $\mathsf{Lin}\, \dot{D}$ is larger than it should be and $k < d - 2$, then one of the children of $\tilde{v}_k(i_0, \ldots, i_{k-1})$ will satisfy the hypotheses of the theorem. Then of course one just goes as far down in the tree as necessary in order to get the result.

In symbols, let $S := \mathsf{L}\{y(i_0), \ldots, y(i_{k-1})\}$. Let R be any subspace such that $X = R \oplus S$. Recall the lower dimensional problem as defined in (5.2.3). Let $\dot{D} := P[C(\pi, M) \mid R, S]$. Note that

$$\dot{D}^+ = (C(\pi, M)^+ \cap S^\perp)\big|_R \neq \langle 0 \rangle$$

since $\tilde{v}_k(i_0, \ldots, i_{k-1})\big|_R \in \dot{D}^+$. Note also that $\mathsf{Lin}\, \dot{D} \supset \mathsf{C}\{z_i,\, i \in M\}$. There are three cases:

Case 1: $\text{Lin } \dot{D} = \mathrm{C}\{z_i, i \in M\}$.

In this case, it can be seen that $\dot{D}^+ \neq \langle 0 \rangle$ is a hill in the $\{z_i, i \in I\}$ problem. In the event that $S \subset C_M$, this follows by (5.2.8). In the event that $S \cap C_M = \langle 0 \rangle$, this follows by (5.2.9).

Suppose then that $J_2 = \{j \in \{0, \ldots, k-1\} : y(i_j) \notin C_M\} \neq \emptyset$ and

$$J_3 = \{j \in \{0, \ldots, k-1\} : y(i_j) \in C_M\} \neq \emptyset.$$

Let $S_2 := \mathrm{L}\{y(i_j) : j \in J_2\}$ and $S_3 := \mathrm{L}\{y(i_j) : j \in J_3\}$. Then $S = S_2 \oplus S_3$ and $X = R \oplus S_2 \oplus S_3$. ($S_2 \cap S_3 = \langle 0 \rangle$ because $\{y(i_j)\}_0^{k-1}$ is linearly independent).

First, observe that since for all $x \in X$, $x = r + s_2 + s_3$ for some $r \in R$, $s_2 \in S_2$, and $s_3 \in S_3$, $P[x \mid R, S] = P[P[x \mid R \oplus S_2, S_3] \mid R, S_2]$.

Now, for all $i \in I$, let $a_i := P[y_i \mid R \oplus S_2, S_3]$. By (5.2.8), $P[C(\pi, M) \mid R \oplus S_2, S_3]^+ = \mathrm{C}\{\pi_i a_i, i \in I \text{ w.o. } M, a_i, i \in M\}^+$ is a hill. Next, for all $i \in I$, let $b_i := P[a_i \mid R, S_2]$. Of course, $b_i = z_i$. Since $d - 2 \geqslant k = d - \dim R$, (5.2.9) can be used to discover that $\dot{D}^+ = \mathrm{C}\{\pi_i z_i, i \in I \text{ w.o. } M, z_i, i \in M\}^+$ is a hill since

$$\text{Lin } \mathrm{C}\{\pi_i b_i, i \in I \text{ w.o. } M, b_i, i \in M\} = \mathrm{C}\{b_i, i \in M\}.$$

Next, following the same argument as given in (5.2.13), if each linear functional in $\mathrm{Tr}(\tilde{v}_k(i_0, \ldots, i_{k-1}))$ is restricted to have domain R, then the resulting tree is a tree which satisfies the hypotheses of (5.2.13) for this lower dimensional problem. Consequently the "restricted" tree contains a nonzero vector in $\dot{D}^+$ and so, using the same kind of argument used in (5.2.13), $\mathrm{Tr}(\tilde{v}_k(i_0, \ldots, i_{k-1}))$ contains a nonzero vector in $C(\pi, M)^+$.

Case 2: $\text{Lin } \dot{D} \neq \mathrm{C}\{z_i, i \in M\}$ and $k = d - 2$.

By the hypothesis, there exists nonzero $\tilde{a} \in S^{\perp} \cap C(\pi, M)^{+}$. $\tilde{a}|_R \neq 0$ and is an element of $\dot{D}^{+} \subset \tilde{R}$.

Since $\dim R = 2$ and $\text{Lin } \dot{D} \neq \text{C}\{z_i, i \in M\} \supset \langle 0 \rangle$, then $\dot{D}$ is not pointed and either $\dim \text{Lin } \dot{D} = 1$ or $\dot{D} = R$. If $\dot{D} = R$, then $\dot{D}^{+} = \langle 0 \rangle$, which is a contradiction. So, $\text{Lin } \dot{D}$ is a one-dimensional subspace and

$$\text{C}\{z_i, i \in M\} = P[\text{Lin } C(\pi, M) \,|\, R, S] = \langle 0 \rangle$$

which implies that $C_M \subset S$.

Let T be some subspace such that $T \oplus C_M = S$. It follows that $X = R \oplus T \oplus C_M$. Let $U = R \oplus T$ and $u_i := P[y_i \,|\, U, C_M]$ for all $i \in I$. Since $C(\pi, M)^{+}$ is a hill, there exists $K^{+}(\pi) \subset I$ w.o. M such that $\{\langle 0 \rangle, \langle u_i \rangle, i \in K^{+}(\pi)\}$ is a frame for $P[C(\pi, M) \,|\, U, C_M]$. Let $b_i := P[u_i \,|\, R, T]$. It follows then that $\{\langle 0 \rangle, \langle b_i \rangle, i \in K^{+}(\pi)\}$ is a spanning set for $\dot{D} = P[P[C(\pi, M) \,|\, U, C_M] \,|\, R, T]$. Hence $\text{C}\{z_i, i \in K^{+}(\pi)\} = \dot{D}$.

Consequently, $\text{Lin } \dot{D} = \text{C}\{z_i : i \in K^{+}(\pi) \text{ and } z_i \in \text{Lin } \dot{D}\}$. So there exist $p, q \in K^{+}(\pi)$ such that $\langle z_p \rangle = \langle -z_q \rangle \neq \langle 0 \rangle$, $\text{Lin } \dot{D} = \langle z_p \rangle \cup \langle 0 \rangle \cup \langle z_q \rangle$, and $\text{L}(\dot{D}^{+}) = z_p^{\perp}$. Now if $[z_p, \tilde{v}_{d-2}(i_0, \ldots, i_{d-3})|_R] = 0$, then $[y_p, \tilde{v}_{d-2}(i_0, \ldots, i_{d-3})] = 0$. But this cannot happen by the hypothesis since $z_p \neq 0$ implies $y_p \notin \text{L}\{y(i_0), \ldots, y(i_{d-3})\}$. As a result, either

$$[z_p, \tilde{v}_{d-2}(i_0, \ldots, i_{d-3})|_R] < 0 \text{ or } [z_q, \tilde{v}_{d-2}(i_0, \ldots, i_{d-3})|_R] < 0.$$

Suppose, without loss of generality, that $[y_p, \tilde{v}_{d-2}(i_0, \ldots, i_{d-3})] < 0$. Algorithm (5.2.11) will then obtain a conical spanning set for the one-dimensional subspace $y(i_0)^{\perp} \cap \cdots \cap y(i_{d-3})^{\perp} \cap y_p^{\perp}$. But inasmuch as $\tilde{a}|_R \in \dot{D}^{+} \subset z_p^{\perp}$, it is clear that $\tilde{a} \in y_p^{\perp}$ and so the algorithm will find $\langle \tilde{a} \rangle$.

Case 3: $\text{Lin } \dot{D} \neq \text{C}\{z_i, i \in M\}$ and $k < d - 2$.

Claim: There exists p such that $z_p \in \text{Lin } \dot{D}$ and $[z_p, \tilde{v}_k(i_0, \ldots, i_{k-1})|_R] < 0$.

To see this, suppose first that $\mathsf{C}\{z_i, i \in M\} = \langle 0 \rangle$. Then as in Case 2, there exists $K^+(\pi)$ such that $\mathsf{C}\{z_i, i \in K^+(\pi)\} = \dot{D}$. Let $J = \{i \in K^+(\pi): z_i \in \mathsf{Lin}\, \dot{D}$ and $z_i \neq 0\}$. Suppose for all $j \in J$, $[z_j, \tilde{v}_k(i_0, \ldots, i_{k-1})|_R] \geqslant 0$. However, since $z_j \neq 0$ implies $y_j \notin S$ for any j, it follows that $[z_j, \tilde{v}_k(i_0, \ldots, i_{k-1})|_R] > 0$ for all $j \in J$. This implies that $\mathsf{C}\{z_i, i \in J\}$ is pointed which is impossible for a nontrivial subspace.

On the other hand, suppose $\mathsf{C}\{z_i, i \in M\} \neq \langle 0 \rangle$. Let $J = \{i \in M: z_i \neq 0\}$. Once again the supposition that for all $i \in J$, $[z_j, \tilde{v}_k(i_0, \ldots, i_{k-1})|_R] \geqslant 0$ leads to a contradiction.

So, there exists p such that $[y_p, \tilde{v}_k(i_0, \ldots, i_{k-1})] < 0$ and consequently, $\tilde{v}_{k+1}(i_0, \ldots, i_{k-1}, p)$ is in the tree. There also exists a nonzero $\tilde{a} \in S^\perp \cap C(\pi, M)^+$ which yields $\tilde{a}|_R \in \dot{D}^+$. Since $z_p \in \mathsf{Lin}\, \dot{D}$, $\tilde{a}|_R \in z_p^\perp$ and so $\tilde{a} \in y(i_0) \cap \cdots \cap y(i_{k-1})^\perp \cap y(p)^\perp \cap C(\pi, M)^+$. Now refer to Case 1, 2, or 3 as appropriate. Since d is finite and the cases are exhaustive, this process will stop after a while and the result will follow. □

Section 5.4:

Displacing Relative Boundary Vectors Into The Relative Interiors Of Hills

The first phase of the general tree algorithm is the relative boundary vector collection algorithm (5.2.11). This section describes the second phase of the general tree algorithm, namely, that procedure which operates on the tree of relative boundary vectors to produce a relative interior vector for each hill. This displacement of relative boundary vectors into the relative interiors of cones is necessary since finding relative interior vectors for all of the max-hills is effectively equivalent to identifying all equivalence classes of solutions to Problem (5.1.1) (cf., (5.1.11)).

As in Chapter 3, the displacement (i.e., second) phase of the tree algorithm will at times find it necessary to call the entire tree algorithm (i.e., both phases) recursively in order to solve a lower dimensional problem.

-- another way to propagate hills to lower dimensions --

The next theorem shows that a vector in a relative boundary face of a hill identifies a certain hill in a certain lower dimensional problem it generates. This theorem makes the above mentioned recursion possible as will be shown towards the end of this section.

(5.4.1) Theorem: Let $C(\pi,M)^+$ be a hill and suppose $F_J(C(\pi,M)^+) \neq \varnothing$ where $J \neq I_0$. Then

$$\mathrm{C}\{\pi_i y_i,\ i \in J \text{ w.o. } M,\ y_i,\ i \in M\}^+$$

is a hill for the $\{y_i,\ i \in J\}$ problem.

This hill is $\langle 0 \rangle$ if and only if $J = M$.

Proof: (a) For all $i \in I$, $y_i \in C_M$ if and only if $i \in M$. Consequently, for all $i \in J$, $y_i \in C_M$ if and only if $i \in M$.

(b) C_M is a subspace.

(c) If $J = M$, then the result follows.

Suppose $J \neq M$. Let T be any subspace such that $T \oplus C_M = \mathsf{L}\{y_i, i \in J\} =: R$ and define $t_i := P[y_i | T, C_M]$ for $i \in J$ w.o. M. It must be shown that $\mathsf{C}\{\pi_i t_i, i \in J \text{ w.o. } M\}$ is pointed.

Let S be some subspace such that $S \oplus R = X = S \oplus C_M \oplus T$. Let $U = S \oplus T$. Then $\mathsf{C}\{\pi_i u_i, i \in I \text{ w.o. } M\}$ is pointed which implies that $\mathsf{C}\{\pi_i u_i, i \in J \text{ w.o. } M\}$ is pointed. But observe that for $i \in J$ w.o. M, $y_i = s_i + x_i + b_i$ and $u_i = s_i + x_i$ for some $s_i \in S$, $x_i \in T$, and $b_i \in C_M$. Also observe $y_i = t_i + e_i$ for some $e_i \in C_M$. Consequently,

$$s_i + x_i - t_i = e_i - b_i \quad \Rightarrow \quad s_i = t_i - x_i$$
$$\Rightarrow \quad s_i = 0 \text{ and } t_i = x_i = u_i$$

So, $\mathsf{C}\{\pi_i t_i, i \in J \text{ w.o. } M\}$ is pointed.

(d) Next, it must be shown that there exists $K^+ \subset J$ w.o. M such that $\mathsf{C}\{t_i, i \in K^+\} = \mathsf{C}\{\pi_i t_i, i \in J \text{ w.o. } M\}$.

Claim: If $\langle \pi_j t_j \rangle$ is an isolated ray of $\mathsf{C}\{\pi_i t_i, i \in J \text{ w.o. } M\}$, then $\langle \pi_j t_j \rangle$ is an isolated ray of $\mathsf{C}\{\pi_i u_i, i \in I \text{ w.o. } M\}$.

The first thing to do is to show that for any $j \in J$ w.o. M,

$$I_j(\pi) := \{i \in I \text{ w.o. } M: \langle \pi_i u_i \rangle = \langle \pi_j u_j \rangle\} \subset J \text{ w.o. } M.$$

(This definition of $I_j(\pi)$ is used only in this proof.) Suppose for some $i \in I$ w.o. J, there exists $\alpha > 0$ such that $\alpha \pi_i u_i = \pi_j u_j$. Then

$$0 < \alpha[\pi_i y_i, \tilde{v}] = \alpha[\pi_i u_i, \tilde{v}] = [\pi_j u_j, \tilde{v}] = [\pi_j y_j, \tilde{v}] = 0$$

for any $\tilde{v} \in F_J(C(\pi, M)^+) \subset C_M^\perp$.

Now assume there exist $\lambda_i \geqslant 0$, not all 0, such that

$$\pi_j u_j = \sum_{I \text{ w.o. } J} \lambda_i \pi_i u_i + \sum_{J \text{ w.o. } (I_j(\pi) \cup M)} \lambda_i \pi_i u_i$$

For any $\tilde{v} \in F_J(C(\pi, M)^+)$,

$$0 = [\pi_j u_j, \tilde{v}] = \sum_{I \text{ w.o. } J} \lambda_i [\pi_i u_i, \tilde{v}] = \sum_{I \text{ w.o. } J} \lambda_i [\pi_i y_i, \tilde{v}]$$

Since $[\pi_i y_i, \tilde{v}] > 0$ for all $i \in I$ w.o. J, $\lambda_i = 0$ for all $i \in I$ w.o. J. Consequently, there exist $\lambda_i \geqslant 0$, not all 0, such that $\pi_j t_j = \sum_{J \text{ w.o. } (I_j(\pi) \cup M)} \lambda_i \pi_i t_i$ which implies the contradiction that $\langle \pi_j t_j \rangle$ is not an isolated ray of $\mathrm{C}\{\pi_i t_i, i \in J \text{ w.o. } M\}$.

The final step is easy. Take $\langle \pi_j t_j \rangle$ in the frame of $\mathrm{C}\{\pi_i t_i, i \in J \text{ w.o. } M\}$. Then $\langle \pi_j t_j \rangle$ is an isolated ray of $\mathrm{C}\{\pi_i u_i, i \in I \text{ w.o. } M\}$. Hence there exists $k \in I$ w.o. M such that $\langle u_k \rangle = \langle \pi_j t_j \rangle$. Since $I_j(\pi) \subset J$ w.o. M, $k \in J$ and $\langle t_k \rangle = \langle \pi_j t_j \rangle$.

Use (f) of (5.1.4) for the rest. □

As was shown in (3.5.6), the converse to Theorem (5.4.1) does not hold in that lower dimensional hills do not necessarily generate hills in the original problem.

-- *the goal is to identify relative interiors of containing hills* --

As was intimated before, the fundamental problem solved in this section is:

Given $\tilde{v}$ and $\pi_i \in \{-1, 1\}$ for $i \in I$ w.o. $J \neq \varnothing$ such that $[\pi_i y_i, \tilde{v}] > 0$ for $i \in I$ w.o. J and $[y_i, \tilde{v}] = 0$ for $i \in J$, identify every hill $\tilde{v}$ is in and obtain from $\tilde{v}$ a relative interior vector for each such hill. In other words, given $\tilde{v}$ as above, for every hill $\tilde{v}$ is in, determine M, π_i for $i \in J$ w.o. M, and $\tilde{u} \in \tilde{X}$ such that $C(\pi, M)^+$ is the hill in question, $\tilde{v} \in F_J(C(\pi, M)^+)$, and $\tilde{u} \in F_M(C(\pi, M)^+) = \text{rel int } C(\pi, M)^+$.

-- *using pointedness to discover hills* --

The next theorem uses the preceding one to show how this may be done in one special case.

(5.4.2) Theorem: Let $C(\pi, M)^+$ be a hill and suppose $F_J(C(\pi, M)^+) \neq \varnothing$ where $J \neq I_0$.

Suppose $C_J := \mathrm{C}\{y_i, i \in J\}$ is pointed. Then:

(a) $M = I_0$

(b) $\pi_i = 1$ for $i \in J$ w.o. M

Proof: (a): Since $C_M \subset C_J$, if C_M were a nonzero subspace, C_J would not be pointed.

(b): By Theorem (5.4.1), $\mathrm{C}\{\pi_i y_i, i \in J \text{ w.o. } I_0\}$ is a hill in the $\{y_i, i \in J\}$ problem. But there exists $\tilde{z} \in \tilde{R}$ for $R := \mathrm{L}\{y_i, i \in J\}$ such that $[y_i, \tilde{z}] > 0$ for all $i \in J$ w.o. I_0. Hence by Theorem (5.2.2), $\tilde{z}$ is in every hill in the $\{y_i, i \in J\}$ problem.

It is now argued that the only hill in $\{y_i, i \in J\}$ problem is $\mathrm{C}\{y_i, i \in J\}^+$. Suppose that $\mathrm{C}\{\theta_i y_i, i \in J \text{ w.o. } L, y_i, i \in L\}^+$ is a hill. L must be I_0 since if it is not, then $\mathrm{C}\{y_i, i \in J\} \supset \mathrm{C}\{y_i, i \in L\}$ is not pointed. Now all θ_i must be 1 since $\theta_i[y_i, \tilde{z}] > 0$ for all $i \in J$ w.o. I_0.

So, since $\mathrm{C}\{y_i, i \in J\}^+$ is the only hill in the lower dimensional problem, $\pi_i = 1$ for $i \in J$ w.o. I_0. □

-- *using dimensional degeneracy to discover hills* --

In another special case, there are at most three candidates for hills containing $\tilde{v}$ in their relative boundaries.

(5.4.3) Theorem: Let $C(\pi, M)^+$ be a hill and suppose $F_J(C(\pi, M)^+) \neq \varnothing$ where $J \neq I_0$. Suppose further that $\dim \mathrm{L}\{y_i, i \in J\} = 1$. Let $p \in J$ w.o. I_0, $J_1 := \{i \in J\colon \langle y_i \rangle = \langle y_p \rangle\}$, and

$J_2 := \{i \in J\colon \langle y_i \rangle = -\langle y_p \rangle\}$. Observe that $J = J_1 \cup J_2 \cup I_0$ and $J_1 \neq \emptyset$.

If $J_2 = \emptyset$, then $M = I_0$ and $\pi_i = 1$ for $i \in J_1$.

If $J_2 \neq \emptyset$, then one of the following holds:

(a) $M = J$

(b) $M = I_0$, $\pi_i = 1$, for $i \in J_1$, and $\pi_i = -1$ for $i \in J_2$

(c) $M = I_0$, $\pi_i = -1$ for $i \in J_1$, and $\pi_i = 1$ for $i \in J_2$.

Proof: If $J_2 = \emptyset$, then $\mathsf{C}\{y_i, i \in J\}$ is pointed and (5.4.2) applies.

Suppose $J_2 \neq \emptyset$. If $M \neq J$, then there exists $q \in J$ w.o. M. By definition of M, $y_q \notin C_M$. If dim $C_M = 1$, then $C_M = C_J$ and a contradiction results. Hence $C_M = 0$ and $M = I_0$. Let $\tilde{u} \in$ rel int $C(\pi, M)^+$. Now either $\pi_p = 1$ or $\pi_p = -1$. Suppose $\pi_p = 1$. Take any $j \in J_1$. Then there exists $\alpha > 0$ such that $y_j = \alpha y_p$ and since $0 < [\pi_j y_j, \tilde{u}] = \alpha\pi_j[y_p, \tilde{u}]$, $\pi_j = 1$. Now take $j \in J_2$. Then $y_j = -\alpha y_p$ for some $\alpha > 0$ and since $0 < [\pi_j y_j, \tilde{u}] = -\alpha\pi_j[y_p, \tilde{u}]$, $\pi_j = -1$. The other case is handled similarly. □

-- *the mechanics of displacement* --

Now of course it is one thing to know M and the π_i defining a candidate hill $C(\pi, M)^+$ and it is another thing to actually have a vector in rel int $C(\pi, M)^+$. The next theorem will show how to produce a relative interior vector from a relative boundary vector and an additional auxiliary vector. In symbols, suppose $\tilde{v} \in F_J(C(\pi, M)^+)$ where $J \neq M$ and suppose $\tilde{z} \in \tilde{X}$ is given such that $[\pi_i y_i, \tilde{z}] > 0$ for $i \in J$ w.o. M. It will be shown that there exists $\alpha \in (0, 1)$ such that $(1-\alpha)\tilde{v} + \alpha\tilde{z} \in$ rel int $C(\pi, M)^+$. Note that the next theorem actually applies to situations more general than this one.

(5.4.4) Theorem: Suppose for $I_0 \neq J \subset I$, $\pi_i \in \{-1, 1\}$ for $i \in I$ w.o. J and some $\tilde{v} \in \tilde{X}$, $[\pi_i y_i, \tilde{v}] > 0$ for $i \in I$ w.o. J and

$[y_i, \tilde{v}] = 0$ for $i \in J$.

Let M be an index set such that $I_0 \subset M \subset J$ and J w.o. $M \neq \emptyset$. Suppose there is $\tilde{z} \in \tilde{X}$ and $\theta_i \in \{-1, 1\}$ for $i \in J$ w.o. M such that $[\theta_i y_i, \tilde{z}] > 0$ for $i \in J$ w.o. M and $[y_i, \tilde{z}] = 0$ for $i \in M$.

Let $K := \{i \in I \text{ w.o. } J: [\pi_i y_i, \tilde{z}] < 0\}$. If $K = \emptyset$, then set $\alpha = \frac{1}{2}$. If $K \neq \emptyset$, then choose α such that

$$0 < \alpha < \min_{i \in K} \frac{[y_i, \tilde{v}]}{[y_i, \tilde{v}] - [y_i, \tilde{z}]} < 1 .$$

Then

$$[\pi_i y_i, (1-\alpha)\tilde{v}+\alpha\tilde{z}] > 0 \text{ for } i \in I \text{ w.o. } J,$$

$$[\theta_i y_i, (1-\alpha)\tilde{v}+\alpha\tilde{z}] > 0 \text{ for } i \in J \text{ w.o. } M,$$

$$\text{and } [y_i, (1-\alpha)\tilde{v}+\alpha\tilde{z}] = 0 \text{ for } i \in M.$$

Proof: When $i \in M$, $i \in J$ w.o. M, or when $i \in I$ w.o. J and $[\pi_i y_i, \tilde{z}] \geqslant 0$, it suffices to have $\alpha \in (0, 1)$. If $K \neq \emptyset$, then it is necessary to have for all $i \in K$, $\alpha < \dfrac{[\pi_i y_i, \tilde{v}]}{[\pi_i y_i, \tilde{v}] - [\pi_i y_i, \tilde{z}]}$. □

-- *the general displacement algorithm* --

The main theorem of this section shows that the following algorithm takes the tree of relative boundary vectors provided by the first phase of the tree algorithm and generates a collection of vectors containing a relative interior vector for each nonzero hill.

(5.4.5) Algorithm: Suppose a tree of relative boundary vectors is provided by Algorithm (5.2.11). For $k \geqslant 0$, each $\tilde{v}_k(i_0, \ldots, i_{k-1})$, and $q = 1, 2$, let $\tilde{w}_k(i_0, \ldots, i_{k-1}) = (-1)^q \, \tilde{v}_k(i_0, \ldots, i_{k-1})$ and determine the index set $J \subset I$ such that $\tilde{w}_k(i_0, \ldots, i_{k-1}) \in F_J(C(\pi)^+)$ for some ambiguous

$C(\pi)^+$, (i.e., where $\tilde{w}_k(i_0, \ldots, i_{k-1}) \in C_J^\perp$ and for $i \in I$ w.o. J, $[\pi_i y_i, \tilde{w}_k(i_0, \ldots, i_{k-1})] > 0)$.

1) If C_J is a subspace of any nonnegative dimension, then save $\tilde{w}_k(i_0, \ldots, i_{k-1})$. (The LP described in (2.3.39) will determine whether or not any $C_J \neq \langle 0 \rangle$ is a subspace).

2) If $J \neq I_0$, then select the appropriate case from the following:

Case 1: $\dim \mathsf{L}\{y_i, i \in J\} = 1$

Here $k = 0$ or 1. Let $p \in J$ w.o. I_0, $J_1 := \{i \in J: \langle y_i \rangle = \langle y_p \rangle\}$, and $J_2 := \{i \in J: \langle y_i \rangle = -\langle y_p \rangle\}$.

If $J_2 = \varnothing$, then let $\tilde{z} = y_p$, $\theta_i = 1$ for $i \in J$ w.o. I_0 and displace $\tilde{w}_k(i_0, \ldots, i_{k-1})$ using (5.4.4). Save the displaced vector.

Suppose $J_2 \neq \varnothing$. Let $\tilde{z} = y_p$, $\theta_i = 1$ for $i \in J_1$, $\theta_i = -1$ for $i \in J_2$ and use (5.4.4) to displace $\tilde{w}_k(i_0, \ldots, i_{k-1})$. Next, let $\tilde{z} = -y_p$, $\theta_i = -1$ for $i \in J_1$, and $\theta_i = 1$ for $i \in J_2$ and use (5.4.4) to displace $\tilde{w}_k(i_0, \ldots, i_{k-1})$. Save the displaced vectors.

Case 2: $\dim \mathsf{L}\{y_i, i \in J\} > 1$ and $\mathsf{C}\{y_i, i \in J\}$ is pointed.

The linear program of (2.3.33) will provide $\tilde{z}$ such that $[y_i, \tilde{z}] > 0$ for $i \in J$ w.o. I_0. Let $\theta_i = 1$ for $i \in J$ w.o. I_0 and use (5.4.4) to displace $\tilde{w}_k(i_0, \ldots, i_{k-1})$. Save the displaced vector. Note that the LP of (2.3.33) also determines whether or not $\mathsf{C}\{y_i, i \in J\}$ is pointed.

Case 3: $\dim \mathsf{L}\{y_i, i \in J\} > 1$ and $\mathsf{C}\{y_i, i \in J\}$ is not pointed.

In this case, recursion is necessary. Call Algorithm (5.2.11) and then Algorithm (5.4.5) to provide a set of vectors containing a relative interior vector for each nonzero hill in the lower dimensional $\{y_i, i \in J\}$ problem. For each vector $\tilde{r}_0 \in \tilde{R}$ in this set where $R = \mathsf{L}\{y_i, i \in J\}$, there exists $I_0 \subset M \subset J$ with J w.o. $M \neq \varnothing$ and $\theta_i \in \{-1, 1\}$ such that $[y_i, \tilde{r}_0] = 0$ for $i \in M$ and $[\theta_i y_i, \tilde{r}_0] > 0$ for $i \in J$ w.o. M. Let $\tilde{z} = \psi^{-1}(\tilde{r}_0)$ and use (5.4.4) to displace $\tilde{w}_k(i_0, \ldots, i_{k-1})$. Save all of the displaced vectors generated by this step. (Recall ψ is the isomorphism from $S^{\perp}$ to $\tilde{R} = S^{\perp}|_R$ where S is any subspace such that $R \oplus S = X$ (cf., (2.1.27)).)

-- *the displacement algorithm (5.4.5) works* --

(5.4.6) Theorem: In order to find a relative interior vector for each nonzero hill, it is sufficient to displace using Algorithm (5.4.5) all of the relative boundary vectors produced by the first phase of the tree algorithm. In order to determine if a given displaced boundary vector in rel int $C(\pi,M)^+$ is in the relative interior of a hill, it is sufficient to determine the frame of $\mathrm{C}\{\pi_i u_i,\ i \in I \text{ w.o. } M\}$ where $u_i := P[y_i | U, C_M]$ for any U such that $U \oplus C_M = X$. (cf., (2.3.22)).

Proof: By (5.2.13), the first phase of the tree algorithm constructs a tree of vectors which contains for each hill at least one vector in some nonzero face of that hill. Since Algorithm (5.4.5) displaces all of these vectors, it suffices to show that if $C(\pi,M)^+$ is a nonzero hill and $\tilde{w}_k(i_0, \ldots, i_{k-1}) \in C(\pi,M)^+$, then Algorithm (5.4.5) will displace $\tilde{w}_k(i_0, \ldots, i_{k-1})$ into rel int $C(\pi,M)^+$.

The proof proceeds by induction.

Suppose $\dim X = 2$. It is known that $\tilde{w}_k(i_0, \ldots, i_{k-1})$ is in some nonzero face of $C(\pi,M)^+$, i.e., there exists J such that

$$\tilde{w}_k(i_0, \ldots, i_{k-1}) \in F_J(C(\pi,M)^+).$$

If $M = J$, then C_J is a subspace and by Step 1 of Algorithm (5.4.5), $\tilde{w}_k(i_0, \ldots, i_{k-1})$ is saved. This is good since

$$\tilde{w}_k(i_0, \ldots, i_{k-1}) \in F_M(C(\pi,M)^+) = \text{rel int } C(\pi,M)^+.$$

Now suppose $M \neq J$ and hence $J \neq I_0$. Since $\tilde{w}_k(i_0, \ldots, i_{k-1}) \neq 0$, $\dim \mathrm{L}\{y_i,\ i \in J\} \leqslant 1$ and so, since $J \neq I_0$, Case 1 of Step 2 holds. In this event, Theorem (5.4.3) identifies M as I_0 and determines at most two choices for $\{\pi_i,\ i \in J \text{ w.o. } M\}$. One of the at most two displaced vectors produced by

Theorem (5.4.4) is in rel int $C(\pi, M)^+$.

Now suppose dim $X \geqslant 3$ and that Algorithms (5.2.11) and (5.4.5) together produce relative interior vectors for every nonzero hill in all problems of dimension between 2 and dim $X - 1$. Once again, there exists J such that $\tilde{w}_k(i_0, \ldots, i_{k-1}) \in F_J(C(\pi, M)^+)$.

If $M = J$, then C_J is a subspace and $\tilde{w}_k(i_0, \ldots, i_{k-1}) \in$ rel int $C(\pi, M)^+$ is saved.

Now suppose $M \neq J$ and hence $J \neq I_0$. If dim $\mathrm{L}\{y_i, i \in J\} = 1$, then Theorem (5.4.3) shows that a relative interior vector of $C(\pi, M)^+$ is obtained and saved through the displacement operation specified in Case 1 of Step 2. If dim $\mathrm{L}\{y_i, i \in J\} > 1$ and C_J is pointed, then Theorem (5.4.2) shows that Case 2 of Step 2 will produce and save a relative interior vector of $C(\pi, M)^+$.

In the last event that dim $\mathrm{L}\{y_i, i \in J\} > 1$ and C_J is not pointed, Case 3 of Step 2 calls for recursion in that one is asked to provide relative interior vectors for each nonzero hill in the $\{y_i, i \in J\}$ problem. This latter problem is truly a lower dimensional version of the original problem since $\{y_i, i \in J\} \subset \tilde{w}_k(i_0, \ldots, i_{k-1})^\perp$.

If these relative interior vectors can be obtained, then Theorem (5.4.1) guarantees that since $M \neq J$, the nonzero hills of the lower dimensional problem can be used via the displacement operations of Case 3 of Step 2 of (5.4.5) to produce and save at least one relative interior vector of $C(\pi, M)^+$.

That the needed relative interior vectors can be obtained from the lower dimensional problem is guaranteed by the induction hypothesis. □

-- *comments on the displacement algorithm* --

Note that Case 2 of Step 2 which uses a linear program is not essential since the recursive case could be entered whenever dim $\mathrm{L}\{y_i, i \in J\} > 1$. Case 2 of Step 2 is useful since it helps the algorithm avoid recursion.

The computational aspects of implementing Algorithm (5.4.5) on a computer are covered by Theorems (3.5.12) and (3.5.13).

Note that if $\langle 0 \rangle$ is a hill, it is stored in the tree of relative boundary vectors and thus Algorithm (5.4.5) need not worry about finding zero hills.

-- *looking for max-hills saves a lot of work* --

Clearly Algorithms (5.2.11) and (5.4.5) call for a tremendous amount of work to find relative interior vectors for each hill in that one or two linear programs are used for each displacement and very likely most of the vectors displaced won't be in hills anyway; in fact, in order to determine if a given cone is a hill, it appears to be necessary to use entire sets of linear programs (cf., (2.3.25)). It is very important to stress however that when looking for all of the max-hills of Problem (5.1.1), an enormous amount of this work can be dispensed with.

-- *the propagation of max-cones to lower dimensions* --

As will be seen, one of the factors contributing to this is that a boundary face of a max-cone in the original problem generates a max-cone in a suitably defined lower dimensional version of Problem (5.1.1).

(5.4.7) Theorem: If $C(\pi,M)^+$ is a max-cone and $\tilde{v} \in F_J(C(\pi,M)^+)$ where $J \supsetneq M$ and $R = \mathsf{L}\{y_i,\, i \in J\}$ and S is such that $R \oplus S = X$, then

$$\{\tilde{r} \in \tilde{R}: [\pi_i y_i, \tilde{r}] \geqslant 0,\, i \in J \text{ w.o. } M,\, [y_i, \tilde{r}] = 0,\, i \in M\}$$

is a max-cone with respect to $H_{\pi,J}$ where for all $\tilde{r} \in \tilde{R}$,

$$\begin{aligned} H_{\pi,J}(\tilde{r}) := g(\, &\{(i,\, 1\{[y_i, \tilde{v}]\ \ R_i\ \ 0\}): i \in I \text{ w.o. } J\} \\ &\cup \{(i,\, 1\{[y_i, \tilde{r}]\ \ R_i\ \ 0\}): i \in J\}) \end{aligned}$$

where R_i = ">" if $i \in I^{\#}$ and R_i = "$\geqslant$" if $i \in I^{\#\#}$. Also, $\sup\limits_{\tilde{r} \in \tilde{R}} H_{\pi,J}(\tilde{r}) = \sup\limits_{\tilde{x} \in \tilde{X}} H(\tilde{x})$ so that the maximum objective function values for each problem are the same.

Proof: (i): Certainly $\mathrm{C}\{y_i, i \in M\}$ is a subspace.

(ii): It is necessary to show that there exists $\tilde{r}_0 \in \tilde{R} = S^{\perp}\big|_R$ such that

$$[\pi_i y_i, \tilde{r}_0] > 0 \text{ for } i \in J \text{ w.o. } M \text{ and } [y_i, \tilde{r}_0] = 0, i \in M.$$

Now, there exists $\tilde{x}_0 \in \tilde{X}$ such that $[\pi_i y_i, \tilde{x}_0] > 0$ for $i \in I$ w.o. M and $[y_i, \tilde{x}_0] = 0$ for $i \in M$. Since $\tilde{X} = R^{\perp} \oplus S^{\perp}$, $\tilde{x}_0 = \tilde{t}_0 + \tilde{w}_0$ where $\tilde{t}_0 \in R^{\perp}$ and $\tilde{w}_0 \in S^{\perp}$. Observe that for $i \in J$ w.o. M, $[\pi_i y_i, \tilde{x}_0] = [\pi_i y_i, \tilde{w}_0] > 0$ and $0 = [y_i, \tilde{x}_0] = [y_i, \tilde{w}_0]$ for $i \in M$. Take $\tilde{r}_0 = \tilde{w}_0\big|_R$.

(iii): Next, it is shown that $H_{\pi,J}(\tilde{r}_0) = \sup_{\tilde{r} \in \tilde{R}} H_{\pi,J}(\tilde{r})$. Suppose there exists $\tilde{r}_1$ such that $H_{\pi,J}(\tilde{r}_1) > H_{\pi,J}(\tilde{r}_0)$. Consequently, there exists $\beta_1 > 0$ such that for $i \in I$ w.o. J,

$$\mathrm{sgn}[y_i, \tilde{v}] = \mathrm{sgn}[y_i, \tilde{v} + \beta_1 \psi^{-1}(\tilde{r}_1)]$$

and

$$H(\tilde{v} + \beta_1 \psi^{-1}(\tilde{r}_1)) > H_{\pi,J}(\tilde{r}_0).$$

However, since for all $i \in I$ w.o. J, $\mathrm{sgn}[y_i, \tilde{v}] = \pi_i = \mathrm{sgn}[y_i, \tilde{x}_0]$ and for all $i \in J$ w.o. M, $\mathrm{sgn}[y_i, \tilde{r}_0] = \pi_i = \mathrm{sgn}[y_i, \tilde{x}_0]$ and for all $i \in M$, $0 = [y_i, \tilde{r}_0] = [y_i, \tilde{x}_0]$, it must be that $H_{\pi,J}(\tilde{r}_0) = H(\tilde{x}_0)$ which leads directly to a contradiction. □

-- *propagating max-cones from lower dimensions to higher ones* --

Curiously enough, the converse of this theorem holds whereas the analogous converse for the corresponding theorem about hills does not (cf., (3.5.6)). The result is that every max-cone in the lower dimensional problem associated with some face of a max-cone generates a max-cone in the original problem.

(5.4.8) Theorem: Suppose $C(\pi,M)^+$ is a max-cone and $\tilde{v} \in F_J(C(\pi,M)^+)$ where $J \underset{\neq}{\supset} M$ and $R = \mathrm{L}\{y_i, i \in J\}$ and S is such that $R \oplus S = X$. Given $H_{\pi,J} : \tilde{R} \to \mathrm{R}$ via $H_{\pi,J}(\tilde{r}) :=$

$$g(\{(i, 1\{[y_i, \tilde{v}]\ R_i\ 0\}): i \in I \text{ w.o. } J\} \cup \{(i, 1\{[y_i, \tilde{r}]\ R_i\ 0\}): i \in J\})$$

where R_i = ">" if $i \in I^{\#}$ and R_i = "$\geqslant$" if $i \in I^{\#\#}$.

Let L be such that $J \supset L \supset I_0$. If $L \neq J$ then let $\theta_i \in \{-1, 1\}$ for $i \in J$ w.o. L. Suppose

$$\{\tilde{r} \in \tilde{R} = S^{\perp}|_R : [\theta_i y_i, \tilde{r}] \geqslant 0, i \in J \text{ w.o. } L, [y_i, \tilde{r}] = 0, i \in L\}$$

is a max-cone relative to $H_{\pi,J}$.

Then

$$\{\tilde{x} \in \tilde{X}: [\pi_i y_i, \tilde{x}] \geqslant 0 \text{ for } i \in I \text{ w.o. } J,\ [\theta_i y_i, \tilde{x}] \geqslant 0 \text{ for } i \in J \text{ w.o. } L,\ [y_i, \tilde{x}] = 0 \text{ for } i \in L\}$$

is a max-cone in the original problem.

Proof: (i): C_L is a subspace.

(ii): Let $\tilde{x}_1 \in S^{\perp}$ be such that $[\theta_i y_i, \tilde{x}_1|_R] > 0$ for $i \in J$ w.o. L and $[y_i, \tilde{x}_1|_R] = 0$ for $i \in L$. Observe that there exists $\alpha > 0$ such that

$$\begin{aligned} &[\pi_i y_i, \tilde{v} + \alpha\tilde{x}_1] > 0 \quad \text{for} \quad i \in I \text{ w.o. } J \\ &[\theta_i y_i, \tilde{v} + \alpha\tilde{x}_1] > 0 \quad \text{for} \quad i \in J \text{ w.o. } L \\ &[y_i, \tilde{v} + \alpha\tilde{x}_1] = 0 \quad \text{for} \quad i \in L \end{aligned}$$

(iii): Note that

$$H(\tilde{v}+\alpha\tilde{x}_1) = H_{\pi,J}(\tilde{x}_1|_R) = \sup_{\tilde{r}} H_{\pi,J}(\tilde{r}) = H_{\pi,J}(\tilde{r}_0) = H(\tilde{x}_0)$$

where $\tilde{x}_0 \in \text{rel int } C(\pi,M)^+$ and $\tilde{r}_0 = P[\tilde{x}_0|S^{\perp}, R^{\perp}]|_R$ (cf., (5.4.7)). □

-- *simplifications made possible by searching only for max-hills* --

It is now possible to discuss how to modify Algorithms (5.2.11) and (5.4.5) to obtain a relative interior vector for each max-hill.

First, it is only necessary to look for vectors in the relative interiors of max-hills and not hills in general since by Theorem (5.4.7), finding all max-hills in the lower dimensional problems generated during recursion in (5.4.5) suffices to produce all max-hills in the original problem. Consequently, as one is executing the relative boundary vector collection algorithm, it is only necessary to save those $\tilde{w}_k$ which could yield displaced vectors with H-values greater than or equal to the largest guaranteed H-value observed so far in the search sequence.

Second, it is never necessary to use the subspace detecting LP mentioned in Step 1 of the displacement algorithm (5.4.5) since if $\tilde{w}_k(i_0, \ldots, i_{k-1})$ is a vector in the relative interior of a max-hill, it will be apparent from its H-value. Similarly, it is never necessary to actually prove that any displaced vector is in hill (cf. (5.4.6)) since the only cones of interest are max-hills and vectors in the relative interiors of max-hills announce themselves by their H-values.

Finally, the general max-hill finding tree algorithm may be improved using any or all of the improvements listed in Section 5.3 which are themselves direct analogs of the improvements listed in Section 3.4. Some of these improvements can be seen in the workings of Algorithm (5.1.13) which shows in one place how Algorithm (5.4.5) works with Algorithm (5.2.11).

It is interesting to contrast the roles of h_J in (3.2.12) with $H_{\pi,J}$ in (5.1.13). Note that it is the linear character of h_I which enables the influence of problems higher up in the recursion (as manifested by the term $\{(i, 1\{[y_i, \tilde{v}]\ R_i\ 0\} : i \in I \text{ w.o. } J\})$ to be subtracted off in the lower dimensional problems whereas in the general case, $H_{\pi,J}$ must take this influence into account.

This section closes with a comment on a peculiar characteristic of these methods. Notice that the form of H and the nature of $I^{\#}$ and $I^{\#\#}$ have absolutely nothing to do with hills. They make their only appearances in the path to solving Problem (5.1.1) at the beginning when they help to characterize the solution geometry and at the end when they are used to evaluate the "worth" (i.e., the H-value) of relative boundary and interior vectors whose genesis did not require H, $I^{\#}$, and $I^{\#\#}$ in the first place.

Summary For Section 5.4

The development of the displacement phase of the general tree algorithm parallels that corresponding to the WOH tree algorithm. Recall that the objective of the displacement phase is to displace vectors from relative boundary faces of hills into the relative interiors of these hills.

The first theorem in this section shows how nonempty faces of hills generate hills in suitably defined lower dimensional problems. The next two theorems use the first to show how in certain special cases, if one has a relative boundary face vector of an otherwise unknown hill, then at most three candidates can be obtained for explicitly identifying that hill. As before, in the general case, it is necessary to resort to recursion.

The sole difference between the displacement phase of the general tree algorithm and that of the WOH tree algorithm when seeking to identify all hills is that if $\tilde{v}_k(i_0, \ldots, i_{k-1}) \in F_J(C(\pi)^+)$ for some ambiguous $C(\pi)^+$, then $\tilde{v}_k(i_0, \ldots, i_{k-1})$ is itself saved as being potentially a relative interior hill vector if $C\{y_i, i \in J\}$ is a subspace. Otherwise, as before, any $\tilde{v}_k$ in $(d-1)$-dimensional boundary faces are displaced without crossing any other hyperspaces to either or both sides (as appropriate) of the hyperspace defining the face they are in. $\tilde{v}_k$ that are able to be displaced (without crossing over any other hyperspaces) to the positive side of all $y_i^\perp \neq X$ that contain them are so displaced. Otherwise, recursion is necessary to accomplish the desired displacements.

The fact that each nonempty face of a max-cone generates a max-cone in the associated lower dimensional problem enables much of the work necessary to identify all hills to be omitted in the event that only max-hills are of interest. What this amounts to in practice is that it is only necessary to keep track of the H-values of the $\tilde{v}_k$ both during relative boundary vector generation and during displacing. It is not necessary then to use the various linear programs that would otherwise have to be used extensively in order to identify all of the hills.

When keeping track of H-values in seeking to identify all max-hills, another difference contrasting the general tree algorithm with the WOH tree algorithm is that during the recursion part of the displacement phase, it is necessary for the general tree algorithm to make use of certain characteristics of the relative boundary face vector from the problem which initiated the recursive call. These characteristics can be ignored by the WOH tree algorithm by virtue of its linear objective function.

Chapter 6:

The Computational Complexity Of The Tree Algorithm

This section develops upper and lower bounds for the time complexity of a version of the tree algorithm. This complexity is exponential in the dimensionality d of the data. This is unfortunate but apparently it must be lived with since the underlying problem here is NP-complete.

-- the WOH problem is NP-complete --

In particular, Johnson and Preparata (1978) have shown that Problem (6.1.1) is NP-complete:

(6.1.1) Problem: Given positive integers d and k and a set of vectors $\{y_i\}_1^n \subset \mathsf{Q}^d$ where Q is the set of rational numbers, do there exist $a \in \mathsf{R}^d$ such that $\sum_1^n 1\{a^T y_i > 0\} \geqslant k$?

To relate this to tree algorithm type problems, consider the following special case of Problem (3.1.1).

(6.1.2) Problem: Let F_p be the set of all binary computer numbers of precision p. Let $n, d \in \mathsf{F}_p$ be positive integers. Let $\{y_i\}_1^n \subset \mathsf{F}_p^d$. Find all $a \in \mathsf{F}_p^d$ which maximize $h(a) := \sum_1^n 1\{a^T y_i > 0\}$.

Now it is generally accepted (although not mathematically proved) that no NP-complete problem can be solved by a Turing machine (or, equivalently, a present-day computer) in time (i.e., number of basic operations) which grows polynomially in the length of the input. Note that solving Problem (6.1.2) for given n, d, $\{y_i\}_1^n$ solves the corresponding instance of Problem (6.1.1) for every choice of $k = 1, \ldots, n$. Hence if there was a general algorithm for solving

Problem (6.1.2) for arbitrary n, d, $\{y_i\}_1^n$, and p which required a number of basic operations polynomial in the length of the input (which is δndp for some $\delta > 1$), then the general case of Problem (6.1.1) could be solved in polynomial time which would be a great surprise to many computer scientists. Consequently, it appears that exponential time algorithms such as the tree algorithm for solving Problem (6.1.2) are the best that can be hoped for.

The situation is not completely bleak however. As the examples in Chapter 9 will indicate, there are problems of practical and real interest which can be solved by the tree algorithm from anywhere between 26 cents to $69. The tree algorithm will be seen to be dramatically faster than various complete enumeration algorithms which are also of exponential time complexity. Also, more problems will enter the range of computational feasibility as computers become faster in the future although, it is true, the exponential time complexity present here tends to strongly mitigate this effect. In the worst case, of course, it is always possible to settle for an algorithm like the fast tree algorithm of (3.4.20) which quickly produces good vectors that cannot be guaranteed to be optimal.

-- *basics* --

The remainder of this chapter is concerned with developing bounds on the time complexity of certain versions of the tree algorithm. It will be useful at this time to informally formalize the concept of algorithm. For present purposes, A is an algorithm if and only if it is a function which maps inputs from a set Ω to sequences of basic operations. Basic operations are those machine (or assembly language) instructions which computers assemble in predetermined fixed numbers and patterns to perform such tasks as addition, multiplication, comparison, finding the square root or cosine of a number, storing a number in a location, printing a character, etc. So, for example, an algorithm which adds two numbers and prints the result can be thought of as mapping each ordered pair (α, β) in $\Omega = \mathsf{R} \times \mathsf{R}$ to the sequence of operations:

(1) put α into number register 1.

(2) put β into number register 2.

(3) add the contents of number register 1 to those of number register 2.

(4) store the contents of number register 2 in location 6.

(5) print the contents of location 6.

-- *prototypical tree algorithm for complexity analysis* --

Next, in order to provide the notation which will facilitate the derivation of the time complexity of the tree algorithm, it is convenient to abstract its structure and present it as the following prototypical tree algorithm.

(6.1.3) Prototypical Tree Algorithm: The set of inputs Ω to this algorithm is the set of all finite matrices of fixed precision computer numbers which are of full row rank and have at least 2 columns. For each $\omega \in \Omega$, $n(\omega)$ is the number of rows of ω, $d(\omega)$ is the number of columns, and $I(\omega) := \{1, \ldots, n(\omega)\}$. For $k = 1, \ldots, q + 1$, let A_k be an algorithm whose set of inputs is

$$\{(p, \omega) : \omega \in \Omega \text{ and } p = (p_1, \ldots, p_k) \in \mathop{\times}_{1}^{n(\omega)} I(\omega)\}.$$

It is assumed that for any fixed $\omega \in \Omega$ and any $p \in \mathop{\times}_{1}^{n(\omega)} I(\omega)$, the sequence of operations $A_k(p, \omega)$ for $k \leqslant q$ produces a set $N_k(p, \omega) \subset I(\omega)$. Let A_0 be a fixed algorithm such that for each $\omega \in \Omega$, $A_0(\omega)$ is a sequence of operations which produces, among whatever else is desired, a set $N_0(\omega) \subset I(\omega)$. The following is a prototypical tree algorithm.

```
A_0
for each i_0 ∈ N_0 do:
  A_1(i_0)
  for each i_1 ∈ N_1(i_0) do:
    A_2(i_0, i_1)
    for each i_2 ∈ N_2(i_0, i_1) do:
      A_3(i_0, i_1, i_2)
        ⋮
          for each i_q ∈ N_q(i_0, . . . , i_{q-1}) do:
            A_{q+1}(i_0, . . . , i_q)
          next i_q;
        ⋮
    next i_2;
  next i_1;
next i_0;
```

This algorithm can be thought of as executing a depth-first search of a tree. The root of the tree has children labeled by the indices of N_0 and each subtree whose root node is labeled by the vector $(i_0, \ldots, i_{k-1})$ has children labeled by vectors, if any, of the form $(i_0, \ldots, i_{k-1}, q)$ where $q \in N_k(i_0, \ldots, i_{k-1})$. The process of going from node to node of the tree requires the execution of the appropriate A_k.

The next theorem shows how to obtain upper and lower bounds on the time complexity of this prototypical tree algorithm. For present purposes, the time complexity of a sequence of basic operations is its length.

(6.1.4) Theorem: For each input matrix $\omega \in \Omega$, let the number of basic operations in $A_k((i_0, \ldots, i_{k-1}), \omega)$ in Algorithm (6.1.3) be $C_k((i_0, \ldots, i_{k-1}), \omega) = \#A_k((i_0, \ldots, i_{k-1}), \omega)$. $C_k(i_0, \ldots, i_{k-1})$ is thus a map from Ω to the positive integers. Assume for $0 \leqslant k \leqslant q+1$, $i_0 \in N_0$, $i_1 \in N_1(i_0), \ldots, i_{k-1} \in N_{k-1}(i_0, \ldots, i_{k-2})$ that

$$\gamma_k \leqslant C_k(i_0, \ldots, i_{k-1}) \leqslant \delta_k$$

for some γ_k, δ_k possibly depending on ω. Analogous to bounding $C_k(i_0, \ldots, i_{k-1})$, assume

$$\mu_k \leqslant \#N_k(i_0, \ldots, i_{k-1}) \leqslant \nu_k \text{ for some } \mu_k,$$

ν_k possibly depending on ω. Let C^* be the time complexity of Algorithm (6.1.3). Then:

$$C^* = C_0 + \sum_{i_0 \in N_0} \left[C_1(i_0) + \sum_{i_1 \in N_1(i_0)} \left[C_2(i_0, i_1) + \cdots + \sum_{i_q \in N_q(i_0, \ldots, i_{q-1})} C_{q+1}(i_0, \ldots, i_q) \right] \cdots \right]$$

C^* is bounded below by

$$\gamma_0 + \gamma_1\mu_0 + \gamma_2\mu_0\mu_1 + \cdots + \gamma_{q+1} \prod_0^q \mu_i$$

and is bounded above by

$$\delta_0 + \delta_1\nu_0 + \delta_2\nu_0\nu_1 + \cdots + \delta_{q+1} \prod_0^q \nu_i$$

This theorem will be used to determine lower and upper bounds on the time complexity of the computer version of the tree algorithm for maximizing certain functions H of systems of linear relations. The computer version of the algorithm is the one which works with the representations in R^d of the vectors $y_i \in X$ and $\tilde{v} \in \tilde{X}$ with the additional change that these representations are approximated by vectors of fixed precision computer numbers.

-- *the problem and algorithm to be analyzed* --

The problem of interest here is to maximize a function $H = g \circ f$ of a system of linear relations when H is in homogeneous canonical form. Define for each $x \in \mathsf{R}^d$, $f(x) := (1\{\underline{y}_1^T x \; R_1 \; 0\}, \ldots, \; 1\{\underline{y}_n^T x \; R_n \; 0\})$ where

$R_i \in \{ > , \geqslant \}$ and where it is assumed that $g : \bigtimes_1^n \{0, 1\} \rightarrow (0, \infty)$ is nondecreasing. Note, no $y_i = 0$.

The next step is to define a version of the general tree algorithm which will solve this problem and whose complexity will be analyzed subsequently.

(6.1.5) Algorithm: Modify general tree algorithm (5.1.13) as follows. The exploration of the tree is to take place in a depth-first search manner (cf., (3.4.3)). New $\tilde{\underline{v}}_k$ are to be computed using the projection method of (3.4.7).

-- *simplifications* --

The complexities of the displacement phase, the restarting improvement, the trimming improvement, and the initial use of the fast approximate algorithm (3.4.20) are omitted in order to simplify the analysis.

As for the displacement phase, it tends to take only a small fraction of the time needed to explore the boundary vector tree even though in some cases the displacement of most of the saved boundary vectors may require the recursive calling of the first phase of the algorithm. The time complexity of the displacement phase is especially insignificant if the data $\{y_i\}_1^n$ is in pointed position in which case only a few linear programs with constraint matrices of $d+1$ rows and at the very most n columns are needed.

As for the effect of the restarting, it does seem to make for a faster algorithm, although since the tree after the last restart is explored completely, the time complexity of the tree algorithm with the restart option is still exponential in d.

The examples in Chapter 9 will show that, after the last restart, the effect of trimming is that it tends to bring the time complexity of the algorithm down to the lower bound which will be derived for Algorithm (6.1.5) (which has neither restarting nor trimming). In these examples, the final tree produced by trimming and restarting is typically from 1/2 to 1/8 the size of the final tree produced by restarting alone.

In all of the examples the author has seen, it has proven to be very useful to obtain a good $\tilde{v}_0$ for the general tree algorithm to start with by using the fast approximate algorithm of (3.4.20) to visit a limited number of best children at each level while restarting whenever beneficial to do so. It will be seen that typically the time complexity of the fast approximate algorithm is dominated by that of the general tree algorithm which follows it.

-- *time complexity and other notation* --

It is necessary to define some of the notation that will be used shortly. Let $\alpha := \inf\{\#N(\tilde{v}) : \tilde{v} \neq 0\}$ where it is recalled that

$$N(\tilde{v}) := \{i \in \{1, \ldots, n\} : [y_i, \tilde{v}] = \underline{y}_i^T \underline{\tilde{v}} < 0\}.$$

(Note that α is a function of the input matrix ω). To get a feeling for α, it can easily be shown that

$$\alpha \leqslant n - \sup_{\tilde{x}} \sum_{1}^{n} 1\{[y_i, \tilde{x}] > 0\}$$

with equality holding by (4.1.15) when $\{y_i\}_1^n$ is in pointed position. Note that $\sum_1^n 1\{[y_i, \tilde{x}] > 0\}$ is the OH criterion function used when the tree algorithm is seeking to find those open halfspaces through the origin which contain the greatest number of points. So when $\{y_i\}_1^n$ is in pointed position and on the unit sphere then α is the smallest number points that can be kept out of an open hemisphere.

Since the standard "big O" notation for working with complexities is neither well-defined nor general enough for subsequent needs, consider the following definitions and theorems (which are expressed in somewhat more generality than will be used):

(6.1.6) Definition: Let S be an arbitrary set and T a vector space over R with norm $\|\cdot\|$. Let $\Phi := T^S = \{f : f : S \to T\}$.

(a) For all $f_1, f_2 \in \Phi$, $f_1 \lesssim f_2$ if and only if there exists $\alpha > 0$ such that $\|f_1\| \leqslant \alpha\|f_2\|$ (i.e., for all $s \in S$, $\|f_1(s)\| \leqslant \alpha\|f_2(s)\|$).

(b) For all $f_1, f_2 \in \Phi$, $f_1 \approx f_2$ if and only if $f_1 \lesssim f_2$ and $f_2 \lesssim f_1$.

(6.1.7) Theorem: $\lesssim$ is a reflexive and transitive relation on $\Phi \times \Phi$. $\approx$ is an equivalence relation on $\Phi \times \Phi$.

(6.1.8) Definition: For all $g \in \Phi$:

(a) $\Lambda(g) := \{f \in \Phi: g \lesssim f\}$ is the set of all functions f of order at least as great as that of g.

"$\lambda(g)$" is used to denote some otherwise unspecified element of $\Lambda(g)$.

(b) $M(g) := \{f \in \Phi: f \lesssim g\}$ is the set of all functions f of order at most as great as that of g.

"$\mu(g)$" is used to denote some otherwise unspecified element of $M(g)$.

(c) $\Theta(g) := \{f \in \Phi: f \approx g\}$ is the set of all functions f of the same order as g.

"$\theta(g)$" is used to denote some otherwise unspecified element of $\Theta(g)$. As a mnemonic, think of Θ as being constructed from O for order and a symmetric connective piece.

(d) $\Theta(\Phi) := \{\Theta(f): f \in \Phi\}$. For all $\Theta(f_1), \Theta(f_2) \in \Theta(\Phi)$, define $\Theta(f_1) \lesssim \Theta(f_2)$ if and only if for all $\theta(f_1) \in \Theta(f_1)$ and all $\theta(f_2) \in \Theta(f_2)$, $\theta(f_1) \lesssim \theta(f_2)$.

(6.1.9) Theorem:

(a) $\Theta(f_1) \lesssim \Theta(f_2)$ if and only if $f_1 \lesssim f_2$.

(b) $\lesssim$ is a partial order (i.e, a transitive, reflexive, and anti-symmetric relation) on $\Theta(\Phi)$.

The basic geometric idea behind $\lesssim$ lies in seeking to formalize the situation where the graph of the norm of one function mapping $S \to T$ can be bounded from above by a positive scaled multiple of the graph of the norm of another function mapping $S \to T$. (Recall that the graph of $\|f\|$ for $f \in \Phi$ is $\{(s, \|f(s)\|) : s \in S\}$.) $\lesssim$ is similar but not identical to the $\lesssim$ of Hardy (1924).

For the purposes of this monograph, $S = \Omega$, $T = \mathrm{R}$, and $\| \cdot \| = | \cdot |$.

M is similar to the usual big O notation. For real-valued functions f on the positive integers, Knuth (1976) defines $O(f(n))$ to be the set of all $g(n)$ such that there exist positive constants C and n_0 with $|g(n)| \leqslant Cf(n)$ for all $n \geqslant n_0$. When $S = \{1,2,3,\ldots\}$, $T = \mathrm{R}$, and $\| \cdot \| = | \cdot |$, $M(f)$ differs from this definition of $O(f(n))$ in that the absolute value of f is taken and in that n_0 is missing. (Note, if $f > 0$, then n_0 is actually superfluous since if the condition holds for one n_0 when $f > 0$, then it holds for all n).

There are three primary reasons why big O is not used here.

1) Standard statements like $n^2 + n = O(n^2)$ and $H_n = \ln n + \gamma + O(\frac{1}{n})$ do not both make sense whether $O(f(n))$ is interpreted

(i) as being a generic function of order at most $f(n)$ or

(ii) as being a specific but unknown function of order at most $f(n)$ or

(iii) as being Knuth's (1976, 1973a) set of functions and using his notion of one-way equality.

As it turns out, reading Knuth closely, one finds that he does not in fact define what it means for a specific function (e.g., H_n) to be $=$ (i.e., one-way equality) to an $O(f(n))$ expression (e.g., $\ln n + \gamma + O(\frac{1}{n})$). It is possible, of course, to write $\{H_n\} = \ln n + \gamma + O(\frac{1}{n})$ in Knuth's notation, but why talk about sets of functions when you probably only want to talk about one specific function? And more fundamentally, why use an equal sign when you don't mean it (i.e., when the relationship isn't even symmetric)? And furthermore still, how does one say that $O(n^2) + O(n^2)$ is precisely the same set as $O(n^2)$ if the expression "$O(n^2) + O(n^2) = O(n^2)$" is interpreted to mean $O(n^2) + O(n^2) \subset O(n^2)$?

In the notation of this monograph, the ideas behind the above two $O(f(n))$ statements seem to be more clearly and precisely expressed by saying $n^2 + n \in M(n^2)$ and $H_n = \ln n + \gamma + \mu(\frac{1}{n})$ for some $\mu(\frac{1}{n}) \in M(\frac{1}{n})$ (note: the usual confusion is being made here between a function and its value, e.g., the map $n \rightarrow n^2$ and the value n^2).

2) $M(f)$ and $O(f(n))$ are not specific enough for this chapter. For example, $1 \in O(n^2)$. When it is known that an algorithm has time complexity between $6n(\omega)^2$ and $36n(\omega)^2$ for all inputs ω, then it will be said here that the complexity is of order $\Theta(n^2)$, not that it is in $M(n^2)$ or $O(n^2)$.

3) This monograph also needs a notational system which permits working with the orders of real-valued functions on arbitrary sets. In particular, it will be necessary to work with the complexities of algorithms defined on the set of all matrices satisfying certain conditions. The usual big O notation is not defined in this generality. In fact, the usual big O notation is not only not defined for general S but also not defined for general T and $\| \cdot \|$.

As a final remark, many of the standard results for $O(f(n))$ such as $O(f(n))+O(f(n)) = O(f(n))$ in Knuth's notation have the appropriate analogs for Θ and these will be used without comment.

-- *complexity bounds for Algorithm (6.1.5)* --

(6.1.10) Theorem: For the tree algorithm of (6.1.5), suppose additionally that $n > d$ and that it takes $\theta(n)$ operations to evaluate $g(t)$ for $t \in \mathop{\times}\limits_1^n \{0, 1\}$. (Notice that the dependence of n on ω is being notationally suppressed). Let $C^*(\omega)$ be the complexity of tree algorithm (6.1.5) for each $\omega \in \Omega$. Then for all input matrices $\omega = \begin{bmatrix} \underline{y}_1^T \\ \vdots \\ \underline{y}_n^T \end{bmatrix} \in \Omega$:

1. If $\alpha = 0$, then $dn \lesssim C^*$, i.e., $C^* \in \Lambda(dn)$.

2. If $\alpha = 1$, then $d^2 n \lesssim C^*$, i.e., $C^* \in \Lambda(d^2 n)$.

3. If $\alpha > 1$, then $dn\frac{\alpha^d - 1}{\alpha - 1} \lesssim C^*$, i.e., $C^* \in \Lambda\left(dn\frac{\alpha^d - 1}{\alpha - 1}\right)$.

4. $C^* \lesssim dn^d/2^{d-1}$, i.e., $C^* \in M(dn^d/2^{d-1})$.

Proof: Referring to the notation of Algorithm (6.1.3), discover that $q = d - 2$. Note that since $\underline{\tilde{v}}_k$ is replaced by $-\underline{\tilde{v}}_k$ when appropriate, $\#N_k(i_0, \ldots, i_{k-1}) \leqslant (n - k)/2 = \nu_k$ for $k = 0, \ldots, d - 2 = q$. By definition of α, $\mu_k = \alpha \leqslant \#N_k(i_0, \ldots, i_{k-1})$ for $k = 0, \ldots, d - 2$.

The next step is to compute the order of the time complexity of algorithm $A_k(i_0, \ldots, i_{k-1})$ for $k = 1, \ldots, q$ and any $i_0, \ldots, i_{k-1}$ that can be provided by the tree algorithm. $\underline{\tilde{v}}_k$ is computed in $\theta(kd)$ basic operations for some $\theta(kd) \in \Theta(kd)$ (cf., (3.5.12)). Similarly, the signs of $[y_i, \tilde{v}_k]$ are computed in $\theta(nd)$ basic operations, $H(\underline{\tilde{v}}_k)$ and/or $H(-\underline{\tilde{v}}_k)$ are computed in $\theta(n)$ basic operations once the sgn $[y_i, \tilde{v}_k]$ are known, and $\underline{\tilde{v}}_k$ and/or $-\underline{\tilde{v}}_k$ are saved, if appropriate, in $\theta(d)$ basic operations. Consequently for each $k = 1, \ldots, d - 2$, $C_k(i_0, \ldots, i_{k-1}) \in \Theta(nd)$.

Essentially the same argument is used to show that $C_{q+1}(i_0, \ldots, i_{k-1}) \in \Theta(nd)$ and since A_0 computes N_0, $C_0 \in \Theta(nd)$.

So, there are positive constants σ_k and τ_k such that for $k = 0, \ldots, q + 1 = d - 1$, for all $\omega \in \Omega$,

$$\sigma_k n(\omega) d(\omega) \leqslant C_k((i_0, \ldots, i_{k-1}), \omega) \leqslant \tau_k\, n(\omega) d(\omega).$$

By Theorem (6.1.4),

$$\sigma_0 nd + \sigma_1 nd\alpha + \cdots + \sigma_{d-1} nd \prod_0^{d-2} \alpha \leqslant C^*$$

$$C^* \leqslant \tau_0 nd + \tau_1 nd\left(\frac{n}{2}\right) + \cdots + \tau_{d-1} nd \prod_{k=0}^{d-2} \frac{n-k}{2}$$

$$\leqslant nd(\max \tau_i) \left[\sum_{k=0}^{d-1} \left(\frac{n}{2}\right)^k \right]$$

$$\leqslant 3(\max \tau_i) dn^d / 2^{d-1}$$

Note that there is no claim that either one of these bounds is tight since they probably aren't.

Also, note that if trimming is added to Algorithm (6.1.5), then the argument deriving the upper time complexity bound still holds since the time cost of doing the trimming is not that expensive. What is somewhat interesting is that when the previous argument for the lower bound is applied, the resulting lower bound is in $\Theta(nd)$ (which is not very useful) since the trimming operation can deprive a parent of all of its children.

Finally, note that if the restarting improvement were added to (6.1.5), then the *very* worst case upper bound for the resulting algorithm would be in $\left(\frac{n}{2} - \alpha\right) \Theta\left(dn^d / 2^{d-1}\right)$.

-- *time complexity bounds for the fast approximate tree algorithm* --

Next, bounds are derived on the time complexity of the fast approximate tree algorithm of (3.4.20) without the restarting improvement. Also, for simplicity's sake, the number of basic operations employed by the displacement phase will not be counted.

(6.1.11) Theorem: Let C^{**} be the time complexity of the fast approximate tree algorithm of (3.4.20) without the restarting improvement and not counting the displacement operations. For each $\tilde{v}_k(i_0, \ldots, i_{k-1})$ which is one of the "best children" examined by the fast tree algorithm, let

$$N''(\tilde{v}_k(i_0, \ldots, i_{k-1})) := \{q \in I : \tilde{v}_{k+1}(i_0, \ldots, i_{k-1}, q) \text{ is one of the "best children" of } \tilde{v}_k(i_0, \ldots, i_{k-1}) \text{ whose subtrees will be explored}\}.$$

Recall (cf., (3.4.20)) that $\#s$ is such that for $k = 0, \ldots, d-2$,

$$\min(\alpha, \#s(k)) \leqslant \#N''(\tilde{v}_k(i_0, \ldots, i_{k-1})) \leqslant \#s(k).$$

Then, for all input matrices $\omega = \begin{bmatrix} \underline{v}_1^T \\ \vdots \\ \underline{v}_n^T \end{bmatrix} \in \Omega,$

1. If $\alpha = 0$, then $dn \lesssim C^{**}$.

2. Suppose $\alpha \geqslant 1$.

   (a) If $\#s(k) = 1$ for all k or if $\alpha = 1$, then $\alpha d^2 n \lesssim C^{**}$.

   (b) If $\#s$ is a constant greater than 1 and $\alpha \geqslant 2$, then

   $$\alpha dn \left(\frac{[\min(\alpha, \#s)]^{d-1} - 1}{\min(\alpha, \#s) - 1} \right) \lesssim C^{**}.$$

   (c) $\alpha dn (1 + \sum_{i=0}^{d-3} \prod_{k=0}^{i} \min(\alpha, \#s(k))) \lesssim C^{**}$.

3. (a) If $\#s(k) = 1$ for all k, then $C^{**} \lesssim d(d-1)n^2$

   (b) If $\#s$ is a constant greater than 1, then $C^{**} \lesssim dn^2 \left(\frac{(\#s)^{d-1} - 1}{\#s - 1} \right)$

   (c) $C^{**} \lesssim dn^2 \left(1 + \sum_{i=0}^{d-3} \prod_{k=0}^{i} \#s(k) \right)$

Proof: Referring to the notation of Algorithm (6.1.3), discover that $q = d - 2$.

As in (6.1.10), there exists $0 < \sigma_{q+1} \leqslant \tau_{q+1}$ such that for all $\omega \in \Omega$, $\sigma_{q+1} n(\omega) d(\omega) \leqslant C_{q+1}((i_0, \ldots, i_q), \omega) \leqslant \tau_{q+1} n(\omega) d(\omega)$.

The following table lists various upper and lower bounds on the complexity of the component actions making up $A_k(i_0, \ldots, i_{k-1})$ for $0 \leqslant k \leqslant q = d - 2$ where if α appears in an expression, it is assumed that $\alpha \geqslant 1$. Here $\theta(f(\alpha, n, d))$ is some specific element of $\Theta(f(\alpha, n, d))$.

action	lower	upper
compute $\tilde{v}_k(i_0, \ldots, i_{k-1})$	$\theta(kd)$	$\theta(kd)$
compute $N(\tilde{v}_k)$	$\theta(nd)$	$\theta(nd)$
for all $q \in N(\tilde{v}_k)$, compute $\tilde{v}_{k+1}(i_0, \ldots, i_{k-1}, q)$	$\theta(\alpha(k+1)d)$	$\theta(\frac{n-k}{2}(k+1)d)$
for all $q \in N(\tilde{v}_k)$, compute $N(\tilde{v}_{k+1}(i_0, \ldots, i_{k-1}, q))$	$\theta(\alpha n d)$	$\theta(\frac{n-k}{2} nd)$
for all $q \in N(\tilde{v}_k)$, compute the necessary H values for $\tilde{v}_{k+1}(i_0, \ldots, i_{k-1}, q)$	$\theta(\alpha n)$	$\theta(\frac{n-k}{2} n)$
identify the best children of $\tilde{v}_k$ and store in $N''(\tilde{v}_k)$	$\theta(n) + \theta(\min(\alpha, \#s(k)))$	$\theta(n) + \theta(\#s(k)^2)$
compute the necessary H values for $\tilde{v}_k$ using $N(\tilde{v}_k)$	$\theta(n)$	$\theta(n)$
store $\tilde{v}_k$ if desired	0	$\theta(d)$

So, for $k = 0, \ldots, d-2$, there are positive constants σ_k and τ_k such that for all $\omega \in \Omega$,

$$\sigma_k\, \alpha(\omega)n(\omega)d(\omega) \leqslant C_k((i_0, \ldots, i_{k-1}), \omega) \leqslant \tau_k n(\omega)^2 d(\omega).$$

Then by (6.1.4),

$$C^{**} \geqslant dn(\min\sigma_i)\Big[\alpha + \alpha\min(\alpha, \#s(0)) + \cdots$$

$$+\alpha\prod_0^{d-3}\min(\alpha, \#s(k)) + \prod_0^{d-2}\min(\alpha, \#s(k))\Big]$$

The rest follows similarly. □

If a truly worst case complexity is desired for the fast approximate algorithm when including the restarting improvement as well, then the worst case complexities of (6.1.11) should be multiplied by $\left(\frac{n}{2} - \alpha\right)$

The bounds given in (6.1.11) when compared with those of the "standard" tree algorithm in (6.1.10) show how judicious choice of $\#s(k)$ (perhaps $\#s(k) = 1$ for small k) can lead to a version of the fast approximate tree algorithm which is very much faster than the standard one.

A more realistic *estimate* for a lower bound to Algorithm (6.1.5) can be obtained by replacing α with $\tilde{\alpha} := \inf\{\#N(\tilde{v}) : \tilde{v} \notin y_i^{\perp}$ for any $i\}$ in $\Theta(nd\frac{\alpha^d - 1}{\alpha - 1})$ when $\tilde{\alpha} > 1$. (Recall no $y_i = 0$ here). Now, if $\{y_i\}_1^n$ is in pointed position, then $\tilde{\alpha} = \inf\{\sum_1^n 1\{[y_i, \tilde{x}] \leqslant 0\} : \tilde{x} \in \tilde{X}\} = \alpha$, so that the utility of $\tilde{\alpha}$ comes in for sets of $\{y_i\}_1^n$ which are not in pointed position. The motivation behind $\tilde{\alpha}$ is that even when $\{y_i\}_1^n$ is not in pointed position, the typical $\tilde{v}_k$ selected by the tree algorithms will not be so over constrained that it cannot be displaced to the positive sides of all the hyperspaces that contain it; hence $\tilde{\alpha}$ is a better estimate of the minimum number of children such $\tilde{v}_k$ can have.

The examples given in Chapter 9 indicate that the rather sophisticated WOH tree algorithm used there has a time complexity near $\theta\left(dn\frac{\tilde{\alpha}^d - 1}{\tilde{\alpha} - 1}\right)$ for $\tilde{\alpha} > 1$ and rather far from $\theta\left(dn^d/2^{d-1}\right)$. For people who need an answer to the problem of maximizing H, this is good news. On the other hand, $\tilde{\alpha}^{d-1}$ may often times be too much of a price to pay (e.g., $30^5 = 24{,}300{,}000$) and it will become necessary to use only some version of the fast approximate algorithm in order to get a reasonably good answer in a reasonable amount of time.

The complexity bounds derived in this chapter are actually of more than theoretical interest. Incorporated into each bound is a factor counting up the number of $\tilde{v}_k$ examined by the algorithm. Using this and if, from previous computer runs, there is a rough feeling of the CPU time necessary to process each $\tilde{v}_k$ on the average and if there is some idea of what $\tilde{\alpha}$ might be, then one can get rough CPU time estimates for running tree algorithms on a fixed set of y_i.

Summary for Chapter 6

Consider the problem of finding those $x \in \mathrm{R}^d$ which maximize $g(1\{a_1^T x\ R_1\ 0\}, \ldots, 1\{a_n^T x\ R_n\ 0\})$ where g is nondecreasing, $R_i \in \{ >, \geqslant \}$, and all $a_i \neq 0$. Upper and lower time complexity bounds are derived in this chapter for a somewhat unsophisticated tree algorithm variant which will solve this problem; bounds are also obtained for a simplified fast approximate tree algorithm.

The first algorithm is defined by modifying the basic tree algorithm (5.1.13) as follows. The exploration of the tree is to take place in a depth-first manner. New $\underline{\tilde{v}}_k$ are to be computed using the projection method. No trimming or restarting is to take place. The fast tree algorithm is not used to generate $\tilde{v}_0$. The complexity of the displacement phase is not to be considered (in practice, it is relatively insubstantial).

Before proceeding further, it is necessary to develop some useful notation for working with time complexities which is more general and well-defined for the tree algorithm's needs than the big O notation. Hence, for real-valued functions f on some set,

$$\Theta(f) := \{g: \text{ there exist } 0 < \delta_1 \leqslant \delta_2 \text{ such that } \delta_1 |f| \leqslant |g| \leqslant \delta_2 |f| \}$$

and $\theta(f)$ denotes some not necessarily known element of $\Theta(f)$. $\Theta(f)$ is thus the set of all real-valued functions with the same domain as f which are of the same order as f.

Suppose it takes $\theta(n)$ operations to evaluate $g(t)$ for any $t \in \times_1^n \{0, 1\}$. For the fixed set $\{a_i\}_1^n \subset \mathrm{R}^d$, let $\alpha = \inf\{\#\{i : a_i^T x < 0\} : x \neq 0\}$. Then the upper time complexity bound for the algorithm is of $\Theta(dn^d/2^{d-1})$ and the lower bound is of $\Theta\left(dn\dfrac{\alpha^d - 1}{\alpha - 1}\right)$ if $\alpha > 1$.

When fast algorithm generation of $\tilde{v}_0$, trimming, restarting, and displacement are added, the empirical evidence of Chapter 9 indicates that the complexity of the currently most sophisticated WOH tree algorithm is at the moment best estimated by

$$\theta\left(nd\frac{\tilde{\alpha}^d - 1}{\tilde{\alpha} - 1}\right)$$

where $\tilde{\alpha} := \inf\{\#\{i : a_i^T x < 0\} : x \notin a_i^\perp \text{ for any } i\}$.

The fast approximate tree algorithm with no restarting and ignoring the displacement operations has, when $\#s(k)$, the number of best subtrees examined at each level k, is a constant function $\geqslant 2$, a time complexity bounded below by

$$\theta\left(\alpha dn\left(\frac{[\min(\#s, \alpha)]^{d-1} - 1}{\min(\#s, \alpha) - 1}\right)\right)$$

when $\alpha \geqslant 2$ and bounded above by

$$\theta\left(dn^2\left(\frac{(\#s)^{d-1} - 1}{\#s - 1}\right)\right)$$

The exponential character of these complexities in d is unfortunate from the computational standpoint but apparently is inescapable since the basic underlying problem here is NP-complete.

Chapter 7:

Other Methodology For Maximizing Functions Of Systems Of Linear Relations

The material in Chapter 4 appears to be the first which describes a method for extremizing general monotone functions of consistent or inconsistent systems of homogeneous or inhomogeneous linear inequalities subject to various constraints. There have, of course, been procedures proposed previously for solving special cases of this problem. This chapter discusses several of these.

Consider the problem of maximizing

$$H_1(x) := \sum_{J} \sigma_i 1\{a_i^T x > \mu_i\} + \sum_{I \text{ w.o. } J} \sigma_i 1\{a_i^T x \geqslant \mu_i\}$$

over $x \in \mathsf{R}^d$ where σ_i, $\mu_i \in \mathsf{R}$ and I and J are index sets with $J \subset I$. To the author's knowledge, all previous work on solving this problem has fallen into two categories:

(a) when all $\sigma_i > 0$ and the underlying system is consistent, i.e.,

$$\bigcap_{J} \{x : a_i^T x > \mu_i\} \cap \bigcap_{I \text{ w.o. } J} \{x : a_i^T x \geqslant \mu_i\} \neq \varnothing$$

and

(b) when the underlying system is homogeneous (i.e., all $\mu_i = 0$) and inconsistent.

Each will be discussed in turn.

-- *consistent systems of linear equations* --

In the first case, it is easy to see that the object is to find a vector x which satisfies every linear inequality in the system. In the event that $J = \varnothing$ and for every $i \in I$ there exists $j \in I$ such that $a_i = -a_j$ and $\beta_i = -\beta_j$, then it is clear that the problem is to find all vectors x which satisfy the system of linear equations $\{a_i^T x = \beta_i : i \in I\}$. There is, of course, an enormous literature on this problem. Perhaps the most widely used general purpose algorithm for solving systems of linear equations is that of Gaussian elimination. See Stewart (1973) for details. The tree algorithm suggests an interesting and efficient way to solve systems of linear equations as will be seen in Chapter 8.

-- *consistent systems of linear inequalities* --

As for the general problem of maximizing H_1 when all $\sigma_i > 0$ and the underlying system is consistent, probably the oldest algorithm for solving this problem is Fourier elimination which dates back to before 1826. In a way analogous to the Gaussian elimination method for solving systems of linear equations, Fourier elimination constructs a sequence of systems of linear relations, each with one less variable than its predecessor, such that a solution to one system in the sequence yields a solution to the system preceding it. Since the final system is easy to solve, one can follow the sequence back and obtain a solution to the original system. For details, see Stoer and Witzgall (1970).

No doubt, the most popular method for obtaining a solution to the consistent system $\{a_i^T x \geqslant \mu_i, i \in I\}$ of linear inequalities is the Phase I method of linear programming. In this method, a linear objective function is defined in terms of additional so-called artificial variables and then minimized over a certain polytope by a linear program. The solution to this linear program yields a vector which is on the boundary of $\bigcap_I \{x : a_i^T x \geqslant \mu_i\}$. (See Hadley (1962) for details). As a related reference, Gale (1969) describes how a variant of the simplex method can be used to find $x \geqslant 0$ (i.e., $\xi_i \geqslant 0$, all i) which solves the system $\{a_i^T x \geqslant \mu_i, i \in I\}$.

Another way to solve the system $\{a_i^T x \geqslant \mu_i, i \in I\}$ is the Motzkin-Schoenberg relaxation method (1954). Suppose $H_i := \{x : a_i^T x \geqslant \mu_i\}$ and assume $A := \bigcap_I H_i$ is nonempty. Then this procedure begins by specifying

$\lambda \in (0, 2]$ and selecting an initial point x_0. If $x_0 \notin A$, let u be the projection (using the usual Euclidean distance) of x_0 onto the boundary of H_j where $x_0 \notin H_j$ and the Euclidean distance between H_j and x_0 is the greatest of all distances between x_0 and halfspaces H_i which don't contain x_0. Let $x_1 = x_0 + \lambda(u - x_0)$. Now repeat the above steps for x_1 in order to obtain x_2 and so forth. If A has an interior then if $0 < \lambda < 2$, the sequence of points so generated either *converges* to a point on the boundary of A or else terminates with a point in A while if $\lambda = 2$, then the sequence is guaranteed to terminate with a point in A. Eaves (1973) proves a uniform property of this sequence for $\lambda = 2$ and cites recent work in this area.

There are some similarities between the tree algorithm (3.3.13) as modified by the projection method for calculating $\underset{\sim}{v}_k$ (cf., Improvement (3.4.7)) and the relaxation method when $\lambda = 1$ and the inequalities are consistent and homogeneous. The tree algorithm projects an initial vector $\tilde{v}_0$ onto each of the halfspaces not containing it. Each of these projected points is then projected in the subspace it is in onto the halfspaces which don't contain it and so on until the resulting tree can grow no further. The relaxation method projects an initial vector onto the furthest halfspace not containing it and then iterates this process for the projected point without further constraining any of the projected points in any way.

-- *Steele's WOH algorithm* --

Next, turn to the problem of maximizing

$$h_1(x) = \sum_J \sigma_i \, 1\{a_i^T x > 0\} + \sum_{I \text{ w.o. } J} \sigma_i 1\{a_i^T x \geqslant 0\}$$

when the underlying system is inconsistent. Suppose attention is directed to the case where all $\sigma_i > 0$ and $\{a_i, i \in J\}$ is in pointed position. An algorithm suggested in conversation by Mike Steele begins by enumerating the edges (i.e., one-dimensional faces) of all of the cones $\mathrm{C}\{\pi_i a_i, i \in I\}^+$ where $\pi_i \in \{-1, 1\}$. Then if the h_1 value of an edge is sufficiently high, a vector in the edge is displaced to the positive side of all its constraining boundary hyperspaces into the interior of the neighboring cone so determined. (This is where the pointed position assumption is used.) The cones determined in this way contain all solutions to the problem of maximizing h_1.

That this procedure is valid can be seen by the following argument. Theorem (4.1.15) states that all maximizing vectors are contained in the max-sum cones associated with the strict inequality version of the problem. Also by Theorem (4.1.15), the interior vectors of max-sum cones attain the maximum possible value of h_1 for the original problem involving "$>$" and "$\geqslant$". Note that knowledge of the value of h_1 on any edge of a max-sum cone readily yields the value of h_1 on the interior since any vector in the edge may be displaced to the positive side of all its constraining boundary hyperspaces. Consequently, by enumerating the value of h_1 on all of the edges, all of the max-sum cones of the strict inequality version of the problem may be located which themselves contain all vectors maximizing the function h_1 of the original problem.

Since the one-dimensional dimensionality spacc of an edge is defined and annihilated by a not necessarily unique linearly independent set of $d - 1$ a_i, all edges can be discovered by determining all possible linearly independent subsets of $\{a_i, i \in I\}$ of size $d - 1$ and forming the one-dimensional subspace which annihilates them. Each such subspace contains two edges. If $\#I = n$, then there are at most $\binom{n}{d-1}$ of these linearly independent subsets and consequently at most $2\binom{n}{d-1}$ edges. These upper bounds are achieved if the set $\{a_i, i \in I\}$ is in general position.

By using the theory of Chapter 3, this algorithm may be extended to the case of maximizing $\sum_I \sigma_i 1\{a_i^T x > 0\}$ for $\sigma_i > 0$ for arbitrary $\{a_i, i \in I\}$. This algorithm is the WOH tree algorithm's current best competitor for solving this WOH problem and the two will be compared in the examples of Chapter 9. This algorithm computes all of the edges as before. Naturally, there is no need to consider y_j if $\langle y_j \rangle = \langle y_i \rangle$ or $\langle y_j \rangle = -\langle y_i \rangle$ for some i. Hence, it is assumed that all such unnecessary vectors are ignored by the edge computation procedure although of course they will have to be considered when computing the associated value of $\sum_I \sigma_i \, 1\{a_i^T x > 0\}$. So, suppose $n-k$ a_i are left after these deletions. Then the edge enumeration procedure need only check $\binom{n-k}{d-1}$ subsets of the $n-k$ points. Some of these subsets may be linearly dependent in which case they can be ignored. Also, the same edge may be generated by more than one subset. Nonetheless, by checking each subset, one obtains at least one vector in the boundary of each max-sum cone.

Now since $\{a_i, i \in I\}$ is not necessarily in pointed position, it may not be possible to displace an edge vector to the positive side of each of its constraining boundary planes. As seen in Chapter 3 however, the problem of determining how to best displace a boundary vector is a lower dimensional version of the original problem and consequently, the edge enumeration procedure can be applied here in a recursive manner in order to solve this WOH problem.

In summary, this edge enumeration algorithm for solving Problem (3.1.1) eliminates unnecessary a_i at the beginning, computes all of the edges, and, if they have sufficiently high values of $\sum_I \sigma_i 1\{a_i^T x > 0\}$, displaces edge vectors into the interiors of cones using recursion if necessary. The resulting set of cones contains all max-sum cones. Ignoring the recursion, the time complexity of this algorithm is $\theta\left(nd\binom{n-k}{d-1}\right)$ (for the θ-notation, see (6.3.8)).

-- *Johnson and Preparata's weighted hemisphere algorithms* --

Johnson and Preparata (1978) also describe algorithms for maximizing over nonzero x, $h_1(x) := \sum_J \sigma_i 1\{a_i^T x > 0\} + \sum_{I \text{ w.o. } J} \sigma_i 1\{a_i^T x \geqslant 0\}$ where without loss of generality, all $\sigma_i > 0$. When $J = \varnothing$, $J = I$, or $\varnothing \neq J \neq I$, the respective problems are called the Weighted Closed, Open, or Mixed Hemisphere problems.

The algorithm Johnson and Preparata give for solving the Weighted Closed Hemisphere (WCH) problem is a recursive one which at each stage reduces each of the current set of WCH problems to at most $n = \#I$ WCH problems of one lower dimension. A careful look at this algorithm with an eye to removing the recursive character yields the following algorithm which computes all the same quantities as the Johnson-Preparata WCH algorithm but in a non-recursive way.

(7.1.1) Algorithm: Assume for simplicity without loss of generality that $\dim \mathsf{L}\{a_i, i \in I\} = d$. The notation $P[a_i \mid S^\perp]$ for orthogonal projectors is defined in (3.4.8).

For each $i_1 \in \{j \in I : a_j \neq 0\}$ do:

For each $k \in I$, compute $z_k(i_1) := P[a_k \mid a_{i_1}{}^\perp]$.

For each $i_2 \in \{j \in I : z_j(i_1) \neq 0\}$ do:

For each $k \in I$, compute $z_k(i_1, i_2) := P[z_k(i_1) \mid z_{i_2}(i_1)^\perp]$.

For each $i_3 \in \{j \in I : z_j(i_1, i_2) \neq 0\}$ do:

For each $k \in I$, compute $z_k(i_1, i_2, i_3) := P[z_k(i_1, i_2) \mid z_{i_3}(i_1, i_2)^\perp]$.

⋮

For each $i_{d-1} \in \{j \in I : z_j(i_1, \ldots, i_{d-2}) \neq 0\}$ do:

For each $k \in I$, compute

$$z_k(i_1, \ldots, i_{d-1}) := P[z_k(i_1, \ldots, i_{d-2}) \mid z_{i_{d-1}}(i_1, \ldots, i_{d-2})^\perp].$$

Let $h_2(x) := \sum_{k=1}^{n} \sigma_k 1\{z_k(i_1, \ldots, i_{d-1})^T x \geqslant 0\}$.

Find some q such that $z_q(i_1, \ldots, i_{d-1}) \neq 0$ and compute $h_2(z_q(i_1, \ldots, i_{d-1}))$ and $h_2(-z_q(i_1, \ldots, i_{d-1}))$, saving those vectors in the pair which have sufficiently high h values.

next i_{d-1};

⋮

next i_1;

Algorithm (7.1.1), which is the Johnson-Preparata WCH algorithm written non-recursively, is actually an edge enumeration algorithm. To see this, begin by noting (after a little work) that the nested projection operations of (7.1.1) are based on the same idea as the Modified Gram-Schmidt algorithm (cf., the discussion following (3.5.12)). This implies that for each ordered set of indices $(i_1, \ldots, i_{d-1})$ used during the execution of (7.1.1), $\{a_{i_1}, \ldots, a_{i_{d-1}}\}$ is linearly independent and for each $k \in I$, $z_k(i_1, \ldots, i_{d-1}) = P[a_k | \mathsf{L}\{a_{i_1}, \ldots, a_{i_{d-1}}\}^{\perp}]$ and is thus in an edge if it is nonzero. Since orthogonal projectors are symmetric and idempotent, for given $(i_1, \ldots, i_{d-1})$,

$$h_2(z_q(i_1, \ldots, i_{d-1})) = \sum_{k=1}^{n} \sigma_k 1\{a_k^T z_q(i_1, \ldots, i_{d-1}) \geqslant 0\}$$

and so when h_2 is defined and evaluated in the $d - 1$ loop for $z_q(i_1, \ldots, i_{d-1})$ and $-z_q(i_1, \ldots, i_{d-1})$, what is actually happening here is that the original WCH criterion function is being evaluated on the two nonzero edges contained in $\mathsf{L}\{a_{i_1}, \ldots, a_{i_{d-1}}\}^{\perp}$.

It is easy to see that every edge is visited at least once by this procedure and in fact if $\{a_i, i \in I\}$ is in general position, then every edge is visited exactly $(d - 1)!$ times. In this event, it is certain that Steele's algorithm is faster. The Johnson-Preparata WCH algorithm has time complexity $O(dn^d)$ which becomes the exact complexity of the algorithm when $\{a_i, i \in I\}$ is in general position.

Now the algorithm which Johnson and Preparata present for solving the Weighted Open Hemisphere Problem (i.e., Problem (3.1.1)) is the same as their algorithm for solving the Weighted Mixed Hemisphere Problem (WMH). The operations the WMH algorithm performs include all of the operations the corresponding WCH algorithm performs plus on the order of 2^{d-2} as much more. The WOH manifestation of the Johnson-Preparata WMH algorithm not only enumerates the values of the criterion function on all of the edges but also in the interiors of all fully dimensional cones (which is the only thing it should do since this is where all of the solutions are). Interestingly, as will be seen in a few paragraphs, there is a closed form expression for the number of fully dimensional cones when $\{a_i, i \in I\}$ is in general position.

Johnson and Preparata describe an improved densest hemisphere algorithm for two dimensions which can be incorporated into their basic WCH and WMH algorithms to reduce the worst case time complexities from $O(dn^d)$ and $O(2^{d-1}dn^d)$ to $O(dn^{d-1}\log n)$ and $O(2^{d-2}dn^{d-1}\log n)$ respectively. These worst case complexities become the actual complexities when $\{a_i, i \in I\}$ is in general position. The improved WCH algorithm nonetheless remains an edge enumeration algorithm which is incorporated into the improved WMH algorithm.

-- *comparison of the various WOH algorithms* --

So how does the WOH tree algorithm compare with the Steele and Johnson-Preparata WOH algorithms? First of all, the examples of Chapter 9 show dramatically that the WOH tree algorithm (as defined by the improvements of section 3.4) as well as, by implication, the general tree algorithm are not edge enumeration algorithms. The tree algorithm in these examples computes far fewer candidate vectors than would be computed by the algorithms of Steele and Johnson-Preparata.

-- *counting fully dimensional cones* --

A brief digression here will show that these examples also demonstrate that the WOH tree algorithm (and for that matter, the general tree algorithm) are not fully dimensional cone enumeration algorithms either. This follows from a theorem of Schläfli (Cover (1965)). First, two definitions are necessary.

(7.1.2) Definition: Let $\{y_i\}_1^n \subset X$. Define

$$\Delta \{y_i\}_1^n := \#\{\{y_i\}_1^n \cap \{x \in X : [x, \tilde{u}] > 0\} : \tilde{u} \neq 0\} .$$

Observe that $\Delta\{y_i\}_1^n$ is the number of subsets of $\{y_i\}_1^n$ which can be generated from $\{y_i\}_1^n$ by intersecting this set with open halfspaces whose boundaries pass through the origin.

(7.1.3) Definition: For $n, d \geqslant 1$, let

$$C(n, d) := \begin{cases} 2^n & \text{if } n \leqslant d \\ 2\sum_0^{d-1} \binom{n-1}{k} & \text{if } n > d \end{cases}$$

(7.1.4) Theorem: (Schläfli): Let $\{y_i\}_1^n \subset X$ where $\dim X = d$. Then $\Delta\{y_i\}_1^n \leqslant C(n, d)$ with equality holding if $\{y_i\}_1^n$ is in general position.

(7.1.5) Theorem: $\Delta\{y_i\}_1^n$ is greater than or equal to the number of fully dimensional $C(\pi)^+$ with equality holding when $\{y_i\}_1^n$ is in pointed position.

Proof: First, observe that any $\tilde{u}$ such that

$$\tilde{u} \in \operatorname{int} C(\pi)^+ = \{\tilde{x} \in \tilde{X}: [\pi_i y_i, \tilde{x}] > 0 \text{ for all } i \text{ such that } y_i \neq 0\}$$

is such that $\{y_i\}_1^n \cap \{x \in X: [x, \tilde{u}] > 0\} = \{y_i: \pi_i = 1\}$; also, if any two such interior vectors generate in this manner the same subset of $\{y_i\}_1^n$, then they must be in the interior of the same $C(\pi)^+$.

Conversely, take any nonzero $\tilde{v} \in \tilde{X}$ and consider $\{y_i\}_1^n \cap \{x \in X: [x, \tilde{v}] > 0\} =: \{y_i, i \in J\}$. If $\{y_i\}_1^n$ is in pointed position, by using (3.5.1), $\tilde{v}$ may be displaced to some nonzero $\tilde{u}$ such that if for any i, $[y_i, \tilde{v}] > 0$ then $[y_i, \tilde{u}] > 0$ and if $[y_i, \tilde{v}] \leqslant 0$ for $y_i \neq 0$ then $[y_i, \tilde{u}] < 0$. Consequently, $\tilde{u} \in \operatorname{int} C(\pi)^+$ where $\pi_i = 1$ if and only if $i \in J \cup I_0$. Observe also that if two subsets of $\{y_i\}_1^n$ of the form $\{y_i\}_1^n \cap \{x \in X: [x, \tilde{v}] > 0\}$ give rise in this fashion to the same $C(\pi)^+$, the two subsets must be identical.

So if $\{y_i\}_1^n$ is in pointed position, then, since the preceding discussion exhibited one-to-one maps mapping each of the sets of fully dimensional cones and subsets of $\{y_i\}_1^n$ counted by $\Delta\{y_i\}_1^n$ into each other, it may be concluded by the Schroeder-Bernstein Theorem that $\Delta\{y_i\}_1^n$ is precisely the number of fully dimensional $C(\pi)^+$ when $\{y_i\}_1^n$ is in pointed position. □

So, in short, if $\{y_i\}_1^n$ is in general position and $n > d$, then the number of fully dimensional $C(\pi)^+$ is $2\sum_0^{d-1}\binom{n-1}{k} \geqslant 2\binom{n}{d-1}$ and so since the WOH tree algorithm doesn't enumerate edges, it doesn't enumerate fully dimensional cones either.

As commented in Chapter 6, the time complexity of the WOH tree algorithm augmented with all of the improvements of Section 3.4 except (3.4.19) tends to be close to $\theta\left(nd\dfrac{\alpha^d - 1}{\alpha - 1}\right)$ where $\alpha = \inf\{\#N(\vec{v}) : \vec{v} \neq 0\}$ which in general is far removed from $\theta\left(nd\binom{n}{d-1}\right)$ or $O(d2^{d-2}n^{d-1}\log n)$ which are the complexities of the edge enumeration procedure of Steele and the WOH-WMH algorithm of Johnson and Preparata, respectively.

As α increases towards n/2, one would expect tree algorithm (6.1.5) to behave more and more like an edge enumeration procedure. When trimming and restarting are added to the procedure, however, its behavior becomes uncertain since while there may be n/2 children to start with at each node, a large number are likely to be trimmed away. Whether or not the tree algorithm with or without trimming can be induced to compute as many if not more vectors than there are edges is not known with the exception of the information in a theorem of Warmack and Gonzalez (1973) which this author does not understand.

Interestingly enough, the fast two-dimensional enumeration idea of Johnson and Preparata can be used to modify the tree algorithm. First, it is reported that by using the hill concept, the Johnson-Preparata two-dimensional algorithm can be improved so that the new algorithm sorts half as many rays and determines the relative order of pairs of rays in the ordering process by a single comparison instead of the more involved procedure given in Johnson and Preparata (1978). When the improved two dimensional algorithm is incorporated into the tree algorithm, it appears that the unmodified tree algorithm will be faster for fixed small $\alpha = \inf\{\#N(\tilde{v}) : \tilde{v} \neq 0\}$ as n grows whereas if $\alpha = \beta n$ some $\beta > 0$ then the modified tree algorithm will be faster as n becomes large. Space limitations preclude going into more detail at this time.

-- *Ibaraki and Muroga's algorithm* --

Ibaraki and Muroga (1970) formulated the WOH problem as a mixed integer linear programming problem with n integer variables, $2d$ positive continuous variables, and $2n$ constraints. This is a very large mixed integer linear programming problem as indicated by the number of variables and constraints. The procedure Ibaraki and Muroga propose for solving it is to force all the continuous variables to be integers, add n more constraints, and use Gomory's all-integer integer linear programming method. They offer no proof that the two problems are equivalent and it seems doubtful (to this author) that they are. In view of the large number of variables and constraints in the original Ibaraki and Muroga formulation and its all-integer version, it appears unlikely that existing integer programming methods using either of the Ibaraki and Muroga formulations would be faster than the tree algorithm in solving the WOH problem.

-- *Warmack and Gonzalez's algorithm* --

Warmack and Gonzalez (1973) presented an OH tree algorithm for maximizing $\sum_{1}^{n} 1\{y_i^T x > 0\}$ when $\{y_i\}_1^n$ is in general position. Just as with the tree algorithms discussed in this monograph, the Warmack-Gonzalez algorithm has two phases: a boundary vector collection phase and a displacement phase. The Warmack-Gonzalez boundary vector collection phase searches a tree which satisfies the requirements of Algorithm (3.3.13). The Warmack-Gonzalez displacement phase rests heavily on the general position assumption. The Warmack and Gonzalez algorithm will solve the WOH problem under the general position assumption although this fact is not stated in their paper.

When comparing the Warmack-Gonzalez algorithm with the Section 3.4 modified WOH tree algorithm, it must be remembered that they are closely related. The crucial difference between the respective boundary vector collection algorithms is that the Warmack-Gonzalez algorithm constrains all of its $\tilde{v}_k$ in an arbitrary way to be in edges. This has several implications. First, as commented in (3.4.11), these arbitrary constraints raise some question as to whether the Warmack-Gonzalez trimming procedure is actually valid for the Warmack-Gonzalez tree algorithm. Second, in comparison with the Section 3.4 modified WOH tree algorithm, the Warmack and Gonzalez method of arbitrary constrainment would seem to lead to a slower algorithm since in the Section 3.4 modified tree algorithm the initial vector $\tilde{v}_0$ and all other $\tilde{v}_k$ for $k \leqslant d - 2$ are not chosen by some arbitrary condition forcing them to be in edges but rather are constrained by only as many hyperspaces as necessary and are otherwise encouraged to lie in heuristically good regions of the space $\tilde{X}$. In particular, the value of a good $\tilde{v}_0$ and hence the value of restarting and using the fast approximate tree algorithm to obtain good $\tilde{v}_0$ is very considerable in reducing the time complexity of the algorithm by reducing the number of $\tilde{v}_k$ that must be examined. Of course, many of the improvements listed in Section 3.4 can be used by the Warmack-Gonzalez algorithm since it is, after all, a tree algorithm.

The necessity of the general position assumption for both phases of the Warmack-Gonzalez algorithm to work is a significant limitation. It might be thought that one could just perturb various points of a set not in general position around in such a manner as to obtain general position in a set "not too far removed" from the original. However, this presents the serious problems of isolating which points are involved in which degeneracies and then deciding how far and in what direction to perturb the points so that all of the linear degeneracies they are involved in vanish. But most importantly, there is simply the fact that sometimes the linear degeneracies are present in $\{y_i\}_1^n$ because they are supposed to be there by the definition of the problem and so consequently, they shouldn't really be destroyed. As a simple example, if some problem concerning the maximization of functions of systems of linear relations uses $1\{y_1^T x = 4\}$, then once this problem has been reduced to homogeneous canonical form, then one will be working with both the vectors $(y_1, -4)$ and $(-y_1, 4)$; if these vectors are perturbed so that they no longer lie in the same line then the algorithm will not try to satisfy the $1\{y_1^T x = 4\}$ equality. Furthermore, in one of the statistical classification examples of Chapter 9, it will be seen that not only do cross sample ties cause unremovable lack of general position when using certain loss functions but, in addition, the data may come from distributions sufficiently discrete for there to be very high order linear degeneracies: in one Chapter 9 example, there is a hyperplane containing 59 of the 86 data points.

So, in short, the Section 3.4 modified WOH tree algorithm and the general tree algorithm of Chapter 5 are seen to be much more general and (probably) much faster than the Warmack-Gonzalez algorithm. Also, their validity has been established in rigorous terms whereas the main proofs in the Warmack-Gonzalez paper are non-rigorous, incomplete, intuitive, and, in the case of their fundamental algorithm theorem, based on a completely incorrect idea (cf., (3.3.21)).

Chapter 8:

Applications Of The Tree Algorithm

This chapter discusses several situations where some of the problems discussed in Chapter 4 arise.

-- *solving systems of linear equations with the tree algorithm* --

Consider then how the tree algorithm of Chapter 5 would solve a consistent system of linear equations $[a_1 \cdots a_n]^T x = b \neq 0$. It will be seen that the tree algorithm's approach is not only novel but also produces a procedure for solving this problem which has a time complexity of the same order as that of Gaussian elimination.

To begin, the problem is stated as that of seeking to find all $x \in \mathsf{R}^d$ which maximize $\sum_1^n 1\{a_i^T x = \beta_i\}$. This problem is equivalent to an instance of Problem I of Chapter 4, namely, that which seeks to find all $y \in \mathsf{R}^{d+1}$ such that $e_{d+1}^T y = 1$ which maximize

$$\sum_{i=1}^{n} 1\{(a_i, -\beta_i)^T y \geqslant 0\} + 1\{(-a_i, \beta_i)^T y \geqslant 0\}.$$

(Recall $e_{d+1} = (0, \ldots, 0, 1)$.) For each $i = 1, \ldots, n$, let $s_i = (a_i, -\beta_i)$. By Theorem (4.1.9), $y = z \, / \, e_{d+1}^T z$ solves the preceding problem if z maximizes

$$(2n + 1) \cdot 1\{e_{d+1}^T z > 0\} + \sum_{i=1}^{n} 1\{s_i^T z \geqslant 0\} + 1\{-s_i^T z \geqslant 0\} .$$

In order to solve the latter problem, the tree algorithm will first check to see whether $T := \mathsf{L}\{e_{d+1}, s_i, i = 1, \ldots, n\}$ is fully dimensional.

In the event that $p = \dim T < d + 1$, using Procedure (4.1.11), the tree algorithm proceeds to find an orthonormal basis for R^{d+1} such that the first k

components span T. Next, it computes the orthogonal change of basis matrix $\underset{\sim}{B} = \begin{bmatrix} \underset{\sim}{B}_1 \\ \underset{\sim}{B}_2 \end{bmatrix}$ where the rows of the $p \times (d+1)$ matrix $\underset{\sim}{B}_1$ span T. Next, b_i is set to equal $\underset{\sim}{B}_1 s_i$ for $i = 1, \ldots, n$ and b_{n+1} is set to equal $\underset{\sim}{B}_1 e_{d+1}$. From Chapter 4, it is known that each w which maximizes

$$(2n+1)\cdot 1\{b_{n+1}^T w > 0\} + \sum_{i=1}^{n} 1\{b_i^T w \geqslant 0\} + 1\{-b_i^T w \geqslant 0\}$$

generates a linear manifold of solutions $\{\underset{\sim}{B}_1^T w + \underset{\sim}{B}_2^T z : z \in \mathsf{R}^{d+1-p}\}$ to the original problem involving the s_i.

If $p = \dim T = d+1$, then there is no need for the tree algorithm to re-express e_{d+1} and the s_i. However, for the sake of simple notation for the following discussion, when $p = \dim T = d+1$, consider $\underset{\sim}{B}_1 = \underset{\sim}{I}$, $\underset{\sim}{B}_2 = \underset{\sim}{0}$, $b_{n+1} = e_{d+1}$, and $b_i = s_i$ for $i = 1, \ldots, n$.

In short, at this point, $\mathsf{L}\{b_i, i = 1, \ldots, n+1\} = \mathsf{R}^p$ and it is desired to use the tree algorithm to maximize

$$H(w) := (2n+1)\cdot 1\{b_{n+1}^T w > 0\} + \sum_{i=1}^{n} 1\{b_i^T w \geqslant 0\} + 1\{-b_i^T w \geqslant 0\}\,.$$

The tree algorithm that will be used is that of Chapter 5 modified as in section 5.3 but without the trimming and restarting improvements.

The relative boundary vector collection phase of this tree algorithm begins by computing $\underline{\tilde{v}}_0 \in \langle b_{n+1} + \sum_{i=1}^{n}(b_i - b_i)\rangle = \langle b_{n+1}\rangle$. In fact, take $\underline{\tilde{v}}_0 = b_{n+1}$. Observe that $\underline{\tilde{v}}_0^T b_{n+1} > 0$. If it should happen that $b_i^T \underline{\tilde{v}}_0 = 0$ for all $i = 1, \ldots, n$, then this first phase of the tree algorithm would save $\underline{\tilde{v}}_0$ for the second or displacement phase and would quit itself at this point since $\tilde{v}_0$ has no children. Suppose this does not occur, i.e., let $i_0 \leqslant n$ be some index such that either $b(i_0)^T \underline{\tilde{v}}_0 < 0$ or $-b(i_0)^T \underline{\tilde{v}}_0 < 0$. Then the tree algorithm would compute $\underline{\tilde{v}}_1(i_0) = P[b_{n+1} \mid \mathsf{L}\{b(i_0)\}^\perp]$. Note that $b_{n+1}^T \underline{\tilde{v}}_1(i_0) > 0$ since orthogonal projectors are symmetric and idempotent. As before, if $b_i^T \underset{\sim}{\tilde{v}}_1(i_0) = 0$ for $i \leqslant n$, the first phase stores $\tilde{v}_1(i_0)$ for the second phase and quits. But, on the other hand, suppose $i_1 \leqslant n$ is some index such that either $b(i_1)^T \underset{\sim}{\tilde{v}}_1(i_0) < 0$ or $-b(i_1)^T \underset{\sim}{\tilde{v}}_1(i_0) < 0$. Then the tree algorithm would next compute $\underset{\sim}{\tilde{v}}_2(i_0, i_1) = P[b_{n+1} \mid \mathsf{L}\{b(i_0), b(i_1)\}^\perp]$. Continue on in this fashion.

Eventually, if it hasn't stopped before, the first phase will compute $\underset{\sim}{\tilde{v}}_{p-1}(i_0, \ldots, i_{p-2}) = P[b_{n+1} \mid \mathsf{L}\{b(i_0), \ldots, b(i_{p-2})\}^\perp]$. As before, $b_{n+1}^T \underset{\sim}{\tilde{v}}_{p-1}(i_0, \ldots, i_{p-2}) > 0$. At this point, it must be the case that for all $i \leqslant n$, $b_i^T \underset{\sim}{\tilde{v}}_{p-1}(i_0, \ldots, i_{p-2}) = 0$. For suppose to the contrary that for some j, $b_j^T \underset{\sim}{\tilde{v}}_{p-1}(i_0, \ldots, i_{p-2}) \neq 0$. Then $\mathsf{L}\{b(i_0), \ldots, b(i_{p-2})\}^\perp \cap b_j^\perp = \langle 0 \rangle$ and the system is inconsistent. So, in short, it can now be said that the relative boundary vector collection phase of the tree algorithm will explore at most one complete branch of its tree before saving one vector for displacement and then quitting.

So, the displacement phase receives from the first phase a vector $\tilde{v}_k$ such that $b_{n+1}^T \underset{\sim}{\tilde{v}}_k > 0$ and for all $i \leqslant n$, $b_i^T \underset{\sim}{\tilde{v}}_k = 0$. Since $\mathsf{C}\{b_i, -b_i, i = 1, \ldots, n\} = \mathsf{R}^{p-1}$ is a subspace and since $\tilde{v}_k$ has the highest possible H value, this $\tilde{v}_k$ is one of the vectors produced by the displacement phase. Clearly, this $\tilde{v}_k$ is a solution to the problem and for that matter, $\langle \tilde{v}_k \rangle$ is precisely the set of solutions to the problem since (i) $\dim \mathsf{C}\{b_i, -b_i, i = 1, \ldots, n\} = p - 1$ and (ii) any attempt whatsoever to displace $\tilde{v}_k$ off $b_i^\perp$ for any $i \leqslant n$ will result in a decrease in the objective function H.

In summary then, the tree algorithm produces the following solution set to the problem of finding all x such that $a_i^T x = \beta_i$ for $i = 1, \ldots, n$ when not all $\beta_i = 0$:

$$\left\{ x : \begin{pmatrix} x \\ 1 \end{pmatrix} = \frac{\underset{\sim}{B}_1^T \underset{\sim}{\tilde{v}}_k + \underset{\sim}{B}_2^T z}{(\underset{\sim}{B}_1^T \underset{\sim}{\tilde{v}}_k + \underset{\sim}{B}_2^T z)^T e_{d+1}} \text{ for some } z \in \mathsf{R}^{d+1-p} \right\}$$

$$= \left\{ x : \begin{pmatrix} x \\ 1 \end{pmatrix} = \frac{\underset{\sim}{B}_1^T \tilde{v}_k}{\tilde{v}_k^T \underset{\sim}{B}_1 e_{d+1}} + \underset{\sim}{B}_2^T z \text{ for some } z \in \mathrm{R}^{d+1-p} \right\}$$

Clearly, the procedure the tree algorithm uses to solve this problem can be modified so as to produce a yet more efficient algorithm for solving systems of linear equations but this matter will not be pursued further at this time. The point is that for an algorithm whose time complexity in general has an exponential lower bound, the tree algorithm does remarkably well in producing a promising polynomial time algorithm for solving systems of linear equations.

-- *the tree algorithm and statistical classification* --

In the field of statistical classification, the tree algorithm offers a solution to a problem which has been in the literature for over 25 years. This problem, solved in the one-dimensional case by Stoller (1954), is concerned with finding a non-enumerative algorithm for consistently estimating best hyperplane classification rules for the two-class Bayes classification problem. The tree algorithm will solve much more general classification problems than this as will now be briefly described.

In the two-class statistical classification problem, there is an ordered pair (X, T) associated with each individual from each of two classes where $X \in \mathrm{R}^p$ is a vector of measurements on that individual and T is equal to 1 or 2 depending upon which of the two classes the individual belongs to. Usually T is not known for certain individuals and the object is to develop good rules for estimating T given knowledge of X.

For $k = 1, 2$, let F_k be the distribution of X when individuals from class k are being sampled. So, for any Borel-measurable set A, $F_k(A) := P_k[X \in A]$. No assumptions at all are made about the nature of F_1 and F_2. T may or may not be a random variable; if it is, then set $\tau_k := P[T = k]$ and consider $X \mid T = k$ distributed as F_k.

Attention will now be restricted to only those classification rules that belong to certain prespecified classes. In general, for each set $A \subset \mathsf{R}^p$ in a set of sets S, define the classification rule d_A to classify x as class 2 if and only if $x \in A$, i.e., $d_A(x) = 2$ if and only if $x \in A$. Let $\mathsf{D}_S := \{d_A : A \in S\}$.

Classes of rules D_S will be defined by characterizing the set of regions S. The only kinds of rules that will be considered here are those for whom $S = \{\{x \in \mathsf{R}^p : g(x) > 0\} : g \in G\}$ for some vector space G of real-valued functions on R^p which is spanned by a fixed set of functions including the identity. So, for some $d \geqslant 1$, f_1 identically 1, and prespecified $f_i : \mathsf{R}^p \to \mathsf{R}$ for $i = 2, \ldots, d$, $G := \{g = \sum_1^d \alpha_i f_i : \alpha_i \in \mathsf{R}\}$. For an example of such an S, consider the set of all open halfspaces in R^p, namely, $\{\{x \in \mathsf{R}^p : \alpha_1 + \alpha_2 \xi_1 + \cdots + \alpha_{p+1} \xi_p > 0\} : \alpha_i \in \mathsf{R}\}$. Here, for example, f_{p+1} maps x onto its last component ξ_p. The set of all open polynomial regions of any desired but fixed degree is another example of an allowable S.

A loss function L is used to measure for each d_A in D_S the value or worth of using d_A to classify individuals. Three modest requirements are made of L:

(a) L must be a function of d_A through only the misclassification probabilities $F_1(A)$ and $F_2(A^c)$.

(b) L must be a nondecreasing function of the misclassification probabilities separately.

(c) L must be a continuous function of the misclassification probabilities as well as τ_1 (if present).

So, consonant with these assumptions, the loss associated with d_A is thought of as being $L(F_1(A), F_2(A^c), \tau_1)$. The nondecreasing requirement requires that if $\beta_1 \leqslant \beta_2$ and $\gamma_1 \leqslant \gamma_2$, then $L(\beta_1, \gamma_1, \tau_1) \leqslant L(\beta_2, \gamma_2, \tau_1)$.

One of the most frequently used loss functions is the Bayes loss function $\lambda_1 \tau_1 F_1(A) + \lambda_2 \tau_2 F_2(A^c)$ where $\lambda_k > 0$ is the cost of misclassifying an individual from class k. The Bayes loss function measures the expected cost of using d_A to classify individuals sampled from the mixture $\tau_1 F_1 + \tau_2 F_2$ of F_1 and F_2.

The statistical problem here is to find d_A which minimize $L(F_1(A), F_2(A^c), \tau_1)$.

Since F_1, F_2, and τ_1 are not known, information is gathered about them by taking samples. In the Bayes situation, a sample $\{(X_i, T_i)\}_1^n$ is taken where the T_i are known. For $k = 1, 2$, let $N_k := \sum_{i=1}^{n} 1\{T_i = k\}$, $\hat{\tau}_k = \frac{N_k}{n}$, and $\hat{F}_k$ be the appropriate empirical measure, i.e., for measurable A,

$$\hat{F}_k(A) := \frac{1}{N_k} \sum_1^n 1\{X_i \in A, T_i = k\}$$

The empirical version of the problem then becomes that of seeking those d_A in D_S which minimize $L(\hat{F}_1(A), \hat{F}_2(A^c), \tau_1) =$

$$L(\frac{1}{N_1} \sum_1^n 1\{g(X_i) > 0, T_i = 1\}, \frac{1}{N_2} \sum_1^n 1\{g(X_i) \leqslant 0, T_i = 2\}, \frac{N_1}{n})$$

In Greer (1979), it was shown that the d_A which minimize the empirical loss for the Bayes situation (and others as well) are almost surely consistent estimates of best d_A in D_S in the sense that the true loss function values of the estimated best d_A converge almost surely as $n \to \infty$ to the best loss function value possible for the given class D_S.

-- *the tree algorithm solves the empirical minimization problem* --

To see how the tree algorithm solves this empirical minimization problem, suppose for a given sample $\{(X_i, T_i)\}_1^n$, the class 1 points are $\{x_{1i}, i = 1, \ldots, n_1\}$ and the class 2's are $\{x_{2j}, j = 1, \ldots, n_2\}$. Then the empirical minimization problem seeks to identify all $(\alpha_1, \ldots, \alpha_d) =: a$ which minimize

$$L\left(\frac{1}{n_1}\sum_{i=1}^{n_1} 1\{\sum_{q=1}^{d}\alpha_q f_q(x_{1i}) > 0\},\ \frac{1}{n_2}\sum_{j=1}^{n_2} 1\{\sum_{q=1}^{d}\alpha_q f_q(x_{2j}) \leqslant 0\},\ \frac{n_1}{n}\right)$$

Next define $y = (f_1(\cdot), \ldots, f_d(\cdot))$ so that for example, $y_{1i} = (f_1(x_{1i}), \ldots, f_d(x_{1i}))$. Then the problem is to find all $a \in \mathsf{R}^d$ which minimize

$$L\left(\frac{1}{n_1}\sum_{1}^{n_1} 1\{a^T y_{1i} > 0\},\ \frac{1}{n_2}\sum_{1}^{n_2} 1\{a^T y_{2j} \leqslant 0\},\ \frac{n_1}{n}\right)$$

Through a little trickery including replacing y_{1i} with $-y_{1i}$ (see Greer (1979) for details), the above minimization problem can be seen to be equivalent in the sense of Chapter 4 to the problem of finding all $a \in \mathsf{R}^d$ which minimize

$$L\left(\frac{1}{n_1}\sum_{1}^{n_1} 1\{a^T y_{1i} \leqslant 0\},\ \frac{1}{n_2}\sum_{1}^{n_2} 1\{a^T y_{2j} \leqslant 0\},\ \frac{n_1}{n}\right)$$

(As will be seen shortly, the advantage of this trickery is that the solution equivalence classes of this second minimization problem are all the interiors of fully-dimensional cones which is a conceptually nice fact.)

For the Bayes loss function, this objective function becomes

$$\sum_{1}^{n_1} \frac{\lambda_1}{n}\, 1\{a^T y_{1i} \leqslant 0\} + \sum_{1}^{n_2} \frac{\lambda_2}{n}\, 1\{a^T y_{2j} \leqslant 0\}$$

Minimizing the above function is equivalent to maximizing the WOH criterion function

$$\sum_{1}^{n_1} \frac{\lambda_1}{n}\, 1\{a^T y_{1i} > 0\} + \sum_{1}^{n_2} \frac{\lambda_2}{n}\, 1\{a^T y_{2j} > 0\}$$

This minimization, of course, can be done by the WOH tree algorithm.

When $\lambda_1 = \lambda_2$, it is clear that the Bayes empirical minimization problem is that of finding all allowable classification rules which make the fewest number of errors on the data.

To return to the general case, consider once more the problem of finding all $a \in \mathsf{R}^d$ which minimize

$$e(a) := L\left(\sum_{i=1}^{n_1} \sigma_{1i} 1\{a^T y_{1i} \leqslant 0\},\ \sum_{j=1}^{n_2} \sigma_{2j}\, 1\{a^T y_{2j} \leqslant 0\}\right)$$

where $L(\cdot,\cdot)$ is nondecreasing, all $\sigma_{1i}, \sigma_{2j} > 0$, and $\{y_{1i}\}_1^{n_1}, \{y_{2j}\}_1^{n_2} \subset \mathsf{R}^d$ are given. This is actually a version of Problem III of Chapter 4 in disguise. To see this, define f via $f(a) :=$

$$(1\{a^T y_{11} > 0\}, \ldots, 1\{a^T y(1n_1) > 0\}, 1\{a^T y_{21} > 0\}, \ldots, 1\{a^T y(2n_2) > 0\})$$

and define g via

$$g(\tau_{11}, \ldots, \tau(1n_1), \tau_{21}, \ldots, \tau(2n_2)) := -L\left(\sum_1^{n} {}_1\, \sigma_{1i}(1-\tau_{1i}),\ \sum_1^{n_2} \sigma_{2j}(1-\tau_{2j})\right).$$

Note that since g is nondecreasing, the problem of maximizing $g \circ f$ is a version of Problem III of Chapter 4 which, by Theorem (4.1.13) and the subsequent discussion, can be solved by the WOH tree algorithm with a modified displacement phase. Of course, the general tree algorithm of Chapter 5 will also solve it. Finally, observe that any vector a which minimizes e maximizes $g \circ f$ and vice-versa.

The minimax loss function is an example of a loss function from the classification problem which yields a g function which is nondecreasing but not strictly increasing. Here, the problem is to find those A which minimize $\max(\hat{F}_1(A), \hat{F}_2(A^c))$. In this case,

$$g(\tau_{11}, \ldots, \tau(1n_1), \tau_{21}, \ldots, \tau(2n_2)) = -\max\left(\sum_1^{n_1} \sigma_{1i}(1-\tau_{1i}),\ \sum_1^{n_2} \sigma_{2j}(1-\tau_{2j})\right).$$

Consider the empirical minimization problem associated with the Neyman-Pearson loss function where the goal is to find those A which minimize $\hat{F}_2(A^c)$ while keeping $\hat{F}_1(A) < \rho$ for some $\rho > 0$. So, for suitable choices of ρ, $\sigma_{1i} > 0$, $\sigma_{2j} > 0$, a_i, b_j, the object is to find x_0 which minimizes

$\sum_1^{n_2} \sigma_{2j} 1\{b_j^T x \leqslant 0\}$ subject to $\sum_1^{n} {}_1\, \sigma_{1i} 1\{a_i^T x \leqslant 0\} < \rho$. This is equivalent to maximizing $\sum_1^{n_2} \sigma_{2j} 1\{b_j^T x > 0\}$ subject to $\sum_1^{n} {}_1\, \sigma_{1i} 1\{a_i^T x > 0\} > -\rho + \sum_1^{n} {}_1\, \sigma_{1i}$. This is an instance of Problem VI of Chapter 4 which the tree algorithm can solve.

-- *operations research and economics* --

Two fields containing further applications are operations research and economics. In these fields, there is occasion to optimize functions subject to requiring the solution vectors to satisfy certain linear relations. It may happen that there are no vectors which satisfy all of the given linear relations at which point it becomes of interest to ask which vectors satisfy as many of the linear relations as possible. The problem of maximizing over $x \in \mathrm{R}^d$ $\sum_1^m 1\{a_i^T x \ R_i \ \mu_i\}$ where $R_i \in \{<, \leqslant, =, \neq, \geqslant, >\}$ can be solved by the general tree algorithm. Also, observe that weights σ_i can be added and the $\sum_1^m \sigma_i 1\{a_i^T x \ R_i \ \mu_i\}$ maximized using the general tree algorithm.

-- *switching theory* --

The tree algorithm can also be used to solve certain problems in switching theory. One of the objectives in switching theory is to find ways to efficiently implement truth functions. For fixed d, a truth function f is a mapping from $B = \times_1^d \{-1, 1\}$ to the set $\{-1, 1\}$. (Without loss of generality and for notational convenience, the set $\{-1, 1\}$ is being used instead of the more traditional $\{0, 1\}$). Truth functions which are particularly easy to implement are ones which are linearly realizable, i.e., ones for which there exists a vector $w \in \mathrm{R}^d$ and scalar γ such that for all $b \in B$, $f(b) = \mathrm{sgn}(b^T w - \gamma)$ where $\mathrm{sgn}(0) = 0$. In order to determine if a truth function f is linearly realizable, the tree algorithm can be used to maximize over (w, γ), $\sum_B 1\{(\text{sgn}\, f(b))(b, -1)^T(w, \gamma) > 0\}$. In general, for fixed d, only a small fraction of the set of truth functions are linearly realizable. Consequently, an optimal (w, γ) will probably have a criterion function value $< 2^d$, in which case it provides the designer with a linearly realizable truth function which has the same value as the given truth function f on as many points of B as possible. The designer has the option of using optimal (w, γ) in conjunction with other circuit elements to produce configurations which implement f. He or she also has the option of weighting desired input vectors $b \in B$ more than others or not at all in order to get optimal (w, γ) closer to what he wants.

A problem of long standing in switching theory is to develop a way to count and, better still, to enumerate the linearly realizable truth functions for fixed d. The tree algorithm does this when it enumerates the hills corresponding to $\{(b, -1), (-b, 1) : b \in B = \times_1^d \{-1, 1\}\}$. In this configuration, every $(d+1)$-dimensional cone is a hill.

-- linear numeric editing --

Finally, the general tree algorithm will solve the following problem from the field of linear numeric editing: Suppose there is a database consisting of vectors in R^d each of which is known to be incorrect if it fails the consistency test of being in some prespecified polytope $\{x \in \mathsf{R}^d : A\, x \leqslant b\}$. Given a vector y which has failed this set of linear edits by not being in the polytope, it is of interest to find the smallest number of components of y which could be change in order to place the modified vector in the polytope. If $z := (\zeta_1, \ldots, \zeta_d)$, then the associated mathematical programming problem is to minimize $\sum_1^d 1\{\zeta_i \neq 0\}$ in such a way that $A(y + z) \leqslant b$. The general tree algorithm will do this.

Chapter 9:

Examples Of The Behavior Of The Tree Algorithm In Practice

This chapter discusses the performance in practice of a sophisticated composite WOH tree algorithm for computing estimated best linear classification rules (cf., Chapter 8) for each of four sets of data. The opportunity is taken here to compare this algorithm with Steele's edge enumeration procedure modified as in Chapter 7 to work with any data. The results are very favorable for the tree algorithm: to be specific, the ratio of the number of linear rules examined by this WOH tree algorithm to the number examined by the edge enumeration procedure ranges between .000034 to .02 for these data sets. It will also be seen that the version of the fast approximate tree algorithm used here to obtain an initial rule for the standard WOH tree algorithm is very successful in finding quite good if not quite optimal rules after looking at only a small fraction of the rules that the subsequent standard algorithm examines.

The first data set to be considered consists of measurements taken at the Stanford Medical Center from men with and without lymph node involvement in prostate cancer. The next set consists of the linearly inseparable pair of flower measurement samples from Fisher's iris data. The last two data sets are artificially generated scatters of O's and ×'s in two dimensions. Hyperbolas are the minimum-error quadratic surfaces for the first of these two data sets. The second data set is the one shown in Chapter 1; its minimum-error quadratic surfaces are ellipses.

The composite WOH tree algorithm used to work on these data sets is that of Chapter 3 which includes all of the improvements in section 3.4 except (3.4.19). Consequently, this composite algorithm consists of a version of the fast approximate tree algorithm followed by a standard WOH tree algorithm based on (3.3.13) and (3.5.7) which obtains its $\tilde{v}_0$ from the fast algorithm. As described in (3.4.20), fast approximate tree algorithms use a best-first depth-first tree exploration algorithm with restarting in order to find good vectors fast. The fast approximate tree algorithm used here visits at most the two best children at each step (i.e., $\#s(k) = 2$ for all k).

-- *prostate cancer data* --

To begin with the first example, at the Stanford Medical Center, a study has been made of ways to predict on the basis of pre-operative measurements whether or not surgery on a given male with prostate cancer will reveal that the cancer has spread to adjoining lymph nodes. It is so important to know whether or not the lymph nodes are cancerous that such exploratory surgery is customary; the goal of this study is to find a way to avoid surgery if at all possible. The author wishes to thank Bill Brown for making available the data collected during this study.

Four measurements were taken on each individual. The first, grade, is a pathologist's ranking on a 4-10 integral scale of a biopsy of the tumor obtained by needle before surgery. Let the ξ_2 variable be the grade variable. The second measurement (ξ_3) is blood serum level of acid phosphatase (acid, for short); it is pseudo-continuous in nature in that it was recorded to three significant digits between 40 and 300 but there were ties nonetheless. The third measurement, stage or (ξ_4), is an ordinal variable assuming integral values from 0 to 4 which indicates the physician's appraisal of the tumor as obtained by physical examination. The last measurement (ξ_5) is a 0-1 variable summarizing the results of reading a lymph angiogram (LAG). In all of these measurements, increasing values indicate increased severity of the disorder.

All four pre-operative measurements were made on each of 86 men with prostate cancer. Subsequent exploratory surgery revealed that 52 did not have lymph node involvement whereas 34 did. Assuming that the sample proportion

52/34 of those without lymph node involvement to those with is an accurate reflection of the analogous proportion in the population at large, the tree algorithm was used to find those hyperplane rules which minimize the empirical loss associated with the Bayes loss function with equal losses $\lambda_1 = \lambda_2$ and $\hat{\tau}_1 = 52/86$. As commented in Chapter 8, this is equivalent to seeking to identify the minimum-error hyperplanes for the data and thus to estimate hyperplane rules which have the smallest probability of making an error when sampling from the mixture of the two distributions.

In this data set, there were two tied pairs of measurement vectors which reduced the total sample size down to 83 (one pair cancelled itself out). Even without these ties, there was a serious lack of general position in this data as demonstrated by the existence of a hyperplane which contains 57 of the 86 data points.

The tree algorithm found 5 distinct minimum-error dichotomies of the data, each making 14 errors. In the search of the final boundary vector tree, 13 solution vectors were found; the number of representatives of each of the 5 solution cones in this set of 13 vectors ranged from 1 to 5. Figure (9.1.1) displays a single representative from each cone. The associated linear classification rules are defined by classifying an individual as having lymph node involvement if the specified linear combinations of the variables are greater than the cut-off values given (i.e., if $\sum_2^5 \alpha_i \ \xi_i > -\alpha_1$).

-- a metric for measuring how close solution vectors are --

When the tree algorithm produces several solution vectors, it is natural to wonder how different the vectors really are. Any reasonable measure of the physical location difference between two solution vectors a and b should depend only on $\langle a \rangle$ and $\langle b \rangle$ since all elements of the open ray $\langle a \rangle$ dichotomize the y_i in precisely the same way as a does (i.e., all vectors in the same open ray are in the same cone in the solution space). The following brief digression

rule	α_2 grade	α_3 acid	α_4 stage	α_5 LAG	$-\alpha_1$ cut-off	# "without" misclassified	# "with" misclassified
A	.3333	.0028	.1038	.2681	3.0436	3	11
B	.1369	.0073	.0998	.2038	1.8086	5	9
C	.1890	.0073	.1529	.1519	2.2738	6	8
D	.3333	.0034	.2279	.1710	3.3458	2	12
E	.1437	.0073	.1046	.1422	1.8086	7	7

(9.1.1) Figure: Representatives from each of the 5 solution equivalence classes for the prostate cancer data. The last two columns contain the respective number of individuals without and with nodal involvement that were misclassified by the associated rule.

introduces such a metric.

(9.1.2) Definition: Define the function ρ which maps pairs of nonzero open rays in R^d onto $[0, 180]$ via, for all nonzero a, $b \in \mathsf{R}^d$, $\rho(\langle a \rangle, \langle b \rangle) = \text{Arccos}\, \dfrac{a^T b}{\| a \| \cdot \| b \|}$ where $\| a \| = \sqrt{(a^T a)}$.

ρ measures the angle between the rays. It is well-defined since if $\langle a_1 \rangle = \langle a_2 \rangle$ and $\langle b_1 \rangle = \langle b_2 \rangle$, then $\rho(\langle a_1 \rangle, \langle b_1 \rangle) = \rho(\langle a_2 \rangle, \langle b_2 \rangle)$. $\rho(\langle a \rangle, \langle b \rangle)$ is 0 if and only if $\langle a \rangle = \langle b \rangle$ and $\rho(\langle a \rangle, \langle b \rangle) = \rho(\langle b \rangle, \langle a \rangle)$. The next theorem shows that ρ satisfies the triangle inequality and is consequently a metric.

(9.1.3) Theorem: Given ρ as in (9.1.2).

(i) Let $\underset{\sim}{C}$ be an orthogonal change of basis matrix for two orthonormal bases of R^d. Then for all nonzero a, $b \in \mathsf{R}^d$, $\rho(\langle a \rangle, \langle b \rangle) = \rho(\langle \underset{\sim}{C} a \rangle, \langle \underset{\sim}{C} b \rangle)$. In other words, ρ is invariant under orthogonal transformations of R^d.

(ii) ρ is a metric.

Proof: (ii): Consider showing $\rho(\langle a_1 \rangle, \langle a_2 \rangle) \leqslant \rho(\langle a_1 \rangle, \langle a_3 \rangle) + \rho(\langle a_3 \rangle, \langle a_2 \rangle)$ when $\{a_1, a_2, a_3\}$ is linearly independent and consequently $d \geqslant 3$. Construct an orthonormal basis for R^d whose first three vectors span $\mathsf{L}\{a_1, a_2, a_3\}$.

Construct the change of basis matrix $\underset{\sim}{C}$ from the standard basis on R^d to this new orthonormal basis. Note that the 4th through d^{th} components (if present) of $\underset{\sim}{C}a_i$ are 0 and that it is sufficient to show

$$\rho(\langle \underset{\sim}{C}a_1 \rangle, \langle \underset{\sim}{C}a_2 \rangle) \leqslant \rho(\langle \underset{\sim}{C}a_1 \rangle, \langle \underset{\sim}{C}a_3 \rangle) + \rho(\langle \underset{\sim}{C}a_3 \rangle, \langle \underset{\sim}{C}a_2 \rangle),$$

which is precisely an inequality for vectors in R^3. The result follows since the sum of any two face angles of a trihedral angle is greater than the third (Theorem 6T2, Lines (1965)). □

Now there might be some question as to how to interpret a distance of 10° in the ρ metric between two rays in R^d since it is hard to visualize R^d for $d \geqslant 4$. Fortunately, it is clear that, by using the orthogonal transformation invariance property of ρ, two rays in R^d are 10° apart in the ρ metric if and only if in terms of an orthogonal coordinate system in their own two-dimensional dimensionality space the angle between them is 10°. Similarly, the distances between three linearly independent rays in R^d can be visualized by seeing the rays in the context of a suitably chosen three-dimensional orthogonal coordinate system. This information is useful in trying to determine whether a distance of 1° is of any physical significance or not.

Note that $\rho(\langle a \rangle, \langle b \rangle) = 180^\circ$ if and only if $\langle a \rangle = \langle -b \rangle$.

In terms of relating this metric on the space of rays back to classification regions in the context of the classification examples being considered in this chapter, it is tempting to define the distance between two linear classification regions as being the ρ distance between their associated rays. This will in fact be the policy here but a certain amount of care is necessary since the same linear classification region can be associated with more than one distinct ray. This phenomenon is unusual and does not occur in the examples given in this chapter. For more information on this matter, see Greer (1979).

At any rate, the ρ distance matrix for the five rules of (9.1.1) is presented in Figure (9.1.4). None of these 5 rules seems to be particularly physically different from the others in spite of the fact that (9.1.1) shows that there is

ρ	A	B	C	D	E
A	0°	2.66°	2.69°	2.90°	2.24°
B		0°	2.72°	3.80°	1.94°
C			0°	1.30°	.89°
D				0°	2.03°
E					0°

(9.1.4) Figure: The matrix of distances in the ρ metric among the 5 minimum-error rules given in Figure (9.1.1) for the prostate cancer data.

substantial variation within this group in terms of the relative number of individuals with lymph node involvement that are misclassified.

-- *performance terminology* --

Before presenting more information on how the tree algorithm performed on this data, it is necessary to explain what is behind the various performance numbers that are given for this and each subsequent example.

To begin with, although the Bayes loss function is the sole loss function used in these examples, most of the performance characteristics remain valid for arbitrary nondecreasing loss functions since precisely the same number of hypersurfaces will be examined by the tree algorithm independent of the loss function used.

Concerning the various performance measures made below, any count of examined hypersurfaces should be multiplied by 2, if desired, in order to yield the number of rules or edges examined.

In order to arrive at an estimate of the number of nodes trimmed by the trimming procedure, it was necessary to make the assumption that the average number before trimming of children of nodes at the kth level of the trimmed tree is the same as the average number of children of nodes at the kth level in the corresponding untrimmed tree. Armed with this assumption, one can proceed to estimate the size of the subtree generated by a trimmed node in the obvious way using the average numbers of untrimmed children generated at each level of the trimmed tree.

The CPU times given in subsequent tables measure the execution time of the tree algorithm only and so do not include the time for compilation, input data processing, or output. The CPU time estimates for the edge enumeration algorithm should be considered approximate since each is computed by multiplying the appropriate number of hypersurfaces by the average amount of time the tree algorithm spends computing a hypersurface. The cost figures are based on $20/CPU minute.

The computer program was written in PL/I and run under the Optimizing Compiler on an Amdahl 470/V6 computer at Bell Laboratories. The program takes about 36 seconds to compile and is 1650 statements long. The input section which transforms the data into a form suitable for the tree algorithm takes .25 seconds to process 100 five-dimensional data vectors. The linear programming necessary to displace vectors was done through a subroutine call to the program LCLIN written by Margaret Wright of Stanford University and the Stanford Linear Accelerator Center.

-- *returning to the prostate cancer data set* --

The discussion of the composite WOH tree algorithm's behavior on the prostate cancer data set can now be resumed. The fast approximate tree algorithm here began with an initial hyperplane rule which made 22 errors. After 4 restarts, the fast approximate algorithm used recursion often down as far as possible to produce vectors in 3 of the 5 solution equivalence classes, each making 14 errors. The fast algorithm explored a total of 309 hyperplanes in 1.57 seconds of CPU time and completed its displacement phase in .41 seconds.

The standard WOH tree algorithm took one of the fast algorithm's best vectors as its $\tilde{v}_0$ and examined 35,541 hyperplanes in 199.58 seconds in order to produce vectors from each of the 5 solution equivalence classes. An estimated 120,000 nodes were trimmed from this final tree. 29 rules were saved by the first phase of the standard algorithm for displacement. The displacement phase produced 13 solution vectors from the preceding 29 in 3.6 seconds after frequently resorting to recursion, often down as far as possible.

Figure (9.1.5) gives further information. Notice that the fast algorithm did very well in this example in that it actually found several solution dichotomies after examining only .0087 of the number of hyperplanes needed by the standard algorithm to find all of the solution dichotomies (and thereby verify that the fast algorithm did in fact terminate with solutions).

Also, the total number of hypersurfaces examined by the composite WOH tree algorithm (i.e., fast followed by standard) is only .020 of the number that must examined by edge enumeration. In this regard, the number of hyperplanes given for the edge enumeration procedure is $\binom{83}{4}$ since it has been assumed here that the edge enumeration procedure computes a hyperplane for each subset of the data of size 4 even though this hyperplane may be identical to that associated with another subset of size 4.

The remaining table gives the average number by level of children for nodes *in the trimmed tree* both before and after the children of those nodes were trimmed. $\#N(\tilde{v}_0) = 13$, not 14, due to the ties in the data. Consequently, $\tilde{\alpha} = \inf\{\#\{y_i : [y_i, \tilde{v}] < 0\} : \tilde{v} \notin y_i^{\perp} \text{ for any } y_i \neq 0\} = 13$. Observe that the time complexity of the final tree exploration phase of this WOH tree algorithm is close to $\theta\left(nd\frac{\tilde{\alpha}^d - 1}{\tilde{\alpha} - 1}\right)$ for this example since on the order of *nd* operations are performed for each hyperplane in the trimmed tree where the average number of children at each node is approximately 13. Also observe that $\frac{\tilde{\alpha}^d - 1}{\tilde{\alpha} - 1} = 30{,}941$ which is reasonably close to 35,541. Since the total time taken by the fast approximate algorithm and the displacement phase of the standard algorithm is about 6 seconds in comparison to the 200 seconds needed for the exploration of the final tree, for all practical purposes, one may consider the time complexity of the WOH tree algorithm used here to be in

	Tree Algorithm Fast	Tree Algorithm Standard	Complete Enumeration
# of hyperplanes considered	309	35,541	1,837,620
Best rule found	14 errors	14 errors	14 errors
CPU time (Amdahl 470/V6)	2 sec	3.4 min	~2.9 hours
cost ($20/CPU min)	67 cents	$68	~$3480

Final Tree

k	avg # of children of $\tilde{v}_k$ before trimming	avg # of children of $\tilde{v}_k$ after trimming
0	13.00	13.00
1	20.08	15.69
2	22.73	12.81
3	25.24	12.52

estimated number of nodes trimmed = 120,000

$\tilde{\alpha} = 13.00$

$$\frac{\tilde{\alpha}^d - 1}{\tilde{\alpha} - 1} = 30{,}941$$

(9.1.5) Figure: Performance numbers for the tree algorithm's behavior on the prostate cancer data.

$\Theta\left(nd\dfrac{\tilde{\alpha}^d - 1}{\tilde{\alpha} - 1}\right)$ for this example. Although the claims of equivalence that will be made here between $\theta\left(nd\dfrac{\tilde{\alpha}^d - 1}{\tilde{\alpha} - 1}\right)$ and the true complexity of this sophisticated WOH tree algorithm are fairly rough, they are sufficiently close to enable one to determine reasonable maximum CPU time limits when running the tree algorithm's program which is fairly useful considering the NP-complete nature of the problem. All that is needed to produce such CPU time estimates is an estimated average processing time per hyperplane obtained from previous runs and an estimate of $\tilde{\alpha}$ from a run of the fast algorithm.

Note that since $\left(\frac{n}{2}\right)^{d-1} \doteq 2{,}966{,}000 >> 35{,}541$, the standard part of the WOH tree algorithm is seen to have a complexity in this example dramatically far from the upper bound $\theta(dn^d/2^{d-1})$ derived for it in Chapter 6.

-- *the Fisher iris data* --

In the next example, the tree algorithm is 30,000 times faster than edge enumeration. Fisher (1936) contains a data set consisting of four measurements made on 50 iris flowers from each of three species. The measurements were sepal length and width and petal length and width. The species were iris setosa, iris versicolor, and iris virginica. Since zero error hyperplanes exist for separating the setosa sample from either of the other two samples, it was decided to give the tree algorithm more of a challenge by applying it to the pair of samples versicolor and virginica.

When asked to discover all minimum-error dichotomies, the WOH tree algorithm discovered two distinct solution equivalence classes, each making one error. Here are the representative rules found by the algorithm:

	α_2	α_3	α_4	α_5	$-\alpha_1$
rule	sepal length	sepal width	petal length	petal width	cut-off
A	.0068	.0403	−.0474	−.1333	−3.056
B	.0114	.0261	−.0513	−.0631	−2.102

The rules are defined by classifying a flower as being versicolor if the specified linear combinations of the variables are greater than the cut-off values given. Rule A misclassifies only the virginica point (6.3, 2.8, 5.1, 1.5) and Rule B misclassifies only the versicolor point (6.0, 2.7, 5.1, 1.6). The ρ distance between rules A and B is .95°.

As for the WOH tree algorithm's performance on this data set, the fast approximate algorithm began with a hyperplane rule that made 8 errors. After 5 restarts and examining 68 nodes in .5 seconds, it found both minimum-error dichotomies. The displacement phase did not have to resort to recursion and took less than .05 seconds to complete.

The standard part of this composite WOH tree algorithm took as its $\tilde{v}_0$ one of the best vectors produced by the fast algorithm and after examining 60 hyperplanes in .3 seconds, determined that the fast algorithm had in fact discovered all minimum-error dichotomies. An estimated 63 nodes were trimmed. 2 rules were saved by the boundary vector collection phase for displacement and both were displaced into solution cones in under .05 seconds without recursion.

Figure (9.1.6) contains further information. Due to the small $\tilde{\alpha}$ here, the fast algorithm explored about as many hyperplanes as its successor did.

The composite WOH tree algorithm examined only a handful (128) of hyperplanes which is a consequence of the fact that the WOH tree algorithm works fast on easy problems and slower on hard problems; this is something that cannot be said for complete enumeration. In this case, since there was one tie in the data, complete edge enumeration would have had to look at $\binom{99}{4}$ = 3,764,376 hyperplanes. The WOH tree algorithm looked at only .000034 of this number.

Other than the tied data points, there were other linear degeneracies in the data leading to a lack of general position but there was no evidence of a lack of pointed position.

	Tree Algorithm Fast	Tree Algorithm Standard	Complete Enumeration
# of hyperplanes considered	68	60	3,764,376
best rule found	1 error	1 error	1 error
CPU time (Amdahl 470/V6)	.484 sec	.310 sec	~6.5 hours
cost ($20/CPU min)	16 cents	10 cents	~$7800

Final Tree

k	avg # of children of $\tilde{v}_k$ before trimming	avg # of children of $\tilde{v}_k$ after trimming
0	1.00	1.00
1	4.00	4.00
2	4.00	2.50
3	6.30	4.40

estimated number of nodes trimmed = 63

$\tilde{\alpha} = 1$

(9.1.6) Figure: Performance numbers for the tree algorithm's behavior on the Fisher iris data set.

This example seems to indicate that when $\tilde{\alpha} = 1$, the time complexity estimates from Chapter 6 are not particularly good but fortunately when $\tilde{\alpha} = 1$, there seems to be no need for the estimates to be good since the algorithm runs so fast.

Once again since $\left(\frac{n}{2}\right)^{d-1} = 6{,}250{,}000 >> 60$, the standard part of the WOH tree algorithm is seen to be operating dramatically far from its upper time complexity bound.

-- *×'s and O's in two-space* --

Figure (9.1.7) shows one of the two minimum-error quadratic dichotomies for a set of 30 ×'s and 30 O's in R^2. These points were generated using a random number generator according to uniform distributions on the indicated ellipses. Both of the minimum-error linear rules in this example are hyperbola dichotomies, each making 4 errors.

One thing to look for in this and the other figures depicting classification rules amidst data is the manifestation of the hill concept. Note that the decision surface of each of the best rules pictured is constrained, as it must be, by ×'s on the ×-side of the surface and O's on the O-side. (See Greer (1979) for further clarification of this comment.)

For this example, the fast algorithm began with a rule which made 18 errors. After 5 restarts and examining 323 hyperplanes in the transformed space in 1.4 seconds, it produced 3 non-optimal rules, each making 5 errors instead of the smallest possible 4 errors. The displacement phase did not have to resort to recursion and took less than .05 seconds to complete.

Starting with one of the best rules produced by the fast algorithm, the standard half of this composite WOH tree algorithm examined 220 nodes before discovering what later proved to be a solution vector. It restarted using this vector and examined 8558 more nodes before stopping with 3 vectors to send to the displacement phase. This took 34.4 seconds. An estimate 64,000 nodes were trimmed. The displacement phase took less than .05 seconds to displace the 3 candidate solution vectors into the two solution cones without recursion. The ρ distance between two of the representatives from the two classes was 3.6°.

Figure (9.1.8) contains further information. The fast algorithm even though it did not produce an optimal rule came within one error of doing so

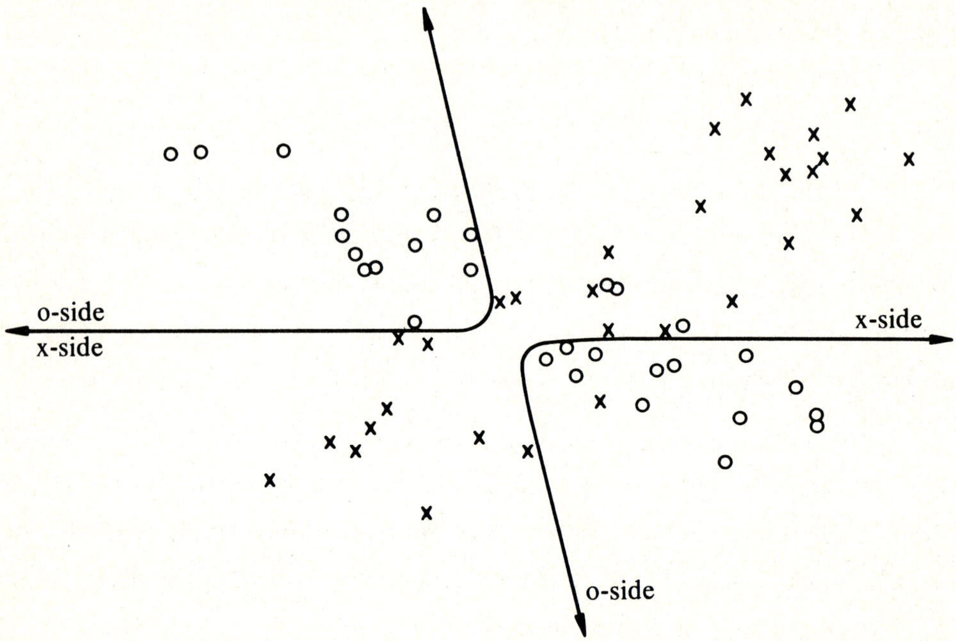

(9.1.7) Figure: The unique minimum-error hyperplane dichotomy for a set of ×'s and O's in R^2.

after examining only 323 vectors or .037 of the number of vectors that the standard algorithm (using the best rule the fast algorithm produced as its $\tilde{v}_0$) needed in order to identify the two solution cones.

The composite WOH tree algorithm examined only .0017 of the number of hyperplanes, namely $5{,}461{,}512 = \binom{60}{5}$, that edge enumeration would have had to check.

There was no evidence of a lack of general position in this data set.

Here $\tilde{\alpha} = 4$ and once past the first level, it can be seen that the effect of trimming is to reduce the average number of children of $\tilde{v}_k$ down to close to 4. Still, $\dfrac{\tilde{\alpha}^d - 1}{\tilde{\alpha} - 1} = 1365$ is not a particularly good estimate of 8558 although it is

	Tree Algorithm Fast	Tree Algorithm Standard	Complete Enumeration
# of hyperplanes considered	323	8778	5,461,512
best rule found	5 errors	4 errors	4 errors
CPU time (Amdahl 470/V6)	1.4 sec	34.4 sec	~6 hours
cost ($20/CPU min)	47 cents	$11.47	~$7200

Final Tree

k	avg # of children of $\tilde{v}_k$ before trimming	avg # of children of $\tilde{v}_k$ after trimming
0	4.0	4.0
1	10.0	9.0
2	9.0	5.3
3	11.1	6.4
4	12.6	5.8

estimated number of nodes trimmed = 64,000

$\tilde{\alpha} = 4$

$$\frac{\tilde{\alpha}^d - 1}{\tilde{\alpha} - 1} = 1365$$

(9.1.8) Figure: Performance numbers for the tree algorithm's behavior on the hyperbola data set.

certainly much better than $\left(\frac{n}{2}\right)^{d-1} = 24{,}300{,}000$ which is the basis for the upper bound on the time complexity of the tree exploration phase developed in Chapter 6.

There is an intuitive explanation here for why the fast algorithm did not find solution vectors and for why the standard algorithm seems to be investigating more nodes than otherwise might have been expected. Recall that the original three dimensional data (remember the class or group variable induces a dimension) is transformed into a 6 dimensional space where quadratic dichotomies in the original space can be found by identifying best separating hyperspaces in the transformed space. Consequently, a set of points which is three dimensional in character is mapped into a three dimensional (nonlinear) manifold in a 6 dimensional space. Reasoning by analogy with a picture of a two dimensional manifold in R^3, it can be surmised that the reason why the tree algorithm had to work harder on this data is because the transformed data points are so concentrated in one area that small changes in a hyperspace passing through the data will result in substantial differences in the resulting dichotomies. This effect seems to be even more pronounced in the next data set.

-- *ellipses as best quadratic separating surfaces* --

The last example is concerned with discussing the behavior of the WOH tree algorithm in estimating minimum-error quadratic rules for the set of 30 ×'s and 30 O's shown in Figure (1.1.1). These points were generated using a random number according to uniform distributions over a sector of a circle and a crescent. Both of the minimum-error classification regions for this data are ellipsoids, each making three errors.

For this example, the fast algorithm began with a rule which made 20 errors. After 8 restarts and examining 403 hyperspaces in the transformed space in 1.9 seconds, it identified the two minimum-error equivalence classes with a total of 3 representatives, each making 3 errors. The displacement phase did not have to resort to recursion and took less than .05 seconds to complete.

The standard algorithm, starting with a best vector from the fast algorithm, examined 28,151 hyperplanes in 114 seconds and sent 2 of them to the displacement phase. These 2 vectors were displaced into the 2 minimum-error equivalence classes in less than .05 seconds without recursion. An estimated 116,000 nodes were trimmed. The ρ distance between the two solution rays produced was 8.7°.

Figure (9.1.9) contains further information. The fast approximate algorithm examined only 403 hyperplanes or .014 of the 28,151 rules examined by the standard algorithm that followed it. Nonetheless, the fast algorithm produced both solution cones.

The composite WOH tree algorithm examined only .0052 of the number, namely $5{,}461{,}512 = \binom{60}{5}$, that edge enumeration would have had to check.

There was no evidence of a lack of general position in this data set.

Even though one might have hoped for the WOH tree algorithm to run faster on this ellipse data set than on the hyperbola data set by virtue of its smaller $\tilde{\alpha} = 3$ and equal n and d, the tree algorithm actually examined over three times as many rules. Perhaps the consequences of the data set being in a lower dimensional manifold of the 6 dimensional transformed space are more severe here.

Further evidence of this can be seen in the "average number of children of $\tilde{v}_k$ after trimming" column where all entries past the first are about three times 3. Here $\dfrac{\tilde{\alpha}^d - 1}{\tilde{\alpha} - 1} = 364$ substantially underestimates 28,151, which is itself nonetheless quite distant from $\left(\dfrac{n}{2}\right)^{d-1} = 24{,}300{,}000$.

	Tree Algorithm Fast	Standard	Complete Enumeration
# of hyperplanes considered	403	28,151	5,461,512
best rule found	3 errors	3 errors	3 errors
CPU time (Amdahl 470/V6)	1.9 sec	1.9 min	~6.2 hours
cost ($20/CPU min)	63 cents	$38	~$7400

Final Tree

k	avg # of children of $\tilde{v}_k$ before trimming	avg # of children of $\tilde{v}_k$ after trimming
0	3.0	3.0
1	9.7	9.3
2	12.6	8.5
3	18.8	10.9
4	19.9	9.7

estimated number of nodes trimmed = 116,000

$\tilde{\alpha} = 3$

$$\frac{\tilde{\alpha}^d - 1}{\tilde{\alpha} - 1} = 364$$

(9.1.9) Figure: Performance numbers for the tree algorithm's behavior on the ellipse data set.

Summary For Chapter 9

These examples suggest the following conclusions. The first is that the tree algorithms presented in this monograph are definitely not algorithms which completely enumerate edges or fully-dimensional cones. (Generalization to the general tree algorithm is possible at this point since it is so similar to the basic WOH tree algorithm (3.2.12).) In none of these examples did the sophisticated WOH tree algorithm used here come even close to enumerating as many hyperplanes as a complete enumeration method would have: in fact, the WOH tree algorithm examined only a small fraction ranging from .000034 to .02 of the number of vectors that would be examined by an edge enumeration procedure.

In every example, the fast approximate WOH tree algorithm used here produced vectors that either were optimal or close to it after examining less than 403 hyperspaces in problems with millions of edges. In these four examples, the fast approximate algorithm was between 6,000 and 55,000 times faster than edge enumeration.

In all cases, the time complexity of the steps necessary to explore the final tree produced by the WOH tree algorithm was dramatically less than the upper bound $\theta(dn^d/2^{d-1})$ established in Chapter 6. $\theta\left(dn\dfrac{\tilde{\alpha}^d - 1}{\tilde{\alpha} - 1}\right)$ where $\tilde{\alpha} := \inf\{\#N(\tilde{v}) : \tilde{v} \notin y_i^{\perp} \text{ for all } i \text{ such that } y_i \neq 0\}$ provided a much better estimate of the time complexity of final tree exploration although at times it too substantially mis-estimated. Since the times needed by the fast approximate algorithm and the displacement phase of the subsequent standard algorithm were seen to be typically only a small fraction of the time needed for final tree exploration, the currently best estimate for the time complexity of the sophisticated WOH tree algorithm used here is $\theta\left(dn\ \dfrac{\tilde{\alpha}^d - 1}{\tilde{\alpha} - 1}\right)$. This at least gives a useful estimate of the maximum CPU time necessary to run the WOH tree algorithm once one has an estimate of $\tilde{\alpha}$ (perhaps obtained from a run of the fast algorithm) and an estimate of the average amount of CPU time necessary to process a single hyperplane (which can be obtained from previous runs).

It is also clear that the trimming improvement is instrumental in reducing the complexity of the algorithm. In the larger examples, from 3 to 7 times the number of nodes in the final trimmed tree were estimated to have been eliminated by the trimming procedure.

In short, the WOH tree algorithm is seen to be much better than complete enumeration for solving the WOH problem and the fast approximate algorithm defined here is very good at producing good if not optimal vectors very quickly.

Chapter 10:

Summary And Conclusion

Without repeating too much of the introduction, this chapter summarizes the nature of the problems that the tree algorithm solves. It then goes on to informally discuss in some detail how and why the tree algorithm solves them.

-- *maximizing functions of systems of linear relations* --

This monograph introduces the problem of extremizing functions of systems of linear relations subject to constraints. As such, it is the first to provide a unifying framework for the research that has been done on finding procedures to produce vectors which satisfy systems of linear relations in certain desired ways. Intuitively, the common denominator of all these problems is that they all seek vectors which satisfy or don't satisfy a set of linear relations in such patterns as will maximize a function of interest.

More formally, H is said to be a function of the system of linear relations $\{a_i^T x \ R_i \ \mu_i\}_1^m$, where $R_i \in \{ <, \leqslant, =, \neq, \geqslant, > \}$ and x is a vector of indeterminates in $\mathbf{R}^d$ if and only if there is a function $g : \times_1^m \{0, 1\} \rightarrow \mathbf{R}$ such that for all $x \in \mathbf{R}^d$,

$$H(x) = g(1\{a_1^T x \ R_1 \ \mu_1\}, \ldots, 1\{a_m^T x \ R_m \ \mu_m\}).$$

The problem is to maximize (or minimize) H over $x \in \mathbf{R}^d$ (i) subject to requiring the maximizing vectors to lie in a designated linear manifold or polyhedral set or (ii) subject to maximizing another function H_2 of a system of linear relations or (iii) subject to maintaining the value of yet another function H_3 of a system of linear relations greater than some preset constant or (iv) any or none of the above constraints.

-- examples of problems in this area --

As has been seen, many problems fall into this general category. For example, the problem of finding solutions to consistent systems of linear inequalities falls into this category. This includes the class of linear programming problems as well as the class of problems of finding solutions to systems of linear equations.

Also in this general category is the problem of maximizing over $x \in \mathrm{R}^d$, $\sum_J \sigma_i \, 1\{a_i^T x > \mu_i\} + \sum_{I \text{ w.o. } J} \sigma_i \, 1\{a_i^T x \geqslant \mu_i\}$ for finite index sets $J \subset I$ when the underlying system of linear inequalities is not necessarily consistent. Note that if all $\sigma_i = 1$, then the object is to find all vectors x which satisfy as many of the linear inequalities as possible.

As another example, when the best linear classification rule estimation problem is transformed out of the measurement space, it becomes one of seeking to minimize over $x \in \mathrm{R}^d$, $L(\sum_{i=1}^{n_1} \sigma_{1i} \, 1\{y_{1i}^T x \leqslant 0\}, \sum_{j=1}^{n_2} \sigma_{2j} \, 1\{y_{2j}^T x \leqslant 0\})$ where $L(\cdot, \cdot)$ is an arbitrary nondecreasing loss function, the σ_{1i}, σ_{2j} are weights, and the y_{1i}, y_{2j} are transformed data points.

-- homogeneous canonical form --

Despite the apparent complexity of the general case, all problems of extremizing functions of systems of linear relations with or without constraints are equivalent to certain other unconstrained problems in a simple homogeneous canonical form. To define this, it is necessary to define nondecreasing and nonincreasing variables. The jth variable of $g : \times_1^m \{0, 1\} \to \mathrm{R}$ is *nondecreasing* if and only if for all choices $\sigma_1, \ldots, \sigma_{j-1}, \sigma_{j+1}, \ldots, \sigma_m \in \{0, 1\}$,

$$g(\sigma_1, \ldots, \sigma_{j-1}, 0, \sigma_{j+1}, \ldots, \sigma_m) \leqslant g(\sigma_1, \ldots, \sigma_{j-1}, 1, \sigma_{j+1}, \ldots, \sigma_m).$$

The jth variable of g is *nonincreasing* if and only if the jth variable $-g$ is nondecreasing. g is a *nondecreasing* function if and only if all of its variables are nondecreasing.

It can be shown that for every problem of extremizing a function H of a system of linear relations subject to constraints, there is both a homogeneous system of linear inequalities $\{b_i^T x \ R_i \ 0\}_1^n$ where $R_i \in \{ > , \geqslant \}$ and a positive function g_2 with no nonincreasing variables such that any vector y which solves the original problem can be obtained from some vector x which maximizes $g_2(1\{b_1^T x \ R_1 \ 0\}, \ldots, 1\{b_n^T x \ R_n \ 0\})$ and vice versa.

Once a problem has been reduced to homogeneous canonical form, then the tree algorithm can solve it if the associated g_2 function is nondecreasing.

-- *the nature of solutions: max-cones and hills* --

As before, the best setting for discussing the tree algorithm is that provided by the arbitrary finite dimensional vector space X over R with its dual space of linear functionals $\tilde{X}$. Consequently, begin by setting $I = \{1, \ldots, n\}$, $b_0 = 0$, and considering $\{b_i, i \in I \cup \{0\}\} \subset X$. Making the convenient but nonessential assumption that $\mathsf{L}\{b_i, i \in I\} = X$, pose the problem of finding all $\tilde{x} \in \tilde{X}$ which maximize

$$H(\tilde{x}) = g_2(1\{[b_1, \tilde{x}] \ R_1 \ 0\}, \ldots, 1\{[b_n, \tilde{x}] \ R_n \ 0\})$$

where g_2 is assumed to be nondecreasing.

Before discussing how the tree algorithm solves this problem, it is necessary to comment on certain structure which is present in the solution space. Observe that for each $\tilde{x} \in \tilde{X}$, by computing $[b_i, \tilde{x}]$ for all $i \in I = \{1, \ldots, n\}$, $J \subset I$ can be determined such that $[b_i, \tilde{x}] = 0$ for all $i \in J$ and, if $J \neq I$, for all $i \in I$ w.o. J, $\pi_i[b_i, \tilde{x}] > 0$ for suitable $\pi_i \in \{-1, 1\}$. Consequently, the solution space can be seen as being partitioned into the union of nonempty level sets of the function H which have the form

$$\{\tilde{x} \in \tilde{X}: [\pi_i b_i, \tilde{x}] > 0 \text{ for } i \in I \text{ w.o. } J, [b_i, \tilde{x}] = 0 \text{ for } i \in J\}.$$

Each level set of this form is the relative interior of a polyhedral convex cone of the form

$$\{\tilde{x} \in \tilde{X}: [\pi_i b_i, \tilde{x}] \geqslant 0 \text{ for } i \in I \text{ w.o. } J,$$
$$[b_i, \tilde{x}] \geqslant 0 \text{ and } [-b_i, \tilde{x}] \geqslant 0 \text{ for } i \in J\}.$$

Since it would clearly be an unhappy situation if it were necessary to enumerate the values of H on the relative interiors of all of these polyhedral convex cones, it is desirable to search for additional structure.

A polyhedral cone is a *max-cone* if and only if the maximum value of H is assumed in its relative interior and $C\{b_i, i \in J\}$ is a subspace. A polyhedral cones is called a *hill* if and only if $C\{b_i, i \in J\}$ is a subspace and there exists $K^+ \subset I$ w.o. J such that the polyhedral cone can be written as

$$\{\tilde{x} \in \tilde{X}: [b_i, \tilde{x}] \geqslant 0 \text{ for } i \in K^+, [b_i, \tilde{x}] = 0 \text{ for } i \in J\}.$$

Any cone which is both a max-cone and a hill is called a *max-hill*.

Consider a vector $\tilde{x}_0$ in the relative interior of a hill $\{\tilde{x} \in \tilde{X}: [b_i, \tilde{x}] \geqslant 0 \text{ for } i \in K^+, [b_i, \tilde{x}] = 0 \text{ for } i \in J\}$. The rationale underlying the choice of the word hill is the following: any attempt by the vector $\tilde{x}_0$ to leave the cone while remaining in the cone's dimensionality space $\{\tilde{x}: [b_i, \tilde{x}] = 0 \text{ for } i \in J\}$ will cause at least one of the indicators $1\{[b_i, \tilde{x}]\ R_i\ 0\}$ for $i \in K^+$ to decrease to 0 as soon as $\tilde{x}_0$ leaves the cone. Typically, but not necessarily always, no other indicators $1\{[b_i, \tilde{x}]\ R_i\ 0\}$ for $i \in I$ w.o. $(J \cup K^+)$ would change their values during this motion and thus the motion would result in an instant decrease in the value of H if g_2 is strictly increasing. Thus hills can be seen to play the role of relative maxima in this problem.

The utility of the max-cone and hill concepts lies in that it can be shown that any vector which maximizes H is in a face of a max-cone which is either a hill or leads through a finite sequence of adjacent max-cones to a max-hill. It can also be shown that:

1) If g_2 is strictly increasing, then every max-cone is a hill.

2) If g_2 is strictly increasing, then, for every lower dimensional max-cone, R_j = "$\geqslant$" for every $j \in J$ where $\mathrm{C}\{b_i, i \in J\}^+$ is the dimensionality space of the cone. Hence, if g_2 is strictly increasing and all R_i = "$>$" then there are no lower dimensional max-cones.

3) If $\{b_i, i \in I\}$ is in pointed position or less generally, if this set is in general position, then all max-cones are either fully dimensional or $\langle 0 \rangle$. (Recall that $\{b_i, i \in I\}$ is in *pointed position* if and only if for all nonempty subsets $K \subset I$, if $\{b_i, i \in K\}^\perp \neq \langle 0 \rangle$, then $\mathrm{C}\{b_i, i \in K\}$ is pointed).

One consequence of requiring $\mathrm{C}\{b_i, i \in J\}$ to be a subspace in the definition of max-cone is that it enables the solution vectors to be organized into larger groups: max-cones may contain several optimal H level sets as boundary faces in addition to the required optimal relative interior face. But the real reason why it is desirable for $\mathrm{C}\{b_i, i \in J\}$ to be a subspace in the max-cone definition is because (i) it is convenient for this to be true for hills and (ii) it is aesthetically pleasing for max-cones to be hills when g_2 is strictly increasing. Why this subspace restriction is desirable for hills will be seen shortly.

One of the major implications of knowing something about the solution geometry is that in order maximize H, it is now clearly sufficient to enumerate the values of H in the relative interiors of all of the hills.

-- *understanding the tree algorithm* --

The philosophy behind the tree algorithm rests on the observation that whenever a nonzero vector is not in a given nonzero hill, then typically that hill will signal this condition by indicating some hyperspace which nontrivially bounds it. More formally, let $\{\tilde{x} \in \tilde{X}: [b_i, \tilde{x}] \geqslant 0$ for $i \in K^+$, $[b_i, \tilde{x}] = 0$

for $i \in J\}$ be a nonzero hill and let $\tilde{v}$ be any nonzero vector in $\tilde{X}$. If $\tilde{v}$ is not in the dimensionality space $\mathsf{C}\{b_i, i \in J\}^+$ for the hill, then there is some $j \in J$ such that $[b_j, \tilde{v}] < 0$. (This is a direct consequence of $\mathsf{C}\{b_i, i \in J\}$ being a subspace.) Observe that since the hill is contained in $\mathsf{C}\{b_i, i \in J\}^+$, $b_j^\perp$ nontrivially bounds the hill. On the other hand, if $[b_i, \tilde{v}] = 0$ for all $i \in J$ and $\tilde{v}$ is not in the hill, then there exists $k \in K^+$ such that $[b_k, \tilde{v}] < 0$. Furthermore, if the hill is at least two dimensional, then $b_k^\perp$ contains nonzero vectors from the hill in this case as well. In any event, for any vector $\tilde{v}$ not in some given nonzero hill, there is some b_i such that $[b_i, \tilde{v}] < 0$; and, what is more important, this $b_i^\perp$ contains some nonzero vector of that hill as long as that hill is not one dimensional with $\tilde{v}$ in its dimensionality space.

Consequently, in order to obtain a nonzero vector in every hill, it is sufficient to select any nonzero vector $\tilde{v}_0$ and then search for nonzero hill vectors first in $\mathsf{L}\{\tilde{v}_0\}$ and then in every $b_j^\perp$ such that $j \in N(\tilde{v}_0) := \{i \in I : [b_i, \tilde{v}_0] < 0\}$. This is the central idea behind the tree algorithm which, as will be seen shortly, applies it in a recursive fashion.

-- *the two phases of the tree algorithm* --

The goal of the tree algorithm is to enumerate the relative interiors of all of the hills. It does this in two steps. The objective of the first step or relative boundary vector collection phase is just to find a set of vectors containing at least one nonzero vector from each nonzero hill. It does this by creating a tree of vectors wherein the desired hill vectors typically lie on the relative boundaries of their hills and not in their relative interiors.

The objective of the second step or displacement phase is to take each vector in this tree of vectors and to perturb or not perturb this vector in such a way that if the vector is in a hill, then it will be displaced into the relative interior of the hill. In this process, one tree vector may result in several displaced vectors, possibly including itself as a result of the null displacement.

In order demonstrate whether or not one of these displaced vectors is actually in the relative interior of a hill, it is necessary to do some linear programming. Fortunately, when seeking to maximize H, it is max-hills, not hills in general, that are of primary interest. In this event, none of this linear programming is necessary; it is sufficient merely to compute the H values of all displaced vectors and save the best.

-- *more on the relative boundary vector collection phase* --

The relative boundary vector collection algorithm constructs a tree of vectors in the following way. An initial nonzero vector $\tilde{v}_0$ is selected and both $\tilde{v}_0$ and $-\tilde{v}_0$ are placed in the root node. $N(\tilde{v}_0) = \{i \in I: [b_i, \tilde{v}_0] < 0\}$ is computed and each $b(i_0)^\perp$ for $i_0 \in N(\tilde{v}_0)$ is explored for nonzero hill vectors in precisely the same way that the search for solutions in X began: namely, for each $i_0 \in N(\tilde{v}_0)$, select some $\tilde{v}_1(i_0) \in b(i_0)^\perp$, compute $N(\tilde{v}_1(i_0))$, and then resolve to explore $b(i_0)^\perp \cap b(i_1)^\perp$ for nonzero hill vectors for each $i_1 \in N(\tilde{v}_1(i_0))$. $b(i_0)^\perp \cap b(i_1)^\perp$ is of course that portion of $b(i_0)^\perp$ which is in $b(i_1)^\perp$ just as $b(i_0)^\perp$ is that portion of $\tilde{X}$ which is in $b(i_0)^\perp$. For each $i_0 \in N(\tilde{v}_0)$, a child node of $\tilde{v}_0$ is created to contain $\tilde{v}_1(i_0)$ and $-\tilde{v}_1(i_0)$. Now in order to explore $b(i_0)^\perp \cap b(i_1)^\perp$ for each $i_0 \in N(\tilde{v}_0)$ and $i_1 \in N(\tilde{v}_1(i_0))$, proceed in a recursive manner treating $b(i_0)^\perp \cap b(i_1)^\perp$ just as $\tilde{X}$ was treated at the start: select some nonzero $\tilde{v}_2(i_0, i_1) \in b(i_0)^\perp \cap b(i_1)^\perp$, compute $N(\tilde{v}_2(i_0, i_1))$, and then resolve to explore $b(i_0)^\perp \cap b(i_1)^\perp \cap b(i_2)^\perp$ in this recursive fashion for all $i_2 \in N(\tilde{v}_2(i_0, i_1))$. For each $i_0 \in N(\tilde{v}_0)$ and $i_1 \in N(\tilde{v}_1(i_0))$, a child node of $\tilde{v}_1(i_0)$ is created to contain $\tilde{v}_2(i_0, i_1)$ and $-\tilde{v}_2(i_0, i_1)$. Continue in this way exploring for nonzero hill vectors and storing the results in the tree.

At some point, one will resolve to explore subspaces $b(i_0)^\perp \cap \cdots \cap b(i_q)^\perp$ which are one-dimensional. Since open rays are the natural generators of cones, there are essentially only two choices for some nonzero $\tilde{v} \in b(i_0)^\perp \cap \cdots \cap b(i_q)^\perp$. So, of course, it suffices to select both but the point is that at this stage, the algorithm must stop since to further constrain $b(i_0)^\perp \cap \cdots \cap b(i_q)^\perp$ in this recursive process of exploration would result at the next level of the tree in an attempt to find a nonzero vector in $\langle 0 \rangle$.

The tree constructed in this manner is guaranteed to contain at least one nonzero vector from every nonzero hill. As for whether or not $\langle 0 \rangle$ is a hill, it is sufficient to determine whether or not $\mathsf{C}\{b_i, i \in I\}$ is a subspace.

-- in what sense recursion? --

The assertion has been made here that the tree algorithm recursively applies the following procedure for finding a nonzero vector from each hill.

1) Given a set $\{b_i, i \in I\}$, select some nonzero $\tilde{v}_0 \in \mathsf{L}\{b_i, i \in I\}^{\sim}$ and explore $\mathsf{L}\{\tilde{v}_0\}$.

2) Search for nonzero hill vectors for the $\{b_i, i \in I\}$ problem by recursively exploring $b_j^\perp$ for every $j \in N(\tilde{v}_0) = \{i \in I : [b_i, \tilde{v}_0] < 0\}$.

On the surface, it may seem that the relative boundary vector collection part of the tree algorithm is not recursively applying this procedure since no lower dimensional problems seem to be generated where one would be exploring $z_i^\perp$ for some z_i; instead, it seems that the algorithm is exploring subspaces of the form $b(i_0)^\perp \cap \cdots \cap b(i_k)^\perp$ instead of subspaces of the original form $b_j^\perp$.

But, in actuality, lower dimensional problems are in effect being generated and the preceding two-step procedure for finding hill vectors *is* being applied recursively to them in order to identify nonzero hill vectors from each of their hills. These lower dimensional problems are not seen because it is both possible and convenient to work exclusively with the b_i.

Consider then what happens if one sets out to obtain a vector from each nonzero hill by recursively applying the above two-step procedure. It is sufficient to fix attention on one nonzero hill C^+ and consider what happens when neither $\tilde{v}_0$ nor $-\tilde{v}_0$ are in C^+. In this event, it can be shown that there is some $k \in N(\tilde{v}_0)$ such that $b_k^\perp \cap C^+ \neq \langle 0 \rangle$ and such that if, for each i, $z_i := P[b_i \mid R, \mathrm{L}\{b_k\}]$ for some R such that $R \oplus \mathrm{L}\{b_k\} = X$, then $b_k^\perp \cap C^+$ is isomorphic to a nonzero hill in the $\{z_i, i \in I\}$ problem.

Consequently, if a nonzero vector in the hill from this lower dimensional problem is found then a nonzero vector from the hill C^+ has effectively been found as well. Since it is not known which $k \in N(\tilde{v}_0)$ will lead to such a useful lower dimensional problem, lower dimensional problems are generated as above for each $j \in N(\tilde{v}_0)$ and the two-step hill vector collection procedure is applied recursively to each; one is thus searching for a nonzero vector from each nonzero hill in each $\{z_i, i \in I\}$ problem. This will call for the examination of yet more problems of even lower dimension until all such lower dimensional problems become two dimensional. At this point, it can be shown directly that the two-step hill vector collection procedure does precisely that: i.e., for each two dimensional problem, it obtains at least one nonzero vector from each nonzero hill. Because of the connection between the hills of one problem and those of certain lower dimensional problems that are generated from it, by recursively obtaining nonzero hill vectors from the nonzero hills of the lower dimensional problems, a nonzero hill vector can be obtained from each nonzero hill in the original problem.

Clearly, the relative boundary hill vector collection algorithm stated earlier does not explicitly compute these lower dimensional problems. However, it does compute precisely the same vectors (upto isomorphic images) that would be produced by the recursive application of the two-step procedure.

-- proving the validity of the boundary vector collection algorithm --

The proof of the validity of the first or relative boundary vector collection phase of the tree algorithm is similar to the preceding discussion of why the recursive application of the two-step procedure works. The proof is inductive in nature. When $\dim \{b_i, i \in I\} = 2$, then the first phase of the tree algorithm is precisely the two-step procedure which explores each $b_j^\perp$ by obtaining a conical spanning set for $b_j^\perp$. When $\dim \{b_i, i \in I\} \geqslant 3$, then one constructs the tree as specified and then focuses attention on a specific nonzero hill C^+. If $\tilde{v}_0, -\tilde{v}_0 \notin C^+$, then it is known that there exists $k \in N(\tilde{v}_0)$ such that $b_k^\perp \cap C^+ \neq \langle 0 \rangle$. As with the two-step procedure, the lower dimensional $\{z_i, i \in I\}$ problem is considered where for all i, $z_i = P[b_i \mid R, \mathsf{L}\{b_k\}]$ where $R \oplus \mathsf{L}\{b_k\} = X$. $b_k^\perp \cap C^+$ is isomorphic to a nonzero hill C_2^+ in this problem. Furthermore, the subtree of the original tree which has root node containing $\tilde{v}_1(k)$ and $-\tilde{v}_1(k)$ is isomorphic to a validly constructed tree for the $\{z_i, i \in I\}$ problem. Consequently, the induction hypothesis asserts that the tree for the $\{z_i, i \in I\}$ problem contains a nonzero vector from C_2^+. Hence, using the isomorphism again, the original tree contains a vector from C^+.

-- more on the displacement phase --

The function of the displacement phase is to take the relative boundary vectors in the tree produced in the first step and to displace or not displace each of these into the relative interiors of cones in such a way as to recover a relative interior vector from each hill. The procedure for displacing a relative boundary vector $\tilde{v} \in \{\tilde{x} \in \tilde{X} : [b_i, \tilde{x}] > 0$ for $i \in I$ w.o. J, $[b_i, \tilde{x}] = 0$ for $i \in J\}$ is the following:

1) If $\mathsf{C}\{b_j, j \in J\}$ is a subspace, then save $\tilde{v}$ since $\tilde{v}$ itself may already be in the relative interior of a hill.

2) If $\mathsf{C}\{b_j, j \in J\}$ is a subspace of dimension 1, then select any $j \in J$ and displace $\tilde{v}$ to both sides of $b_j^\perp$ in turn without crossing over any

other hyperspaces $b_i^{\perp}$ for $i \notin J$.

3) If $\mathrm{C}\{b_j, j \in J\}$ is pointed, then displace $\tilde{v}$ to the positive side of all $b_j^{\perp}$ for $j \in J$ in such a way as to not cross over any other hyperspaces $b_i^{\perp}$ for $i \notin J$.

4) If $\mathrm{C}\{b_j, j \in J\}$ is not pointed and $\dim \mathrm{C}\{b_j, j \in J\} > 1$, then recursion is necessary. Find relative interior vectors for all hills generated by $\{b_j, j \in J\}$ in what is a strictly lower dimensional problem. Each of these hill vectors is mapped under a certain isomorphism to a certain vector back in the original problem. Displace $\tilde{v}$ in turn towards each of these isomorphic image vectors in such a way as to not cross over any other hyperspaces $b_i^{\perp}$ for $i \notin J$. Note that more than one displaced vector may result from this operation.

The resulting set of displaced vectors contains at least one relative interior vector from every hill. The principal reason why the displacement algorithm works is that nonempty faces of hills generate hills in suitably defined lower dimensional problems.

It also turns out that every nonempty, nonzero boundary face of a max-cone generates a max-cone in a suitably defined lower dimensional problem and that, in turn, every max-cone in this lower dimensional problem generates a max-cone back in the original problem. The most important implications of this fact are that it is only necessary to displace relative boundary vectors with sufficiently high H values and that it is sufficient when doing recursion to produce relative interior vectors, not from all of the hills, but rather just from all of the max-hills in the lower dimensional problems. The H function which defines the max-hills in the lower dimensional problems generated by recursion is a version of the original H function modified according to the nature of certain relative boundary vectors in certain higher dimensional problems.

-- *comments* --

It is to be stressed that the tree algorithm just sketched is in some sense just the basic one. This monograph shows how many improvements to this basic algorithm result in a much faster, much more extensively defined and sophisticated tree algorithm.

There are at least two curiosities concerning the nature of tree algorithms for maximizing functions H in homogeneous canonical form with nondecreasing g_2 functions. The first is that the particular form of g_2 and the choice of "$>$" and "$\geqslant$" for R_i in $H(x) = g_2(1\{b_1^T x > 0\}, \ldots, 1\{b_n^T x > 0\})$ have absolutely nothing to do with hills. The nature of g_2 and the R_i make their only appearances in the path towards maximizing H at the beginning when they help to characterize the solution geometry and at the end when they are used to evaluate the "worth" (i.e., the H-value) of relative boundary and interior vectors whose genesis did not require them in the first place. This leads straight into the second curiosity, namely that one single tree of vectors contains all the information necessary for one to maximize any and all functions H in homogeneous canonical form with nondecreasing g_2 functions.

-- *the behavior of the tree algorithm in practice* --

Without encroaching too much upon what was said in the introduction, this chapter concludes with a few more summary remarks.

Chapter 6 developed lower and upper bounds on the order of the time complexity of a version of the tree algorithm for maximizing a function H in homogeneous canonical form. Under the assumption that g_2 is computed in order of n steps and defining $\alpha = \inf\{\#\{i \in I : b_i^T x < 0\} : x \neq 0\}$, the lower bound is $dn\dfrac{\alpha^d - 1}{\alpha - 1}$ when $\alpha \geqslant 2$ and the upper bound is always $dn^d/2^{d-1}$. In practice, the lower bound is much more indicative of the tree algorithm's performance than the upper bound is. The exponential character of the lower bound comes as no surprise since the basic problem the tree algorithm solves is

NP-complete.

The only other algorithms for solving the general class of problems that tree algorithms solve are complete enumeration procedures of one kind or another. Their time complexities are on the order of $dn^{d-1} \log n$ or greater. For the four examples in Chapter 9, a sophisticated WOH tree algorithm was between 50 and 30,000 times faster than the best of the complete enumeration algorithms.

A fast approximate tree algorithm has also been developed. Its modus operandi is to greedily explore subsets of a sequence of trees in an attempt to find good, if not optimal, vectors very quickly. Once it stops, of course, there is no guarantee that the best vectors it produces are optimal. The examples of Chapter 9 indicate that the fast algorithm used there is very successful in producing good, often optimal, vectors 6,000 to 55,000 times faster than complete enumeration.

References

[1] COVER, T. (1965). Geometrical and statistical properties of systems of linear inequalities with applications in pattern recognition. *IEEE Trans. Elec. Comp.* **14** 326-334.

[2] DAVIS, C. (1954). Theory of positive linear dependence. *Amer. J. Math.* **76** 733-746.

[3] EAVES, B. C. (1973). Piecewise linear retractions by reflexion. *Lin. Alg. App.* **7** 93-98.

[4] FISHER, R. A. (1936). The use of multiple measurements in taxonomic problems. *Ann. Eug.* **7** 179-188.

[5] GALE, D. (1960). *Theory of Linear Economic Models.* McGraw-Hill, New York.

[6] GALE, D. (1969). How to solve linear inequalities. *Amer. Math. Monthly* **76** 589-599.

[7] GERSTENHABER, M. (1951). Theory of convex polyhedral cones. In *Activity Analysis of Production and Allocation* (T. C. Koopmans, ed.). John Wiley, New York, 298-316.

[8] GOLDMAN, A. J. and TUCKER, A. W. (1956). Polyhedral convex cones. In *Linear Inequalities and Related Systems*, Annals of Mathematics Study **38** (H. W. Kuhn and A. W. Tucker, eds.). Princeton University Press, Princeton. 41-51.

[9] GREER, R. L. (1979). Consistent nonparametric estimation of best linear classification rules / Solving inconsistent systems of linear inequalities. Ph.D. dissertation, Stanford University.

[10] HADLEY, G. (1962). *Linear Programming.* Addison-Wesley, Reading, Mass.

[11] HALMOS, P. R. (1974). *Finite Dimensional Vector Spaces.* Springer-Verlag, New York.

[12] HARDY, G. H. (1971). *Orders of Infinity.* Hafner, New York.

[13] IBARAKI, T. and MUROGA, S. (1970). Adaptive linear classifier by linear programming. *IEEE Trans. SSC-6* 53-62.

[14] JOHNSON, D. S. and PREPARATA, F. P. (1978). The densest hemisphere problem. *Theor. Comp. Sci.* **6** 93-107.

[15] KELLEY, J. L. (1955). *General Topology.* D. Van Nostrand, New York.

[16] KNUTH, D. E. (1973a). *The Art of Computer Programming: Volume One: Fundamental Algorithms, Second Edition.* Addison-Wesley, Reading, Massachusetts.

[17] KNUTH, D. E. (1973b). *The Art of Computer Programming: Volume Three: Sorting and Searching.* Addison-Wesley, Reading, Massachusetts.

[18] KNUTH, D. E. (1976). Big Omicron and Big Omega and Big Theta. *SIGACT News* **8** 18-24.

[19] LINES, L. (1965). *Solid Geometry.* Dover, New York.

[20] MOTZKIN, T. and SCHOENBERG, I. (1954). The relaxation method for linear inequalities. *Can. J. Math.* **6** 393-404.

[21] NERING, E. (1963). *Linear Algebra and Matrix Theory.* John Wiley, New York.

[22] RUDIN, W. (1973). *Functional Analysis.* McGraw-Hill, New York.

[23] SLATER, M. L. (1951). A note on Motzkin's transposition theorem. *Econometrika* **19** 185-187.

[24] STEWART, G. W. (1973). *Introduction to Matrix Computations.* Academic Press, New York.

[25] STOER, J. and WITZGALL, C. (1970). *Convexity and Optimization in Finite Dimensions.* Springer-Verlag, New York.

[26] STOLLER, D. (1954). Univariate two-population distribution-free discrimination. *J. Amer. Stat. Assoc.* **49** 770-777.

[27] WARMACK, R. and GONZALEZ, R. (1973). An algorithm for the optimal solution of linear inequalities and its application to pattern recognition. *IEEE Trans. Comp.* **22** 1065-1075.

[28] WETS, R. J.-B. and WITZGALL, C. (1967). Algorithms for frames and lineality spaces of cones. *J. Res. Nat. Bur. Standards* **71B** 1-7.

Index